Food Packaging Technology

Packaging Technology Series

Series Editor: Geoff A. Giles, Global Pack Management, GlaxoSmithKline, London.

A series which presents the current state of the art in chosen sectors of the packaging industry. Written at professional and reference level, it is directed at packaging technologists, those involved in the design and development of packaging, users of packaging and those who purchase packaging. The series will also be of interest to manufacturers of packaging machinery.

Titles in the series:

Design and Technology of Packaging Decoration for the Consumer Market
Edited by G.A. Giles

Materials and Development of Plastics Packaging for the Consumer Market
Edited by G.A. Giles and D.R. Bain

Technology of Plastics Packaging for the Consumer Market
Edited by G.A. Giles and D.R. Bain

Canmaking for Can Fillers
Edited by T.A. Turner

PET Packaging Technology
Edited by D.W. Brooks and G.A. Giles

Food Packaging Technology
Edited by R. Coles, D. McDowell and M.J. Kirwan

Packaging Closures and Sealing Systems
Edited by N. Theobald

FOOD PACKAGING TECHNOLOGY

Edited by

RICHARD COLES
Consultant in Food Packaging, London

DEREK McDOWELL
Head of Supply and Packaging Division
Loughry College, Northern Ireland

and

MARK J. KIRWAN
Consultant in Packaging Technology
London

Blackwell
Publishing

CRC Press

© 2003 by Blackwell Publishing Ltd

Editorial Offices:
9600 Garsington Road, Oxford OX4 2DQ
  Tel: +44 (0) 1865 776868
108 Cowley Road, Oxford OX4 1JF, UK
  Tel: +44 (0) 1865 791100
Blackwell Munksgaard, 1 Rosenørns Allè,
P.O. Box 227, DK-1502 Copenhagen V,
Denmark
  Tel: +45 77 33 33 33
Blackwell Publishing Asia Pty Ltd,
550 Swanston Street, Carlton South,
Victoria 3053, Australia
  Tel: +61 (0)3 9347 0300
Blackwell Publishing, 10 rue Casimir
Delavigne, 75006 Paris, France
  Tel: +33 1 53 10 33 10

Published in the USA and Canada (only) by
CRC Press LLC
2000 Corporate Blvd., N.W.
Boca Raton, FL 33431, USA
Orders from the USA and Canada (only) to
CRC Press LLC

USA and Canada only:
ISBN 0-8493-9788-X

The right of the Author to be identified as the Author of this Work has been asserted in accordance with the Copyright, Designs and Patents Act 1988.

All rights reserved. No part of this publication may be reproduced, stored in a retrieval system, or transmitted, in any form or by any means, electronic, mechanical, photocopying, recording or otherwise, except as permitted by the UK Copyright, Designs and Patents Act 1988, without the prior permission of the publisher.

This book contains information obtained from authentic and highly regarded sources. Reprinted material is quoted with permission, and sources are indicated. Reasonable efforts have been made to publish reliable data and information, but the author and the publisher cannot assume responsibility for the validity of all materials or for the consequences of their use.

Trademark Notice: Product or corporate names may be trademarks or registered trademarks, and are used only for identification and explanation, without intent to infringe.

First published 2003

Library of Congress Cataloging in Publication Data
A catalog record for this title is available from the Library of Congress

British Library Cataloguing in Publication Data
A catalogue record for this title is available from the British Library

ISBN 1-84127-221-3
Originated as Sheffield Academic Press

Set in 10.5/12pt Times
by Integra Software Services Pvt Ltd,
Pondicherry, India
Printed and bound in Great Britain,
using acid-free paper by
MPG Books Ltd, Bodmin, Cornwall

For further information on
Blackwell Publishing, visit our website:
www.blackwellpublishing.com

Contents

Contributors	xv
Preface	xvii

1 Introduction   1
RICHARD COLES

1.1	Introduction		1
1.2	Packaging developments – an historical perspective		2
1.3	Food supply and the protective role of packaging		4
1.4	The value of packaging to society		7
1.5	Definitions and basic functions of packaging		8
1.6	Packaging strategy		9
1.7	Packaging design and development		9
	1.7.1	The packaging design and development framework	12
		1.7.1.1 Product needs	13
		1.7.1.2 Distribution needs and wants of packaging	13
		1.7.1.3 Packaging materials, machinery and production processes	16
		1.7.1.4 Consumer needs and wants of packaging	18
		1.7.1.5 Multiple food retail market needs and wants	22
		1.7.1.6 Environmental performance of packaging	26
	1.7.2	Packaging specifications and standards	28
1.8	Conclusion		29
Literature reviewed and sources of information			29

2 Food biodeterioration and methods of preservation   32
GARY S. TUCKER

2.1	Introduction		32
2.2	Agents of food biodeterioration		33
	2.2.1	Enzymes	33
	2.2.2	Microorganisms	34
		2.2.2.1 Bacteria	35
		2.2.2.2 Fungi	38
	2.2.3	Non-enzymic biodeterioration	40
2.3	Food preservation methods		41
	2.3.1	High temperature	41
		2.3.1.1 Blanching	42
		2.3.1.2 Thermal processing	42
		2.3.1.3 Continuous thermal processing (aseptic)	47
		2.3.1.4 Pasteurisation	51
	2.3.2	Low temperature	52
		2.3.2.1 Freezing	52
		2.3.2.2 Chilling and cooling	53

CONTENTS

	2.3.3 Drying and water activity control	54
	2.3.4 Chemical preservation	56
	2.3.4.1 Curing	57
	2.3.4.2 Pickling	58
	2.3.4.3 Smoking	58
	2.3.5 Fermentation	59
	2.3.6 Modifying the atmosphere	60
	2.3.7 Other techniques and developments	61
	2.3.7.1 High pressure processing	61
	2.3.7.2 Ohmic heating	62
	2.3.7.3 Irradiation	62
	2.3.7.4 Membrane processing	62
	2.3.7.5 Microwave processing	63
References		63

3 Packaged product quality and shelf life   65
HELEN BROWN and JAMES WILLIAMS

3.1	Introduction	65
3.2	Factors affecting product quality and shelf life	68
3.3	Chemical/biochemical processes	69
	3.3.1 Oxidation	70
	3.3.2 Enzyme activity	73
3.4	Microbiological processes	74
	3.4.1 Examples where packaging is key to maintaining microbiological shelf life	75
3.5	Physical and physico-chemical processes	77
	3.5.1 Physical damage	77
	3.5.2 Insect damage	78
	3.5.3 Moisture changes	78
	3.5.4 Barrier to odour pick-up	81
	3.5.5 Flavour scalping	81
3.6	Migration from packaging to foods	81
	3.6.1 Migration from plastic packaging	83
	3.6.2 Migration from other packaging materials	86
	3.6.3 Factors affecting migration from food contact materials	88
	3.6.4 Packaging selection to avoid migration and packaging taints	89
	3.6.5 Methods for monitoring migration	89
3.7	Conclusion	91
References		91

4 Logistical packaging for food marketing systems   95
DIANA TWEDE and BRUCE HARTE

4.1	Introduction	95
4.2	Functions of logistical packaging	96
	4.2.1 Protection	97
	4.2.2 Utility/productivity	98
	4.2.3 Communication	99

4.3		Logistics activity-specific and integration issues	100
	4.3.1	Packaging issues in food processing and retailing	100
	4.3.2	Transport issues	101
	4.3.3	Warehousing issues	104
	4.3.4	Retail customer service issues	106
	4.3.5	Waste issues	107
	4.3.6	Supply chain integration issues	108
4.4		Distribution performance testing	109
	4.4.1	Shock and vibration testing	110
	4.4.2	Compression testing	111
4.5		Packaging materials and systems	112
	4.5.1	Corrugated fiberboard boxes	112
	4.5.2	Shrink bundles	115
	4.5.3	Reusable totes	115
	4.5.4	Unitization	116
4.6		Conclusion	119
References		119

5 Metal cans   120
BEV PAGE, MIKE EDWARDS and NICK MAY

5.1		Overview of market for metal cans	120
5.2		Container performance requirements	120
5.3		Container designs	121
5.4		Raw materials for can-making	123
	5.4.1	Steel	123
	5.4.2	Aluminium	124
	5.4.3	Recycling of packaging metal	124
5.5		Can-making processes	124
	5.5.1	Three-piece welded cans	125
	5.5.2	Two-piece single drawn and multiple drawn (DRD) cans	126
	5.5.3	Two-piece drawn and wall ironed (DWI) cans	127
5.6		End-making processes	129
	5.6.1	Plain food can ends and shells for food/drink easy-open ends	130
	5.6.2	Conversion of end shells into easy-open ends	130
5.7		Coatings, film laminates and inks	131
5.8		Processing of food and drinks in metal packages	132
	5.8.1	Can reception at the packer	132
	5.8.2	Filling and exhausting	133
	5.8.3	Seaming	135
	5.8.4	Heat processing	137
	5.8.5	Post-process can cooling, drying and labelling	138
	5.8.6	Container handling	139
	5.8.7	Storage and distribution	140
5.9		Shelf life of canned foods	141
	5.9.1	Interactions between the can and its contents	142
	5.9.2	The role of tin	142
	5.9.3	The dissolution of tin from the can surface	144
	5.9.4	Tin toxicity	145

	5.9.5	Iron	146
	5.9.6	Lead	147
	5.9.7	Aluminium	147
	5.9.8	Lacquers	147
5.10	Internal corrosion	148
5.11	Stress corrosion cracking	148
5.12	Environmental stress cracking corrosion of aluminium alloy beverage can ends	149
5.13	Sulphur staining	149
5.14	External corrosion	149
5.15	Conclusion	150
References and further reading	151

6 Packaging of food in glass containers   152
P.J. GIRLING

6.1	Introduction	152
	6.1.1	Definition of glass	152
	6.1.2	Brief history	152
	6.1.3	Glass packaging	152
	6.1.4	Glass containers market sectors for foods and drinks	153
	6.1.5	Glass composition	153
		6.1.5.1 White flint (clear glass)	153
		6.1.5.2 Pale green (half white)	154
		6.1.5.3 Dark green	154
		6.1.5.4 Amber (brown in various colour densities)	154
		6.1.5.5 Blue	154
6.2	Attributes of food packaged in glass containers	154
	6.2.1	Glass pack integrity and product compatibility	156
		6.2.1.1 Safety	156
		6.2.1.2 Product compatibility	156
	6.2.2	Consumer acceptability	156
6.3	Glass and glass container manufacture	156
	6.3.1	Melting	156
	6.3.2	Container forming	157
	6.3.3	Design parameters	158
	6.3.4	Surface treatments	158
		6.3.4.1 Hot end treatment	158
		6.3.4.2 Cold end treatment	159
		6.3.4.3 Low-cost production tooling	160
		6.3.4.4 Container inspection and quality	161
6.4	Closure selection	163
	6.4.1	Normal seals	164
	6.4.2	Vacuum seals	164
	6.4.3	Pressure seals	164
6.5	Thermal processing of glass packaged foods	165
6.6	Plastic sleeving and decorating possibilities	165
6.7	Strength in theory and practice	166
6.8	Glass pack design and specification	167
	6.8.1	Concept and bottle design	167
6.9	Packing – due diligence in the use of glass containers	169

	6.10	Environmental profile	171
		6.10.1 Reuse	171
		6.10.2 Recycling	171
		6.10.3 Reduction – lightweighting	172
	6.11	Glass as a marketing tool	172
	References		172
	Further reading		173

7 Plastics in food packaging — 174
MARK J. KIRWAN and JOHN W. STRAWBRIDGE

7.1	Introduction		174
	7.1.1	Definition and background	174
	7.1.2	Use of plastics in food packaging	175
	7.1.3	Types of plastics used in food packaging	177
7.2	Manufacture of plastics packaging		178
	7.2.1	Introduction to the manufacture of plastics packaging	178
	7.2.2	Plastic film and sheet for packaging	179
	7.2.3	Pack types based on use of plastic films, laminates etc.	183
	7.2.4	Rigid plastic packaging	186
7.3	Types of plastic used in packaging		189
	7.3.1	Polyethylene	189
	7.3.2	Polypropylene (PP)	191
	7.3.3	Polyethylene terephthalate (PET or PETE)	194
	7.3.4	Polyethylene naphthalene dicarboxylate (PEN)	195
	7.3.5	Polycarbonate (PC)	196
	7.3.6	Ionomers	196
	7.3.7	Ethylene vinyl acetate (EVA)	197
	7.3.8	Polyamide (PA)	197
	7.3.9	Polyvinyl chloride (PVC)	198
	7.3.10	Polyvinylidene chloride (PVdC)	199
	7.3.11	Polystyrene (PS)	200
	7.3.12	Styrene butadiene (SB)	201
	7.3.13	Acrylonitrile butadiene styrene (ABS)	201
	7.3.14	Ethylene vinyl alcohol (EVOH)	201
	7.3.15	Polymethyl pentene (TPX)	202
	7.3.16	High nitrile polymers (HNP)	202
	7.3.17	Fluoropolymers	203
	7.3.18	Cellulose-based materials	203
	7.3.19	Polyvinyl acetate (PVA)	204
7.4	Coating of plastic films – types and properties		205
	7.4.1	Introduction to coating	205
	7.4.2	Acrylic coatings	205
	7.4.3	PVdC coatings	206
	7.4.4	PVOH coatings	206
	7.4.5	Low-temperature sealing coatings (LTSCs)	206
	7.4.6	Metallising with aluminium	207
	7.4.7	SiO_x coatings	207
	7.4.8	DLC (Diamond-like coating)	208
	7.4.9	Extrusion coating with PE	208

7.5	Secondary conversion techniques		208
	7.5.1	Film lamination by adhesive	208
	7.5.2	Extrusion lamination	210
	7.5.3	Thermal lamination	211
7.6	Printing		211
	7.6.1	Introduction to the printing of plastic films	211
	7.6.2	Gravure printing	211
	7.6.3	Flexographic printing	212
	7.6.4	Digital printing	212
7.7	Printing and labelling of rigid plastic containers		212
	7.7.1	In-mould labelling	212
	7.7.2	Labelling	213
	7.7.3	Dry offset printing	213
	7.7.4	Silk screen printing	213
	7.7.5	Heat transfer printing	213
7.8	Food contact and barrier properties		214
	7.8.1	The issues	214
	7.8.2	Migration	214
	7.8.3	Permeation	215
	7.8.4	Changes in flavour	216
7.9	Sealability and closure		217
	7.9.1	Introduction to sealability and closure	217
	7.9.2	Heat sealing	217
		7.9.2.1 Flat jaw sealing	218
		7.9.2.2 Crimp jaw conditions	219
		7.9.2.3 Impulse sealing	220
		7.9.2.4 Hot wheel sealing	220
		7.9.2.5 Hot air sealers	221
		7.9.2.6 Gas flame sealers	221
		7.9.2.7 Induction sealing	221
		7.9.2.8 Ultrasonic sealing	221
	7.9.3	Cold seal	221
	7.9.4	Plastic closures for bottles, jars and tubs	221
	7.9.5	Adhesive systems used with plastics	222
7.10	How to choose		222
7.11	Retort pouch		224
	7.11.1	Packaging innovation	224
	7.11.2	Applications	225
	7.11.3	Advantages and disadvantages	226
	7.11.4	Production of pouches	227
	7.11.5	Filling and sealing	228
	7.11.6	Processing	229
	7.11.7	Process determination	230
	7.11.8	Post retort handling	231
	7.11.9	Outer packaging	231
	7.11.10	Quality assurance	232
	7.11.11	Shelf life	232
7.12	Environmental and waste management issues		233
	7.12.1	Environmental benefit	233
	7.12.2	Sustainable development	233
	7.12.3	Resource minimisation – lightweighting	233

		7.12.4	Plastics manufacturing and life cycle assessment (LCA)	234
		7.12.5	Plastics waste management	235
			7.12.5.1 Introduction to plastics waste management	235
			7.12.5.2 Energy recovery	236
			7.12.5.3 Feedstock recycling	236
			7.12.5.4 Biodegradable plastics	237

Appendices 238
References 239
Further reading 240
Websites 240

8 Paper and paperboard packaging 241
M.J. KIRWAN

8.1	Introduction		241
8.2	Paper and paperboard – fibre sources and fibre separation (pulping)		243
8.3	Paper and paperboard manufacture		245
	8.3.1	Stock preparation	245
	8.3.2	Sheet forming	245
	8.3.3	Pressing	246
	8.3.4	Drying	247
	8.3.5	Coating	248
	8.3.6	Reel-up	248
	8.3.7	Finishing	248
8.4	Packaging papers and paperboards		248
	8.4.1	Wet strength paper	249
	8.4.2	Microcreping	249
	8.4.3	Greaseproof	249
	8.4.4	Glassine	249
	8.4.5	Vegetable parchment	249
	8.4.6	Tissues	250
	8.4.7	Paper labels	250
	8.4.8	Bag papers	250
	8.4.9	Sack kraft	250
	8.4.10	Impregnated papers	250
	8.4.11	Laminating papers	251
	8.4.12	Solid bleached board (SBB)	251
	8.4.13	Solid unbleached board (SUB)	251
	8.4.14	Folding boxboard (FBB)	252
	8.4.15	White lined chipboard (WLC)	253
8.5	Properties of paper and paperboard		254
	8.5.1	Appearance	254
	8.5.2	Performance	254
8.6	Additional functional properties of paper and paperboard		255
	8.6.1	Treatment during manufacture	255
		8.6.1.1 Hard sizing	255
		8.6.1.2 Sizing with wax	255
		8.6.1.3 Acrylic resin dispersion	255
		8.6.1.4 Fluorocarbon dispersion	255
	8.6.2	Lamination	255

xii CONTENTS

		8.6.3	Plastic extrusion coating and laminating	256
		8.6.4	Printing and varnishing	257
		8.6.5	Post-printing roller varnishing/coating/laminating	258
	8.7	Design for paper and paperboard packaging		258
	8.8	Package types		259
		8.8.1	Tea and coffee bags	259
		8.8.2	Paper bags and wrapping paper	259
		8.8.3	Sachets/pouches/overwraps	260
		8.8.4	Multiwall paper sacks	262
		8.8.5	Folding cartons	263
		8.8.6	Liquid packaging cartons	265
		8.8.7	Rigid cartons or boxes	267
		8.8.8	Paper based tubes, tubs and composite containers	268
			8.8.8.1 Tubes	268
			8.8.8.2 Tubs	268
			8.8.8.3 Composite containers	268
		8.8.9	Fibre drums	268
		8.8.10	Corrugated fibreboard packaging	269
		8.8.11	Moulded pulp containers	272
		8.8.12	Labels	273
		8.8.13	Sealing tapes	275
		8.8.14	Cushioning materials	276
		8.8.15	Cap liners (wads) and diaphragms	276
	8.9	Systems		277
	8.10	Environmental profile		277
	Reference			281
	Further reading			281
	Websites			281

9 Active packaging
BRIAN P.F. DAY
282

	9.1	Introduction	282
	9.2	Oxygen scavengers	284
		9.2.1 ZERO2™ oxygen scavenging materials	288
	9.3	Carbon dioxide scavengers/emitters	289
	9.4	Ethylene scavengers	290
	9.5	Ethanol emitters	292
	9.6	Preservative releasers	293
	9.7	Moisture absorbers	295
	9.8	Flavour/odour adsorbers	296
	9.9	Temperature control packaging	297
	9.10	Food safety, consumer acceptability and regulatory issues	298
	9.11	Conclusions	300
	References		300

10 Modified atmosphere packaging
MICHAEL MULLAN and DEREK McDOWELL
303

	Section A MAP gases, packaging materials and equipment	**303**
	10.A1 Introduction	303
	10.A1.1 Historical development	304

10.A2	Gaseous environment			304
	10.A2.1	Gases used in MAP		304
		10.A2.1.1	Carbon dioxide	304
		10.A2.1.2	Oxygen	305
		10.A2.1.3	Nitrogen	305
		10.A2.1.4	Carbon monoxide	305
		10.A2.1.5	Noble gases	306
	10.A2.2	Effect of the gaseous environment on the activity of bacteria, yeasts and moulds		306
		10.A2.2.1	Effect of oxygen	306
		10.A2.2.2	Effect of carbon dioxide	307
		10.A2.2.3	Effect of nitrogen	308
	10.A2.3	Effect of the gaseous environment on the chemical, biochemical and physical properties of foods		308
		10.A2.3.1	Effect of oxygen	309
		10.A2.3.2	Effects of other MAP gases	310
	10.A2.4	Physical spoilage		311
10.A3	Packaging materials			311
	10.A3.1	Main plastics used in MAP		312
		10.A3.1.1	Ethylene vinyl alcohol (EVOH)	312
		10.A3.1.2	Polyethylenes (PE)	312
		10.A3.1.3	Polyamides (PA)	313
		10.A3.1.4	Polyethylene terephthalate (PET)	313
		10.A3.1.5	Polypropylene (PP)	313
		10.A3.1.6	Polystyrene (PS)	314
		10.A3.1.7	Polyvinyl chloride (PVC)	314
		10.A3.1.8	Polyvinylidene chloride (PVdC)	314
	10.A3.2	Selection of plastic packaging materials		315
		10.A3.2.1	Food contact approval	315
		10.A3.2.2	Gas and vapour barrier properties	315
		10.A3.2.3	Optical properties	318
		10.A3.2.4	Antifogging properties	318
		10.A3.2.5	Mechanical properties	318
		10.A3.2.6	Heat sealing properties	319
10.A4	Modified atmosphere packaging machines			319
	10.A4.1	Chamber machines		319
	10.A4.2	Snorkel machines		319
	10.A4.3	Form-fill-seal tray machines		320
		10.A4.3.1	Negative forming	320
		10.A4.3.2	Negative forming with plug assistance	321
		10.A4.3.3	Positive forming with plug assistance	321
	10.A4.4	Pre-formed trays		323
		10.A4.4.1	Pre-formed trays versus thermoformed trays	323
	10.A4.5	Modification of the pack atmosphere		324
		10.A4.5.1	Gas flushing	324
		10.A4.5.2	Compensated vacuum gas flushing	324
	10.A4.6	Sealing		325
	10.A4.7	Cutting		325
	10.A4.8	Additional operations		325
10.A5	Quality assurance of MAP			326
	10.A5.1	Heat seal integrity		326
		10.A5.1.1	Nondestructive pack testing equipment	328

		10.A5.1.2	Destructive pack testing equipment	328
	10.A5.2	\multicolumn{2}{l	}{Measurement of transmission rate and permeability in packaging films}	329

		10.A5.1.2 Destructive pack testing equipment	328
	10.A5.2	Measurement of transmission rate and permeability in packaging films	329
		10.A5.2.1 Water vapour transmission rate and measurement	329
		10.A5.2.2 Measurement of oxygen transmission rate	331
		10.A5.2.3 Measurement of carbon dioxide transmission rate	331
	10.A5.3	Determination of headspace gas composition	331
		10.A5.3.1 Oxygen determination	331
		10.A5.3.2 Carbon dioxide determination	331

Section B Main food types   **331**

10.B1	Raw red meat	331
10.B2	Raw poultry	332
10.B3	Cooked, cured and processed meat products	333
10.B4	Fish and fish products	334
10.B5	Fruits and vegetables	335
10.B6	Dairy products	338
References		338

Index   **340**

Contributors

Helen Brown — Biochemistry Section Manager, Campden & Chorleywood Food Research Association, Chipping Campden, Gloucestershire, GL55 6LD, UK

Richard Coles — Consultant in Food Packaging, Packaging Consultancy and Training, 20 Albert Reed Gardens, Tovil, Maidstone, Kent ME15 6JY, UK

Brian P.F. Day — Research Section Leader, Food Packaging & Coatings, Food Science Australia, 671 Sneydes Road (Private Bag 16), Werribee, Victoria 3030, Australia

Mike Edwards — Microscopy Section Manager, Chemistry & Biochemistry Department, Campden & Chorleywood Food Research Association, Chipping Campden, Gloucestershire, GL55 6LD, UK

Patrick J. Girling — Consultant in Glass Packaging, Doncaster, UK (formerly with Rockware Glass)

Bruce Harte — Director, Michigan State University, School of Packaging, East Lansing, Michigan, 48824-1223, USA

Mark J. Kirwan — Consultant in Packaging Technology, London, UK (formerly with Iggesund Paperboard)

Nick May — Senior Research Officer, Process and Product Development Department, Campden & Chorleywood Food Research Association, Chipping Campden, Gloucestershire, GL55 6LD, UK

Derek McDowell — Head of Supply and Packaging Division, Loughry College, The Food Centre, Cookstown, Co. Tyrone, BT80 9AA, Northern Ireland

Michael Mullan — Head of Food Education and Training Division, Loughry College, The Food Centre, Cookstown, Co. Tyrone, BT80 9AA and Department of Food Science, The Queen's University of Belfast, Newforge Lane, Belfast, BT9 5PX, Northern Ireland

Bev Page	Packaging Consultant, Oak Shade, 121 Nottingham Road, Ravenshead, Nottingham NG15 9HJ, UK
John W. Strawbridge	Consultant in Plastics Packaging, Welwyn, UK (formerly with Exxon-Mobil)
Gary S. Tucker	Process Development Section Leader, Department of Process and Product Development, Campden & Chorleywood Food Research, Association Chipping Campden, Gloucestershire, GL55 6LD, UK
Diana Twede	Associate Professor, Michigan State University, School of Packaging, East Lansing, Michigan, 48824-1223, USA
James Williams	Flavour Research and Taint Investigations Manager, Campden & Chorleywood Food Research Association, Chipping Campden, Gloucestershire, GL55 6LD, UK

Preface

This volume informs the reader about food preservation processes and techniques, product quality and shelf life, and the logistical packaging, packaging materials, machinery and processes, necessary for a wide range of packaging presentations.

It is essential that those involved in food packaging innovation have a thorough technical understanding of the requirements of a product for protection and preservation, together with a broad appreciation of the multi-dimensional role of packaging. Business objectives may be:

- the launch of new products or the re-launch of existing products
- the provision of added value to existing products or services
- cost reduction in the supply chain.

This book sets out to assist in the attainment of these objectives by informing designers, technologists and others in the packaging chain about key food packaging technologies and processes. To achieve this, the following five principal subject areas are covered:

1. food packaging strategy, design and development (chapter 1)
2. food bio-deterioration and methods of preservation (chapter 2)
3. packaged product quality and shelf life (chapter 3)
4. logistical packaging for food marketing systems (chapter 4)
5. packaging materials and processes (chapters 5–10).

Chapter 1 introduces the subject of food packaging and its design and development. Food packaging is an important source of competitive advantage for retailers and product manufacturers. Chapter 2 discusses bio-deterioration and methods of food preservation that are fundamental to conserving the integrity of a product and protecting the health of the consumer. Chapter 3 discussess packaged product quality and shelf life issues that are the main concerns for product stability and consumer acceptability. Chapter 4 discusses logistical packaging for food marketing systems – it considers supply chain efficiency, distribution hazards, opportunities for cost reduction and added value, communication, pack protection and performance evaluation. Chapters 5, 6, 7 and 8 consider metal cans, glass, plastics and paper and paperboard, respectively. Chapters 9 and 10 discuss active packaging and modified atmosphere packaging (MAP) respectively – these techniques are used to extend the shelf life and/or guarantee quality attributes such as nutritional content, taste and the colour of many types of fresh, processed and prepared foods.

The editors are grateful for the support of authors who are close to the latest developments in their technologies, and for their efforts in making this knowledge available.

We also wish to extend a word of gratitude to others who have contributed to this endeavour: Andy Hartley, Marketing Manager, and Sharon Crayton, Product Manager of Rockware Glass, UK; Nick Starke, formerly Head of Research & Development, Nampak, South Africa; Frank Paine, Adjunct Professor, School of Packaging, Michigan State University; and Susan Campbell.

Richard Coles
Derek McDowell
Mark Kirwan

1 Introduction
Richard Coles

1.1 Introduction

This chapter provides a context for considering the many types of packaging technology available. It includes an historical perspective of some packaging developments over the past 200 years and outlines the value of food packaging to society. It highlights the protective and logistical roles of packaging and introduces packaging strategy, design and development.

Packaging technology can be of strategic importance to a company, as it can be a key to competitive advantage in the food industry. This may be achieved by catering to the needs and wants of the end user, opening up new distribution channels, providing a better quality of presentation, enabling lower costs, increasing margins, enhancing product/brand differentiation, and improving the logistics service to customers.

The business drive to reduce costs in the supply chain must be carefully balanced against the fundamental technical requirements for food safety and product integrity, as well as the need to ensure an efficient logistics service. In addition, there is a requirement to meet the aims of marketing to protect and project brand image through value-added pack design. The latter may involve design inputs that communicate distinctive, aesthetically pleasing, ergonomic, functional and/or environmentally aware attributes.

Thus, there is a continual challenge to provide cost effective pack performance that satisfies the needs and wants of the user, with health and safety being of paramount importance. At the same time, it is important to minimise the environmental impact of products and the services required to deliver them. This challenge is continually stimulated by a number of key drivers – most notably, legislation and political pressure. In particular, there is a drive to reduce the amount of packaging used and packaging waste to be disposed of.

The growing importance of logistics in food supply means that manufacturing and distribution systems and, by implication, packaging systems, have become key interfaces of supplier–distributor relationships. Thus, the role of the market and the supply chain has increasing significance in the area of packaging innovation and design.

Arising from the above discussion is the need for those involved in packaging design and development to take account of technological, marketing, legal, logistical and environmental requirements that are continually changing. Consequently, it is asserted that those involved in packaging need to develop an integrated view of the effect on packaging of a wide range of influences, including quality, production, engineering, marketing, food technology R&D, purchasing, legal issues, finance, the supply chain and environmental management.

1.2 Packaging developments – an historical perspective

The last 200 years have seen the pack evolve from being a container for the product to becoming an important element of total product design – for example, the extension from packing tomato ketchup in glass bottles to squeezable co-extruded multi-layer plastic bottles with oxygen barrier material for long shelf life.

Military requirements have helped to accelerate or precipitate some key packaging developments. These include the invention of food canning in Napoleonic France and the increased use of paper-based containers in marketing various products, including soft cheeses and malted milk, due to the shortage of tinplate for steel cans during the First World War. The quantum growth in demand for pre-packaged foods and food service packaging since the Second World War has dramatically diversified the range of materials and packs used. These have all been made possible by developments in food science and technology, packaging materials and machine technology. An overview of some developments in packaging during the past 200 years is given below.

- *1800–1850s*. In 1809 in France, Nicolas Appert produced the means of thermally preserving food in hermetically sealed glass jars. In 1810, Peter Durand designed the soldered tinplate canister and commercialised the use of heat preserved food containers. In England, handmade cans of 'patent preserved meats' were produced for the Admiralty (Davis, 1967). In 1852, Francis Wolle of Pennsylvania, USA, developed the paper bag-making machine (Davis, 1967).
- *1870s*. In 1871, Albert L. Jones in the USA patented (no. 122,023) the use of corrugated materials for packaging. In 1874, Oliver Long patented (no. 9,948) the use of lined corrugated materials (Maltenfort, 1988). In 1879, Robert Gair of New York produced the first machine-made folding carton (Davis, 1967).
- *1880s*. In 1884, Quaker Oats packaged the first cereal in a folding box (Hine, 1995).
- *1890s*. In 1892, William Painter in Baltimore, USA, patented the Crown cap for glass bottles (Opie, 1989). In 1899, Michael J. Owens of Ohio conceived the idea of fully automatic bottle making. By 1903, Owens had commercialised the industrial process for the Owens Bottle Machine Company (Davis, 1967).
- *1900s*. In 1906, paraffin wax coated paper milk containers were being sold by G.W. Maxwell in San Francisco and Los Angeles (Robertson, 2002).

- *1910s.* Waxed paperboard cartons were used as containers for cream. In 1912, regenerated cellulose film (RCF) was developed. In 1915, John Van Wormer of Toledo, Ohio, commercialised *the paper bottle*, a folded blank box called Pure-Pak, which was delivered flat for subsequent folding, gluing, paraffin wax coating, filling with milk and sealing at the dairy (Robertson, 2002).
- *1920s.* In 1923, Clarence Birdseye founded *Birdseye Seafoods* in New York and commercialised the use of frozen foods in retail packs using cartons with waxed paper wrappers. In 1927, Du Pont perfected the cellulose casting process and introduced their product, Cellophane.
- *1930s.* In 1935, a number of American brewers began selling canned beer. In 1939, ethylene was first polymerised commercially by Imperial Chemical Industries (ICI) Ltd.. Later, polyethylene (PE) was produced by ICI in association with Du Pont. PE has been extensively used in packaging since the 1960s.
- *1940s.* During the Second World War, aerosol containers were used by the US military to dispense pesticides. Later, the aerosol can was developed and it became an immediate postwar success for dispensing food products such as pasteurised processed cheese and spray dessert toppings. In 1946, polyvinylidene chloride (PVdC) – often referred to as Saran – was used as a moisture barrier resin.
- *1950s.* The retort pouch for heat-processed foods was developed originally for the US military. Commercially, the pouch has been most used in Japan. Aluminium trays for frozen foods, aluminium cans and squeezable plastic bottles were introduced e.g. in 1956, the *Jif* squeezable lemon-shaped plastic pack of lemon juice was launched by Colman's of Norwich, England. In 1956, Tetra Pak launched its tetrahedral milk carton that was constructed from low-density polyethylene extrusion coated paperboard.
- *1960s.* The two-piece drawn and wall-ironed (DWI) can was developed in the USA for carbonated drinks and beers; the Soudronic welded side-seam was developed for the tinplate food can; tamper evident bottle neck shrink-sleeve was developed by Fuji Seal, Japan – this was the precursor to the shrink-sleeve label; aluminium roll-on pilfer-proof (ROPP) cap was used in the spirits market; tin-free steel can was developed. In 1967, the ring-pull opener was developed for canned drinks by the Metal Box Company; Tetra Pak launched its rectangular Tetra Brik Aseptic (TBA) carton system for long-life ultra-heat treated (UHT) milk. The TBA carton has become one of the world's major pack forms for a wide range of liquid foods and beverages.
- *1970s.* The bar code system for retail packaging was introduced in the USA; methods were introduced to make food packaging tamper evident; boil-in-the-bag frozen meals were introduced in the UK; MAP retail packs were introduced to the US, Scandinavia and Europe; PVC was used for beverage bottles; frozen foods in microwaveable plastic containers, bag-in-box systems and a range of aseptic form, fill and seal (FFS) flexible packaging systems were developed. In 1973, Du Pont developed the injection stretch blow-moulded PET bottle which was used for colas and other carbonated drinks.

- *1980s*. Co-extruded plastics incorporating oxygen barrier plastic materials for squeezable sauce bottles, and retortable plastic containers for ambient foods that could be microwave heated. PET-coated dual-ovenable paperboard for ready meals. The widget for canned draught beers was commercialised – there are now many types of widget available to form a foamy head in canned and glass bottled beers. In 1988, Japan's longest surviving brand of beer, Sapporo, launched the contoured can for its lager beer with a ring-pull that removed the entire lid to transform the pack into a handy drinking vessel.
- *1990s*. Digital printing of graphics on carton sleeves and labels for food packaging was introduced in the UK; shrink-sleeve plastic labels for glass bottles were rapidly adopted by the drinks industry; shaped can technology became more widely adopted in the USA and Europe as drinks companies sought ways of better differentiating their brands.

Since the advent of the food can in the 19th century, protection, hygiene, product quality and convenience have been major drivers of food technology and packaging innovation. In recent years, there has been a rising demand for packaging that offers both ease of use and high quality food to consumers with busy lifestyles. The 1980s, in particular, saw the widespread adoption by the grocery trade of innovations such as gas barrier plastic materials utilised in aseptic FFS plastic containers for desserts, soups and sauces; plastic retail tray packs of premium meat cuts in a modified atmosphere; and retortable plastic containers for ambient storage ready meals that can be microwave heated.

Technological developments often need to converge in order for a packaging innovation to be adopted. These have included developments in transportation, transport infrastructures, post-harvest technology, new retail formats and domestic appliances such as refrigerators, freezers and microwave ovens. For example, the development of the microwave oven precipitated the development of convenience packaging for a wide range of foods. In addition, the socio-cultural and demographic trends, consumer lifestyles and economic climate must generate sufficient market demand for an innovation to succeed.

1.3 Food supply and the protective role of packaging

Packaging for consumer products is an area where supply and demand is continuously changing due to the development of an international food market and adaptation to consumer, distribution, legal and technological requirements. Broad external influences on packaging for fast-moving consumer products may be summarised as follows:

- technological
- political/legal

- socio-cultural
- demographic
- ecological
- raw material availability
- economic.

The world's total food production has more than doubled over the past fifty years due to improved methods in animal husbandry, the use of advanced seed varieties and crop protection products that boost crop yields and quality. Mass production of packaged food has been enabled by technological innovations in food production, processing and logistics with packaging playing a key role. The economies of scale involved and the intense industrial competition have made many products more affordable.

Consumer demand for pre-packaged food continues to increase in advanced economies and a growing global population is also fuelling the demand. This is increasingly the case in newly industrialised countries experiencing rapid urbanisation.

In response to changing consumer lifestyles, large retail groups and food service industries have evolved. Their success has involved a highly competitive mix of logistical, trading, marketing and customer service expertise, all of which is dependent on quality packaging. They have partly driven the dramatic expansion in the range of products available, enabled by technological innovations, including those in packaging.

The retailing, food manufacturing and packaging supply industries are continuing to expand their operations internationally. The sourcing of products from around the world is increasingly assisted by a reduction in trade barriers. The effect has been an increase in competition and a downward pressure on prices. Increased competition has led to a rationalisation in industry structure, often in the form of mergers and takeovers. For packaging, it has meant the adoption of new materials and shapes, increased automation, extension of pack size ranges and a reduction in unit cost. Another effect of mergers among manufacturers and retailing groups on packaging is the reappraisal of brands and their pack designs.

Increasing market segmentation and the development of global food supply chains have spurred the adoption of sophisticated logistical packaging systems. Packaging is an integral part of the logistical system and plays an important role in preventing or reducing the generation of waste in the supply of food. Figure 1.1 illustrates the distribution flows of food from the farm to the consumer. It should be noted, however, that some parts of the chain permit the use of returnable packages.

Packaging assists the preservation of the world's resources through the prevention of product spoilage and wastage, and by protecting products until they have performed their function. The principal roles of packaging are to contain, protect/preserve food and inform the user. Thereby, food waste may be minimised and the health of the consumer safeguarded.

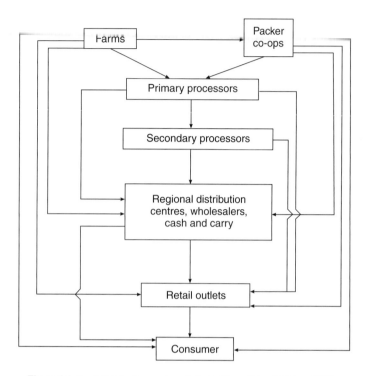

Figure 1.1 Food distribution systems (adapted from Paine & Paine, 1983).

Packaging combined with developments in food science, processing and preservation techniques, has been applied in a variety of ways to ensure the safety of the consumer and integrity of the product. The success of both packaging and food technology in this regard is reflected by the fact that the contents of billions of packs are being safely consumed every day.

In order to help minimise food waste throughout the supply chain and save cost, an optimum level of packaging is required. Significant food wastage occurs in many less developed countries – between 30% and 50% of food produced is wasted due to inadequate means of preservation, protection, storage and transportation (World Health Organisation). In developed countries, where modern processing, packaging and distribution systems are commonplace, food wastage before it reaches the consumer is only 2–3%.

> Less than 1% of packaged food goes to waste, compared with between 10% and 20% of unpackaged food.
> – Industry Council for Packaging and the Environment (INCPEN)

Food wastage can represent a much greater financial loss than just the cost of spoilt product. For example, there may be costs associated with salvage, disposal, administration, replacement, insurance and litigation. There is the potential loss of customer goodwill, which is an important consideration in today's highly competitive marketplace.

A Tetra Pak motto is that *a package should save more than it costs.*

1.4 The value of packaging to society

The value of food packaging to society has never been greater nor, paradoxically, has packaging attracted so much adverse media publicity and political attention. In response, stakeholders in the food industries need to fully appreciate and actively promote the positive contributions that their packaging makes to the quality of life. Food packaging is governed by a mass of laws, regulations, codes of practice and guidelines.

The societal benefits of packaging may include the following:

- prevents or reduces product damage and food spoilage, thereby saving energy and vital nutrients, and protecting the health of the consumer
- requires *less* municipal solid waste disposal since it promotes processed food residue recycling for use as animal feed or compost. For example, from 454 g (1 lb) of fresh corn-on-the-cob purchased at the supermarket, the customer eats approximately only 170 g (six ounces), and the rest ends up in the trash can and, ultimately, in the local landfill (Institute of Packaging Professionals, IOPP, USA). This same amount of edible frozen corn can be packed in a polyethylene bag weighing less than 5 g (less than 0.18 ounce)
- lowers the cost of many foods through economies of scale in mass production and efficiency in bulk distribution. Savings are also derived from reduced product damage
- reduces or eliminates the risk of tampering and adulteration
- presents food in an hygienic and often aesthetically attractive way
- communicates important information about the food and helps consumers make informed purchases
- provides functional convenience in use or preparation, freeing up more time
- promotes goods in a competitive marketplace and increases consumer choice
- facilitates the development of modern retail formats that offer consumers the convenience of the one-stop shop and the availability of food from around the world throughout the year
- extends the shelf life with the benefit of prolonged product use, thereby reducing wastage
- saves energy through the use of ambient packs that do not require refrigeration or frozen distribution and storage.

The food industry is aware of current public concerns related to packaging which include:

- packaging litter and the volume of packaging waste in municipal waste
- cost of disposal and recovery of discarded packaging in municipal waste
- pollution associated with methods of disposal, i.e. landfill and incineration
- ease of opening
- perception of over-packaging due to apparently excessive ullage (free space) resulting from product settlement
- legibility of labels
- integrity of information on labels
- contamination of food due to the packaging itself
- accidents involving packaging.

1.5 Definitions and basic functions of packaging

There are many ways of defining packaging reflecting different emphases. For example:

- A means of ensuring safe delivery to the ultimate consumer in sound condition at optimum cost.
- A coordinated system of preparing goods for transport, distribution, storage, retailing and end-use.
- A techno-commercial function aimed at optimising the costs of delivery while maximising sales (and hence profits).

However, the basic functions of packaging are more specifically stated:

- *Containment*: depends on the product's physical form and nature. For example, a hygroscopic free-flowing powder or a viscous and acidic tomato concentrate
- *Protection*: prevention of mechanical damage due to the hazards of distribution
- *Preservation*: prevention or inhibition of chemical changes, biochemical changes and microbiological spoilage
- *Information about the product*: legal requirements, product ingredients, use etc.
- *Convenience*: for the pack handlers and user(s) throughout the packaging chain
- *Presentation*: material type, shape, size, colour, merchandising display units etc.
- *Brand communication*: e.g. pack *persona* by the use of typography, symbols, illustrations, advertising and colour, thereby creating visual impact
- *Promotion (Selling)*: *free* extra product, *new* product, money off etc.

- *Economy*: for example, efficiency in distribution, production and storage
- *Environmental responsibility*: in manufacture, use, reuse, or recycling and final disposal.

1.6 Packaging strategy

Packaging may also be defined as: *a means of safely and cost effectively delivering products to the consumer in accordance with the marketing strategy of the organisation.* A packaging strategy is a plan that addresses all aspects and all activities involved in delivering the packaged product to the consumer. Packaging strategy should be allied to clearly defined marketing and manufacturing strategies that are consistent with the corporate strategy or mission of the business. Key stakeholders in the strategic development process include management from technical/quality, manufacturing, procurement, marketing, supply chain, legal and finance functions.

Packaging is both strategically and tactically important in the exercise of the marketing function. Where brands compete, distinctive or innovative packaging is often a key to the competitive edge companies seek. In the UK, for example, the development of the famous widget for canned draught beers opened up marketing opportunities and new distribution channels for large breweries. The packaging strategy of a food manufacturer should take into consideration the factors listed in Table 1.1.

1.7 Packaging design and development

Marketing *pull* is a pre-requisite to successful innovation in packaging materials, forms, designs or processes. The most ingenious technological innovation has little chance of success unless there is a market demand. Sometimes, an

Table 1.1 Framework for a packaging strategy

Technical requirements of the product and its packaging to ensure pack functionality and product protection/preservation throughout the pack's shelf life during distribution and storage until its consumption
Customer's valued packaging and product characteristics, for example, aesthetic, flavour, convenience, functional and environmental performance
Marketing requirements for packaging and product innovation to establish a distinct (product/service) brand proposition; protect brand integrity and satisfy anticipated demand at an acceptable profit in accordance with marketing strategy
Supply chain considerations such as compatibility with existing pack range and/or manufacturing system
Legislation and its operational/financial impacts, for example, regulations regarding food hygiene, labelling, weights and measures, food contact materials, due diligence etc.
Environmental requirements or pressures and their impacts, for example, light-weighting to reduce impact of taxes or levies on amount of packaging used

innovation is ahead of its time but may be later adopted when favoured by a change in market conditions. Specialist technical research, marketing research and consumer research agencies are employed to identify opportunities and minimise the financial costs and risks involved in the development, manufacture and marketing of a new product.

For example, the radical redesign of tea bag packs in the UK was based on focus group consumer research. The result was a rigid upright carton with an integral easy tear-off board strip but without the traditional film over-wrap that was difficult to open. Nitrogen gas-flushed metallised polyester pouches are used to contain 40 tea bags for convenient tea caddy or cupboard storage. Carton designs may contain either a single pouch or multiple pouches. The pouch prevents spillage of tea dust, provides freshness and conveys a fresh image. The carton shape, label and colour combinations were also redesigned for extra on shelf impact. This packaging innovation has been widely adopted by retailers and other manufacturers for their branded teas.

Generally, more successful new product developments are those that are implemented as a total concept with packaging forming an integral part of the whole. An example of the application of the *total product concept* is the distinctive white bottle for the rum-based spirit drink *Malibu* which reflects the coconut ingredient. There are many examples such as cartons with susceptors for microwave heating of frozen chips, pizzas and popcorn, and dispensing packs for mints.

Ideally, package design and distribution should be considered at the product concept stage. Insufficient communication may exist between marketing and distribution functions; a new product is manufactured and pack materials, shape and design are formulated to fulfil the market requirements. It is only then that handling and distribution are considered. Product failure in the marketplace due to inadequate protective packaging can be very costly to rectify. Marketing departments should be aware of distribution constraints when designing a *total product concept*. With high distribution costs, increased profitability from product and pack innovation can be wiped out if new packaging units do not fit in easily with existing distribution systems. It is necessary to consider whether packs are produced for their marketability or, for their physical distribution practicability. This would not necessarily be important if it were not for the significance of distribution costs, in particular those for refrigerated products.

The development of packs is frequently a time-consuming and creative endeavour. There may be communication difficulties between business functions and resource issues that impede pack development. The use of multi-disciplinary teams may expedite the packaging development process. This has the effect of improving the quality of the final product by minimising problems caused by design consequences that can result from sequential development. Computer assisted design (CAD) and rapid prototyping facilities for design and physical modelling of packs give packaging development teams the ability to accelerate the initial design process.

In packaging development, thorough project planning is essential. In particular, order lead times for packaging components need to be carefully planned with suppliers at an early stage in order to ensure a realistic time plan. For example, the development of a plastic bottle pack for a juice drink may involve typical stages listed in Table 1.2. There may be issues such as a supplier's availability of injection stretch blow-moulding machines due to seasonal demand for drinks containers and consequent lack of spare production capacity.

With reference to the definition: *Packaging in product distribution is aimed at maximising sales (and repeat sales, and so profits), while minimising the total overall cost of distribution from the point of pack filling onwards.* Packaging is regarded as a benefit to be optimised rather than merely a cost to be minimised (Paine & Paine, 1983).

Packaging optimisation is a main concern of the packaging development function. The aim is to achieve an optimal balance between performance, quality and cost, i.e. value for money. It involves a detailed examination of each cost element in the packaging system and an evaluation of the contribution of each item to the *functionality* of the system (Melis, 1989).

Packaging should be considered as part of the process of product manufacturing and distribution, and the economics of the supply chain should take into account all those operations – including packaging – involved in the delivery of the product to the final user. In certain cases, this may be extended to take account of the costs involved in reuse or waste collection, sorting, recovery

Table 1.2 Typical stages in the design and development of a new plastic bottle pack

Define packaging strategy
Prepare packaging brief and search for pack design concepts: functional and graphical
Concept costing, screening and approval by cross-functional packaging team
Pack component supplier selection through liaison with purchasing
Cost tooling; design and engineer new moulds for bottles and caps with suppliers
Test pack prototype: dimensional, drop impact, leak, compression, cap fit etc.
Commission artwork for labels
Shelf life testing; barrier performance evaluation
Model and sample production: filling system; labelling; casing etc.
Market test prototype
Design, cost and evaluate transit pack performance for prototype: drop, compression etc.
Determine case arrangement on pallets and assess influence of factors affecting stacking performance: brick or column stacking, relative humidity, moisture, pallet design etc.
Define quality standards and packaging specifications
Conduct production and machine trials: efficiency and productivity performance
Plan line change-overs
Develop inspection methods and introduce a quality assurance service
Commission production line for new or changed packaging systems
Fine-tune packaging operations and specifications

Table 1.3 Typical handling operations for an ambient storage retail pack

Production line container forming, de-palletising or de-nesting
Container transfer on conveyor system and container inspection (cleaning)
Filling, sealing (processing) and labelling
Casing, case sealing and coding
Palletising and stretch-wrapping
Plant storage
Transport to warehouse
Lorry transport to retail regional distribution centre (RDC)
RDC storage
Pallet break-bulk and product order pick for stores at RDC
Mixed product load on pallets or roll cages to RDC dispatch
Loaded pallets or roll cages delivered by lorry to retail stores
Loads moved to back of store storage area for a short period
Load retail cabinet or fill shelf merchandising display

and disposal. The overall or total packaging system cost stems from a number of different components including materials utilisation, machinery and production line efficiency, movement in distribution, management and manpower. They may include some of the operations listed in Table 1.3.

Adopting a systems approach to packaging can yield significant benefits other than just cost. Savings can be functionally derived by, perhaps, even increasing packaging costs for better pack performance and recouping savings in other areas such as more productive plant operations or cheaper handling, storage/transportation. This is known as a *total systems approach to packaging optimisation* (Melis, 1989).

1.7.1 The packaging design and development framework

The framework presented in Table 1.4 ideally models the information requirements for packaging design and development. It considers all the tasks a pack has to perform during production and in distribution from the producer to the consumer, taking into account the effect on the environment.

Each of the aspects listed in Table 1.4 is discussed and a checklist of factors for each aspect presented. The market selected for discussion here is the

Table 1.4 The packaging design and development framework (developed from Paine, 1981)

Product needs
Distribution needs and wants
Packaging materials, machinery and production processes
Consumer needs and wants
Market needs and wants
Environmental performance

multiple retail market that dominates the food supply system in the UK grocery trade.

1.7.1.1 Product needs

The product and its package should be considered together i.e. *the total product concept*. A thorough understanding of a product's characteristics, the intrinsic mechanism(s) by which it can deteriorate, its fragility in distribution and possible interactions with packaging materials – i.e. compatibility – is essential to the design and development of appropriate packaging. These characteristics concern the physical, chemical, biochemical and microbiological nature of the product (see Table 1.5). The greater the value of the product, the higher is the likely investment in packaging to limit product damage or spoilage i.e. there is an optimum level of packaging.

1.7.1.2 Distribution needs and wants of packaging

A thorough understanding of the distribution system is fundamental for designing cost-effective packaging that provides the appropriate degree of protection to the product and is acceptable to the user(s). Distribution may be defined as the journey of the pack from the point of filling to the point of end use. In some instances, this definition may be extended to include packaging reuse, waste recovery and disposal. The three distribution environments are climatic, physical and biological (Robertson, 1990). Failure to properly consider these distribution environments

Table 1.5 Product needs

Nature of the product	
Physical nature	Gas, viscous liquid, solid blocks, granules, free-flowing powders, emulsions, pastes etc.
Chemical or biochemical nature	Ingredients, chemical composition, nutritional value, corrosive, sticky, volatile, perishable, odorous etc.
Dimensions	Size and shape
Volume, weight & density	Method of fill, dispense, accuracy, legal obligation etc.
Damage sensitivity	Mechanical strength properties or fragility/weaknesses
Product deterioration: Intrinsic mechanism(s) including changes in	
Organoleptic qualities	Taste, smell, colour, sound and texture
Chemical breakdown	For example, vitamin C breakdown in canned guavas
Chemical changes	For example, staling of bread
Biochemical changes	For example, enzymatic, respiration
Microbiological status	For example, bacterial count
Product shelf life requirement	
Average shelf life needed	
Use-life needed	
Technical shelf life	For example, is migration within legal limits?

will result in poorly designed packages, increased costs, consumer complaints and even avoidance by the customer.

Climatic environment is the environment that can cause damage to the product as a result of gases, water and water vapour, light (particularly UV), dust, pressure and the effects of heat or cold. The appropriate application of technology will help prevent or delay such deleterious effects during processing, distribution and storage (see Table 1.6).

Physical environment is the environment in which physical damage can be caused to the product during warehouse storage and distribution that may involve one or more modes of transportation (road, rail, sea or air) and a variety of handling operations (pallet movement, case opening, order picking etc.). These movements subject packs to a range of mechanical hazards such as impacts, vibrations, compression, piercing, puncturing etc. (see Table 1.7). In general, the more break-bulk stages there are, the greater is the opportunity for manual handling and the greater is the risk of product damage due to drops. In the retail environment, the ideal is a through-distribution merchandising unit – for example, the roll cage for cartons of fresh pasteurised milk.

Biological environment is the environment in which the package interacts with pests – such as rodents, birds, mites and insects – and microbes. For pests,

Table 1.6 The climatic environment

Protection requirement against the climatic environment includes:	
High/low temperature	Small or extreme variations
Moisture	Ingress or egress
Relative humidity	Condensation, moisture loss or gain
Light	Visible, infra-red and UV
Gases and vapour	Ingress/egress: oxygen, moisture etc.
Volatiles and odours	Ingress or egress – aromas, taints
Liquid moisture	For example, corrosion due to salt laden sea spray
Low pressure	External pressure/internal pack pressure variation due to change in altitude or aircraft pressurisation failure
Dust	Exposure to wind driven particles of sand, grit etc.

Table 1.7 The physical environment

Protection against mechanical hazards of storage and transportation by	
Shocks	Vertical and horizontal impacts, e.g. from drops, falls, throwing
Vibration	Low frequency vibrations from interactions of road or rail surfaces with vehicle suspension and engines; handling equipment; machinery vibration on ships
	High frequency aerodynamic vibration on aircraft
Compression/crushing	Dynamic or static loading; duration of stacking; restraint etc.
Abrasion	Contact with rough surfaces
Puncture	Contact with sharp objects, e.g. hooks
Racking or deformation	Uneven support due to poor floors, pallet design, pallet support
Tearing	Wrong method of handling

an understanding of their survival needs, sensory perceptions, strength, capabilities and limitations is required. For microbes, an understanding of microbiology and methods of preservation is necessary (see Table 1.8).

Other factors that need to be considered when designing packaging for distribution purposes include, convenience in storage and display, ease of handling, clearly identifiable and secure. There are trade-offs among these factors. These trade-offs concern the product and distribution system itself. For distributors, the package is the product and they need characteristics that help the distribution process (see Table 1.9). Any change in distribution requirements for certain products affects the total performance of the pack.

Identifying the optimum design of a packaging system requires a cost–benefit trade-off analysis of the performance of the three levels of packaging:

- primary pack: in direct contact with the food or beverage, e.g. bottle and cap, carton
- secondary or transit package: contains and collates primary packs – for example, a shrink-wrapped corrugated fibreboard tray or case
- tertiary package, e.g. pallet, roll cage, stretch-wrap.

An example is the multi-pack made from solid unbleached board (unbleached sulphate or Kraft board) used to collate 12 cans of beer. It can offer benefits such as enhanced promotional capability, more effective use of graphics, better shelf display appearance (no discarded trays), significant saving in board usage, increased primary package protection, better print flexibility during production, improved handling efficiency in retail operations (for example, faster shelf fill), tamper evidence, stackability, ease of handling by the consumer, faster product scanning at the store retail checkout, thereby improving store efficiency and/or customer service.

In terms of the physical nature of a product, it is generally not presented to the distribution function in its primary form, but in the form of a package or unit load. These two elements are relevant to any discussion concerned with the relationship of the product and its package. The physical characteristics of a product, any specific packaging requirements and the type of unit load are

Table 1.8 The biological environment

Microbes	Bacteria, fungi, moulds, yeasts and viruses
Pests	Rodents, insects, mites and birds

Table 1.9 Special packaging features for distribution to enable:

Ease of distribution: handling, stocking and shipment
Protection against soiling, stains, leaks, paint flakes, grease or oil and polluted water
Security in distribution for protection against pilferage, tampering and counterfeiting
Protection from contamination or leakage of material from adjacent packs

all-important factors in the trade-off with other elements of distribution when trying to seek least cost systems at given service levels (Rushton & Oxley, 1989) For example, individual two-pint cartons of milk may be assembled in shrink-wrapped collations of eight cartons, which in turn are loaded onto pallets, stretch-wrapped and trans-shipped on lorries capable of carrying a given number of pallet loads. At the dairy depot, the shrink-wrapped multi-packs may be order picked for onward delivery to small shops. In the case of large retail stores, the individual cartons of milk may be automatically loaded at the dairy into roll cages that are delivered to the retailer's merchandising cabinet display area without an intermediate break-bulk stage.

1.7.1.3 Packaging materials, machinery and production processes
Packaging is constantly changing with the introduction of new materials, technology and processes. These may be due to the need for improved product quality, productivity, logistics service, environmental performance and profitability. A change in packaging materials, however, may have implications for consumer acceptance. The aim is a fitness for purpose approach to packaging design and development that involves selection of the most appropriate materials, machinery and production processes for safe, environmentally sound and cost effective performance of the packaging system.

For example, there is the case of a packaging innovation for a well-known brand of a milk chocolate covered wafer biscuit. The aluminium foil wrap and printed-paper label band were replaced by a printed and coated oriented polypropylene (OPP) film flow-wrap with good gas and moisture barrier properties. Significant cost savings in pack materials and production operations were achieved. For example, only one wrapping operation is now required instead of the two previously used, and production speeds are much higher on account of the high tensile properties of OPP. There is also a lower risk of damage to the plastic wrapper in distribution and a net environmental benefit from using minimal material and energy resource. However, initial consumer research revealed a degree of resistance to this packaging change by those consumers who enjoyed the traditional ceremony of carefully unwrapping the foil pack and their ability to snap-off bars through the foil. The company promoted the new pack to the consumer on the basis of product freshness and the offer of a free extra bar.

Some key properties of the main packaging media are listed in Tables 1.10, 1.11, 1.12 and 1.13, though it should be remembered that, in the majority of primary packaging applications, they are used in combination with each other in order to best exploit their functional and/or aesthetic properties.

Most packaging operations in food manufacturing businesses are automatic or semi-automatic operations. Such operations require packaging materials that can run effectively and efficiently on machinery. Packaging

Table 1.10 Key properties of glass

Inert with respect to foods
Transparent to light and may be coloured
Impermeable to gases and vapours
Rigid
Can be easily returned and reused
Brittle and breakable
Needs a separate closure
Widely in use for both single and multi-trip packaging

Table 1.11 Key properties of tinplate and aluminium

Rigid material with a high density for steel and a low density for aluminium
Good tensile strength
An excellent barrier to light, liquids and foods
Needs closures, seams and crimps to form packs
Used in many packaging applications: food and beverage cans, aerosols, tubes, trays and drums
Can react with product causing dissolution of the metal

Table 1.12 Key properties of paper and paperboard

Low-density materials
Poor barriers to light without coatings or laminations
Poor barriers to liquids, gases and vapours unless they are coated, laminated or wrapped
Good stiffness
Can be grease resistant
Absorbent to liquids and moisture vapour
Can be creased, folded and glued
Tear easily
Not brittle, but not so high in tensile as metal
Excellent substrates for inexpensive printing

Table 1.13 Key properties of plastics

Wide range of barrier properties
Permeable to gases and vapours to varying degrees
Low density materials with a wide range of physical and optical properties
Usually have low stiffness
Tensile and tear strengths are variable
Can be transparent
Functional over a wide range of temperatures depending on the type of plastic
Flexible and, in certain cases, can be creased

needs to be of the specified dimensions, type and format within specified tolerances. The properties of the material will need to take account of the requirements of the packing and food processing operations. They will, therefore, need to have the required properties such as tensile strength and stiffness, appropriate for each container and type of material. For example, a horizontal form/fill/seal machine producing flow wrapped product will require roll stock film of a particular width and core diameter, with a heat or cold sealing layer of a particular plastic material of a defined gauge, and film surfaces possessing appropriate frictional, anti-static and anti-blocking properties to provide optimum machine performance.

Packaging machinery is set up to run with a particular type of packaging material and even minor changes in the material can lead to problems with machine performance. The introduction of new packaging materials and new designs must be managed with care. Materials should be selected after machine trials have shown that the required machine efficiency and productivity can be realised. New designs may require minor or major machine modification that will add direct costs in retooled parts. Indirect costs may result from machine downtime, prolonged changeover times and additional training costs for operators. Design changes in primary packs can have a knock-on effect on secondary packs and volume (cube) efficiencies during distribution and storage that results in height and diameter modifications. For example, a minor change in container profile can impact on machine operations from depalletising through conveying, rinsing, filling, sealing, labelling, casing and palletising. Depalletisers will need adjustment to cope with the new profile of containers. Conveyor guide rails may require resetting. Filler and labeller in-feed and out-feed star-wheels spacing screws may need replacing or modification. Fill head height may require adjustment and new filler tubes and cups may be required. Closure diameter may be affected having an effect on sealer heads that might necessitate adjustment or modification. New labels may be needed which will require modifications and possibly new components such as label pads and pickers. Casing machines may need readjustment to match the new position of containers. A redesigned case may be required and a new pallet stacking plan needed to optimise pallet stability.

The direct costs of new package design and machine modification and the indirect costs of reduced productivity prior to packaging lines settling down can be significant. It is important to bring machine and material suppliers into the design project and keep line operations informed at all stages of implementation.

Packaging machinery has developed into a wide range of equipment and integrated systems, to achieve a complete range of operational, filling and sealing techniques steered by computerised micro-electronic systems. Technical considerations in packaging materials, machinery and production processes are listed in Table 1.14.

1.7.1.4 Consumer needs and wants of packaging
The overall implications of social and economic trends relating to nutrition, diet and health can be summarised concisely as quality, information, convenience, variety,

Table 1.14 Packaging materials, machinery and production processes

Product/packaging compatibility
Identify any packaging material incompatibilities, e.g. migration and environmental stress cracking of plastics
Is there a need to be compatible during all conditions of distribution and use?
Must the package allow gaseous exchange? For example, to allow respiration of fruits and vegetables

Method of processing the product either in the package or independent of it

Elevated thermal treatment	E.g. Retort sterilisation and pasteurisation, cooking, hot filling, drying, blanching, UHT aseptic, ohmic heating, microwave processing
Low temperature treatments	Freezing, chilling and cooling
Gas change or flush	Modified atmosphere gassing
Removal of air	Vacuumising
Chemical	Smoking, sugaring, salting, curing, pickling etc.
Fermentation	E.g. Bacterial fermentation of carbohydrates for yoghurt production
Irradiation	E.g. Gamma rays to kill pathogens in poultry, herbs and spices

Others: Electron beam pasteurisation and sterilisation, gas sterilisation, high pressure processing and membrane processing

Closure performance
Does the seal need to provide the same degree of integrity as the packaging materials?
Re-closure requirement to protect or contain unused portion?
Degree of protection required against leakage or sifting?
Degree of seal strength and type of seal testing method employed?
Application torque and opening torque requirement of caps and closures

Performance requirements of packaging in production may concern
Machinery for container forming
Materials handling
Filling, check-weighing and metal detection
Sealing, capping or seaming
Food processing treatments
Labelling/coding
Casing
Shrink-wrapping; stretch-wrapping
Palletisation
Labour requirements

---

product availability, health, safety and the environment. Consequently, the food processing and packaging systems employed need to be continuously fine-tuned to meet the balance of consumer needs in particular product areas (see Table 1.15).

A branded product is a product sold carrying the product manufacturer's or retailer's label and generally used by purchasers as a guide in assessing quality. Sometimes, the qualities of competing branded products are almost indistinguishable and it is packaging which makes the sale. An interesting or visually attractive pack can give the crucial marketing edge and persuade the impulsive consumer. Packaging should, however, accurately reflect product quality/brand values in order to avoid consumer disappointment, encourage repeat purchase and build brand loyalty. Ideally, the product should exceed customer expectations.

Table 1.15 Consumer needs and wants of packaging

Quality	Processing and packaging for flavour, nutrition, texture, colour, freshness, acceptability etc.
Information	Product information, legibility, brand, use etc.
Convenience	Ease of access, opening and disposal; shelf life, microwaveable etc.
Product availability	Product available at all times
Variety	A wide range of products in variety of pack sizes, designs and pack types
Health	E.g. Enables the provision of extended or long shelf life foods, without the use of preservatives
Safety	The prevention of product contamination and tampering
Environment	Environmental compatibility

Packaging is critical to a consumer's first impression of a product, communicating *desirability, acceptability, healthy eating image* etc. Food is available in a wide range of product and pack combinations that convey their own *processed image* perception to the consumer e.g. freshly packed/prepared, chilled, frozen, ultra-heat treated (UHT) aseptic, in-can sterilised and dried products.

One of the most important quality attributes of food, affecting human sensory perception, is its flavour, i.e. taste and smell. Flavour can be significantly degraded by processing and/or extended storage. Other quality attributes that may also be affected include colour, texture and nutritional content. The quality of a food depends not only on the quality of raw ingredients, additives, methods of processing and packaging, but also on distribution and storage conditions encountered during its expected shelf life. Increasing competition amongst food producers, retailers and packaging suppliers; and quality audits of suppliers have resulted in significant improvements in food quality as well as a dramatic increase in the choice of packaged food. These improvements have also been aided by tighter temperature control in the cold chain and a more discerning consumer.

One definition of shelf life is: *the time during which a combination of food processing and packaging can maintain satisfactory eating quality under the particular system by which the food is distributed in the containers and the conditions at the point of sale*. Shelf life can be used as a marketing tool for promoting the concept of *freshness*. Extended or long shelf life products also provide the consumer and/or retailer with the time convenience of product use as well as a reduced risk of food wastage. The subject of *Packaged product quality and shelf life* is discussed in detail in Chapter 3.

Packaging provides the consumer with important information about the product and, in many cases, use of the pack and/or product. These include facts such as weight, volume, ingredients, the manufacturer's details, nutritional value, cooking and opening instructions. In addition to legal guidelines on the minimum size of lettering and numbers, there are definitions for the various types of product. Consumers are seeking more detailed information about products and, at the same time, many labels have become multilingual. Legibility of labels is an issue for the visually impaired and this is likely to become more important with an increasingly elderly population.

A major driver of food choice and packaging innovation is the consumer demand for convenience. There are many convenience attributes offered by modern packaging. These include ease of access and opening, disposal and handling, product visibility, resealability, microwaveability, prolonged shelf life etc. Demographic trends in the age profile of the UK and other advanced economies reveal a declining birth rate and rapid growth of a relatively affluent elderly population. They, along with a more demanding young consumer, will require and expect improved pack functionality, such as ease of pack opening (The Institute of Grocery Distribution, IGD).

There is a high cost to supplying and servicing the retailer's shelf. Failure to stock a sufficient variety of product or replenish stock in time, especially for staple foods such as fresh milk, can lead to customer dissatisfaction and defection to a competitor's store, where product availability is assured. Modern distribution and packaging systems allow consumers to buy food when and where they want them. Consumer choice has expanded dramatically in recent years. In the UK, for example, between the 1960s and 1990s the number of product lines in the average supermarket rose from around 2000 to over 18 000 (INCPEN).

Since the 1970s, food health and safety have become increasingly major concerns and drivers of food choices. Media attention has alerted consumers to a range of issues such as the use of chemical additives and food contamination incidents. These incidents have been both deliberate, by malicious tampering, and accidental, occurring during the production process. However, many consumers are not fully aware of the importance of packaging in maintaining food safety and quality. One effect has been the rapid introduction of tamper evident closures for many pre-packaged foods in order to not only protect the consumer but also the brand. Another impact has been to motivate consumers to give more attention to the criteria of freshness/shelf life, minimum processing and the product's origin (OECD).

Consumers have direct environmental impact through the way they purchase and the packaging waste they generate. Consumers purchase packaging as part of the product and, over the years, the weight of packaging has declined relative to that of the product contained. However, consumption patterns have generated larger volumes of packaging due to changing demographics and lifestyles. It is the volume of packaging rather than the weight of packaging that is attracting critical public attention. In addition, the trend toward increased pre-packaged foods and food service packaging has increased the amount of plastics packaging waste entering the solid waste stream.

One of the marketing tactics used by retailers and manufacturers is environmental compatibility. However, consumers are often confused or find it difficult to define what is environmentally responsible or *friendly* packaging. It is this lack of clarity that has so far prevented retailers and packaging companies from taking advantage to gain a competitive edge. Consumers need clear information and guidance on which of their actions make the most difference. Each sector of the packaging chain takes responsibility for explaining the functions and benefits of its own packaging. The manufacturers sell the virtues of their packaging to their customers, the product manufacturers, but relatively little of this specific information reaches the ultimate customer.

1.7.1.5 Multiple food retail market needs and wants
Packaging has been a key to the evolution of modern fast-moving consumer goods retailing that in turn has spurred on packaging developments to meet its requirements. The most significant development for the food packaging supply industries has been the emergence of large retail groups. These groups exert enormous influence and control over what is produced, how products are presented and how they are distributed to stores. The large retailers handle a major share of the packaged grocery market and exert considerable influence on food manufacturers and associated packaging suppliers. It is, therefore, important for packaging suppliers to be fully aware of market demand and respond quickly to changes. In addition, the concentration of buyer power at the retail level means that manufacturers may have to modify their distribution and packaging operations in response to structural changes in retailing.

Packaging for fast-moving consumer goods (f.m.c.g.) has been referred to as part of the food retail marketing mix and thus closely affects all the other marketing variables i.e. product, price, promotion, and place (Nickels & Jolsen, 1976; see Fig. 1.2).

The discussion on packaging in the multiple food retail environment may be considered in terms of its role in brand competition and retail logistics.

The role of packaging in brand competition. Packaging plays a vital role in food marketing representing a significant key to a brand's success or mere survival in a highly competitive marketplace. Packaging innovation and design are in the front line of competition between the brands of both major retailers and product manufacturers, having been driven in recent years by dramatic retail growth, intense industry competition and an increasingly demanding and sophisticated consumer. On an individual product/brand basis, success is dependent on the product manufacturer's rapid innovative response to major trends. One of the most effective ways to respond is through distinctive packaging, and this has become one key factor in the success of a brand. The retailers' own brand products compete intensely with manufacturers' brands in virtually every product category. Brand differentiation can be enhanced by innovative packaging designs that confer aesthetic and/or functional attributes. Table 1.16 lists factors influencing retail trade acceptability of packaging.

Packaging plays an important supporting role in projecting the image of the retailer to gain competitive advantage. The general purpose of the image of retailers' own brands is to support the overall message such as *high quality*, *healthy eating*, *freshness*, *environmentally aware* or *value for money*. For example, retailers who are keen to be seen as environmentally aware in part

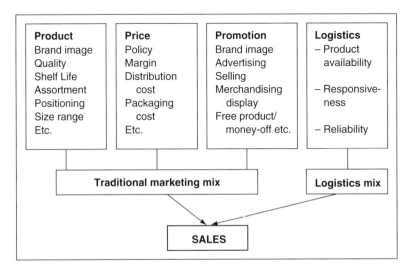

Figure 1.2 Model of the "Marketing Mix" for f.m.c.g. products (adapted from Darden, 1989).

Table 1.16 Factors influencing retail trade acceptability of f.m.c.g. packaging design may include

Sales appeal to target customer	Consumer profile: demographics and psychographics; product usage and perceptions
Retail competition	Local, regional retail formats and offerings
Retail environment	Lighting, aisle, shelf depth/spacing etc.
Brand competition	Retailer's own brands vs. manufacturers' brands
Brand image/positioning	Quality, price, value, healthy, modern, ethical etc.
Brand 'persona'	Combined design elements match the psychographic/demographic profile of the targeted customer
Brand impact/differentation	Aesthetic: colour, shape, material type etc. Functional: dispensing, pouring, opening etc.
Brand promotion	Character merchandising, money-off, free extra product, competitions etc.
Brand communication/presentation	Advertising, merchandising, labelling, typography, logos, symbols etc.
Consumer and brand protection	Tamper-evident/resistant features
Retail customer service	E.g. efficient bar code scanning and pack unitisation for fast service at check-out, hygiene, ease of access to pack units etc.
Retailer's margin	E.g. packaging design to increase display area on shelf for a minimum turnover of money per unit length of shelf space

drive the growing niche market for biodegradable and compostable packaging. They are using it as a point of communication with their customers.

Packaging is closely linked to advertising but it is far more focused than advertising because it presents the product to the consumer daily in the home and on the retail shelf. Merchandising displays that present the pack design in an attractive or interesting way and media advertising consistent with the pack's image also serve to promote the brand. The brand owner is frequently responsible for the merchandising operation. A key to promotional activities is through effective use of packaging and there exist many kinds of on-pack promotions such as *free extra product, money-off, special edition, new improved product, foil packed for freshness.*

Bar code scanning information linked to the use of retailers' loyalty card schemes has made a big impact on buying and marketing decision-making by retailers. Their task is to make better use of this information on consumer behaviour for promotional purposes and to build store brand loyalty. Retailers can also use this information to evaluate the effectiveness of new pack designs, on-pack promotions and the sales appeal of new products.

The role of packaging in multiple retail logistics. There are tight constraints on physical distribution and in-store merchandising. The retailer is receptive to packaging that reduces operating costs, increases inventory turnover, transforms to attractive merchandising displays – such as pre-assembled or easy-to-assemble aisle displays – and satisfies logistics service levels (reliability, responsiveness and product availability). For example, combined transit and point-of-sale packaging saves store labour through faster shelf loading, provides ease of access to product thereby obviating the need to use potentially dangerous unsafe cutting tools, and presents an opportunity for source reduction.

The total distribution cost affects the total volume of demand through its influence on price (McKinnon, 1989). For some fast-moving commodity type products, such as pasteurised milk, the cost of distribution and retail merchandising is usually a sizeable proportion of total product cost representing up to 50 per cent or more of the sales price. The cost of packaging materials and containers also adds slightly to the cost but design of the optimal packaging system can significantly reduce cost in the retail distribution chain. The development of global food supply chains has meant that many points of production have located further away from the points of consumption, often resulting in higher distribution cost.

Controlling distribution cost through improved operational efficiency in the supply chain is a key to competitive advantage for a retailer. The retailer must maximise operational efficiency in the distribution channel (West, 1989). The goal of distribution is to deliver the requisite level of service to customers at the least cost. The identification of the most cost-effective logistical packaging is becoming more crucial. Cost areas in distribution include storage, inventory, transport, administration and packaging. Storage, inventory, transport and store labour are major cost areas for the retailer while transport, storage and packaging are the main cost areas for the food manufacturer.

The efficiency of the multiple retail food supply chain relies on close communication between retailers, food manufacturers and packaging suppliers. It also relies on accurate order forecasting of likely demand. Massive investment in information technology has enabled closer integration of the supply chain and, through electronic data interchange (EDI), has ensured that stock moves to stores on a just-in-time (JIT) basis, and is sold well before the expiry date. The bar code is a code that allows the industry-wide identification of retail product units by means of a unique reference number, the major application being the electronic point of sale (EPoS) system at the retail checkout. The use of the bar code for identification of primary, secondary and tertiary packaging has enabled efficient distribution management and stock control.

Packaging is a means of ensuring the safe delivery of a product to the consumer at the right time in sound condition at optimum cost. The need for *safe delivery* of products at *the right time* and *optimum cost* demands cost-effective protective packaging that facilitates high logistics performance. The objective is to arrive at the optimum protection level that will meet the customer's service requirements at minimal expense. Most organisations build in an expected damage rate, which may be as high as 10% or more for inexpensive and readily replaceable items. Other issues that concern the distributor and store manager are cleanliness and hygiene.

The retailer's challenge is to make the most profitable use of shelf space. There is a need to maintain availability of a wide range of high turnover goods on the retail shelf with good shelf life or freshness. This often conflicts with the requirement to minimise product inventory in the retail distribution channel. Consequently, effective supply chain management and a well-integrated food packaging chain are necessary. For packaging material suppliers and converters, the implications of an increasing range of products, often involving shorter runs and lead times, are higher stock-holding of materials at extra cost, more frequent deliveries or developing just-in-time (JIT) techniques.

A key to competitive edge for a product manufacturer may depend on how quickly and effectively it responds to the retailer's need for:

- minimal stockholding
- high product turnover
- optimal level of fill on shelf
- efficient handling practice
- product integrity.

Increasingly, the manufacturer's packaging line must respond rapidly to promotional needs and shorter order lead times while ensuring minimal downtime. Packaging systems may need to be not only *reliable* but also *flexible*, to change the shape, volume, design and message with relative ease. Flexibility is equally important when there is regional marketing need, sudden seasonal demand or where there has been product failure in the marketplace.

Modular or standardised packaging systems enhance the logistical value of products. Modular systems allow pallets, roll containers and transport containers to be better utilised and enable packs to be bundled in trays and outer cases to fit supermarket shelves more efficiently. Outer packaging is being minimised for the direct transfer of the product from lorry to shelf display. The consequent requirement for increased quality of primary packaging presents innovation opportunities.

Food packaging and the shelf life issue can be strategically important in logistics because of the new distribution channels it can open up and their impact on industry structure. Any process that can extend shelf life – even by only one or two days – can bring about effective rationalisation in distribution and finished goods stock levels.

Retailers are striving to adopt packaging systems that integrate the requirements of improved environmental performance along with marketing and operational efficiency. Examples include the use of returnable plastic trays, refillable packs and the collection for recycling of corrugated cases with increased recycled fibre content. The significant moves towards centralised warehouses controlled by retailers, temperature-controlled delivery systems and just-in-time manufacture/delivery, all contribute to the potential for reducing the amount of packaging used. Table 1.17 lists some packaging characteristics valued in multiple retail logistics and distribution.

1.7.1.6 Environmental performance of packaging

An important strategic issue facing the food industry is the political and public pressure over the environment, particularly in relation to concerns over the amount of packaging and packaging waste. For over a decade, packaging and packaging waste have been the focus of attention from politicians as a result of pressure from the public, media and environmental pressure groups. The industry's options to improve the environmental performance of packaging are summarised in Table 1.18.

Packaging forms only a small part of the total solid waste (TSW) stream. In the UK, packaging waste represents approximately 6% of non-agricultural land-filled TSW which amounts to 120 million tonnes per year. This packaging waste comprises 3% of packaging from the household waste stream and 3% of packaging from commercial/industrial waste streams (INCPEN).

For many years, the packaging sectors of the food industry have made significant efforts, for both commercial and environmental reasons, to reduce the

Table 1.17 Packaging characteristics valued in multiple retail logistics and distribution include:

Meets the retailer's guidelines for acceptable transit packaging	Pallet type, size and security; pack stability, handling, opening features, bar code scan, ease of read, minimum pack damage, hygiene etc.
Minimise overall distribution cost	Storage, inventory, transport, store labour costs etc.
Facilitates logistics service requirements to be met	Product availability, reliability and responsiveness, e.g. efficient consumer response (ECR), just-in-time (JIT) delivery; and modular packaging systems for efficient distribution, retail shelf space utilisation and ease of merchandising operations
Returnable packaging systems	Waste minimisation, e.g. plastic tray systems for fruit, vegetables, meat and baked products
Shelf life extension	For example, perishable product availability, reduced spoilage, expansion of chilled product range, stock rationalisation and reduced inventory costs

Table 1.18 Environmental performance of packaging

Resource efficiency in pack manufacture	Energy, water, materials, source reduction and product redesign
Waste minimisation in production and distribution	Energy, water, materials, pollutants
Waste recovery	Reuse, recycle, composting, incineration with energy recovery; collection and sorting systems
Waste disposal	Landfill, incineration: pollutants; biodegradation, photo-degradation

amount of packaging and its environmental impact through light-weighting and packaging redesign. Examples of light-weighting are given below:

- Food cans – 50% lighter than 50 years ago
- Yoghurt pots – 60% lighter than 30 years ago
- PET bottles for carbonated drinks – 33% lighter than almost 30 years ago
- Drinks cartons – 16% lighter than 10 years ago

Source: INCPEN

Environmental policy on packaging should focus on resource-efficiency and not just waste and recycling. A full strategic response to the environmental issue would include:

- minimising energy and raw material use
- minimising the impact on the waste stream
- not causing environmental damage.

There are many alternative routes to achieve these objectives but the key possibility for gaining a competitive edge for a retailer or manufacturer is repositioning all products to satisfy a comprehensive environmental audit. The risk and uncertainty involves the relative strength of environmental concerns and other key consumer attributes.

There are management tools to reduce or compare the environmental impacts of industrial systems and these include life-cycle inventory analysis (LCI) and life-cycle assessment (LCA). The International Organization for Standardization (ISO) has responded to the need for an internationally recognised methodology for LCI and LCA.

1.7.2 Packaging specifications and standards

The packaging assessment must include a definition of the optimum quality standards and these standards should not be compromised by cost. Ideally, packaging supplier selection is a techno-commercial decision agreed during discussions between the purchasing function and packaging technologists. Buyers are becoming more discerning and now expect suppliers to have quality assurance schemes that are accredited by a third party. Widely used quality management systems are those based on ISO 9000.

Total quality management (TQM) is a technique that examines the overall quality image as perceived by suppliers and customers. There has been a change in focus from inspecting production outputs to monitoring process variations. TQM can be regarded as a source of competitive advantage, especially where quality is perceived to be a problem by the customer (Christopher *et al.*, 1993).

Quality assurance on production and packaging lines has been facilitated by the use of integrated computerised micro-electronic control systems that can detect a range of defects and automatically eject reject packs. For example, there are automatic check-weighers, metal detectors, fill-level sensors, pack-leak detectors, label detectors, pack-dimension sensors, light-transmission sensors and odour detectors.

A retailer's own brand products, delivered by the manufacturer to the retailer's centralised distribution system, may have packaging specifications that are tailored to meet the rigours of that retailer's specific system. The packaging specifications for a national brand, however, may require packaging that needs to provide a higher degree of protection due to the wide variation in storage conditions and distribution hazards experienced during a pack's delivery through several retailers' distribution channels. Thus, the rigours of the distribution system and a lack of control over it by food manufacturers often lead to specifications that generate extra cost and use up resources. Packaging specifications are geared to ensuring that a very high percentage of products arrive in pristine and safe condition, despite the rigours of the journey. However, this approach to packaging is in conflict with the significant pressure to minimise packaging.

The main testing methods for materials are available from a wide range of sources including ISO, American Society for Testing and Materials (ASTM), British Standards Institute (BSI), DIN etc.

1.8 Conclusion

This chapter has introduced the subject of packaging design and innovation which is driven by the need of industry to create and sustain competitive advantage, changes in consumer behaviour, industry's cost sensitivity, shorter product life cycles, new legislation, growing environmental pressures, availability of new materials and technological processes.

A framework has been used to analyse and describe key considerations in the design and development of packaging. It has emphasised the importance of an integrated approach to packaging by those involved in product design and development, materials and machinery development, production processes, logistics and distribution, quality assurance, marketing and purchasing. An integrated approach to packaging design and innovation requires consideration of key issues and values throughout the packaging chain.

Literature reviewed and sources of information

References

Christopher, M., Payne, A. and Ballantine, D. (1993) Relationship marketing: Bringing quality, customer service and marketing together (CIM Professional Development), 2nd edn, Butterworth-Heinemann.

Darden, W.R. (1989) The impact of logistics on the demand for mature industrial products, *European Journal of Marketing*, **23**(2), 47–57.

Davis Alec (1967) Package & print – the development of container and label design, Faber & Faber Ltd, London, ISBN 0571085903.

Hine, T. (1995) The total package – the evolution and secret meanings of boxes, bottles, cans and tubes, 1st edn, Little, Brown & company (Canada) Ltd, p. 62, ISBN 0–316–36480–0.

IGD, Grocery packaging openability and the elderly consumer – consumer concern programme.

INCPEN: 'UK code for good retail packaging'; Booklets: 'Consumer attitudes in packaging' 'Environmental impact of packaging in the UK food supply system'; 'Packaging reduction doing more with less'; 'Pack facts'. Refer website.

IOPP (1993) USA: 'Thirteen important benefits that packaging offers society' in '71 Reasons why packaging matters!' by the Institute of Packaging Professionals.

Maltenfort, G.G. (1988) Corrugated shipping containers – an engineering approach, Jelmar Publishing Co., Inc, New York, pp. 2–3, ISBN 0–9616302–1–3.

McKinnon, A.C. (1989) Physical distribution systems, London: Routledge, ISBN 0415004381.

Melis, T. (1989) Packaging Manager, Unilever NV, 'The European Challenge', NEDO/CBI Conference: 'Wrapping it up in the 1990s', London, 18 July.

Nickel, W.G. and Jolsen, M.A. (1976) Packaging – the fifth 'P' in the marketing mix? *S.A.M. Advanced Management Journal*, Winter, pp. 13–21.

OECD (2001) Sustainable consumption: Sector case study series – Household food consumption: trends, environmental impacts and policy responses.

Opie, R. (1989) Packaging source book, Macdonald & Co. (Publishers) Ltd, London, ISBN 0–356–17665–7.

Paine, F.A. (ed.) (1981) Fundamentals of packaging, revised edn, Institute of Packaging, ISBN 0950756709.
Paine, F.A. and Paine, H.Y. (1983) A handbook of food packaging, Blackie, Glasgow, ISBN 0–249–44164–0.
Robertson, G.L. (1990) Good and bad packaging. Who decides? in *Packaging – A major influence on competitive advantage in the 1990s* (ed. J.L. Gattorna). *International Journal of Physical Distribution & Logistics Management*, **20**(8), 5–20.
Robertson, G.L. (2002) The paper beverage carton: past and future, *Food Technology*, **56**(7), 46–52.
Rushton, A. and Oxley, J. (1989) The product, packaging and unitisation; in *Handbook of Logistics and Distribution Management*, Chapter 22, Kogan Page Ltd, London, pp. 283–298, ISBN 1–85091–443–5.
West, A. (1989) Managing distribution and change: the total distribution concept, New York, Wiley, ISBN 0471922609.

Further reading

Anon (1987) CAD/CAM system aids folding carton design, Boxboard Containers, **94**(9), 42.
Anthony, S. (1988) Toward better packaging design research, *Prep. Foods,* **157**(1), 111–112.
Bakker, M. (ed.) (1986) Wiley Encyclopaedia of packaging technology, New York, Wiley, ISBN 0 471 80940 3.
Barrasch, M.J. (1987) Process of developing a new package, *Brew Dig.*, **62**(7), 16–17.
Besk, F.C. (1986) Application of product fragility information in package design. *Packaging Tech.*, 10–12.
Croner, E. and Paine, F.A. (2002) Market motivators – the special worlds of packaging and marketing, CIM Publishing, ISBN 090213082X.
Cullis, R. and Dawson, A. (1994) Environmental impact management – achievements in transit, IGD, ISBN 1 898044 04 X.
DTI (1992) Managing product design projects – a management overview, HMSO.
DTI (1992) Managing packaging design projects – a management overview, HMSO.
Evans, S., Lettice, F.E. and Smart, P.K. (1997) Using concurrent engineering for better product development, in CIM Institute Report, Cranfield University.
Feder, M. (1987) Packaging magic. *Food Eng. Int.* **12**(9), 30–31, 33, 36–38, 41 (Du Pont Design Technologies).
Gattorna, J.L. (1996) Handbook of logistics and distribution management, 4th edn, Gower Pub. Co., ISBN 0566 09009 0.
Gomez, F. (1988) Ten tips for a successful package design project, *Food and Drug Packaging*, 24.
Griffin, R.C. and Sacharow, S. (1972) Principles of package development. Westport Conn., USA: AVI Publishing, ISBN 0 87055 1183.
Hindle, A. (1994) Environmental impact management – better for business, IGD, ISBN 0 90731675 1.
Hohmann, H.J. (1988) Filling characteristics and dosing accuracy of volumetric dosing devices in VFFS machines, *Pkg. Tech. and Science*, **1**(3), 123–137.
Jonson, G. (1996) LCA – a tool for measuring environmental performance. Leatherhead: Pira International, ISBN 12 85802 128 6.
Leonard, E.A. (1986) Economics of packaging, in *Wiley Encyclopaedia of Packaging Technology* (ed. M. Bapper), New York, Wiley, pp. 25–26, ISBN 0 471 80940 3.
Leonard, E.A. (1996) Packaging specifications, purchasing and quality control, 4th edn, Marcel Dekker, Inc., ISBN 0 8247 9755 8.
Levans, U.I. (1975) The effect of warehouse mishandling and stacking patterns on the compression strength of corrugated boxes, Tappi, **58**(8), 108–111.
Levy, Geoffrey, M. (ed.) (1993) Packaging in the environment, Kluwer Academic/Plenum, ISBN 0834213478.
Lox, F. (1992) Packaging and ecology, Pira International, ISBN 1 85802 0131.
Paine, F.A. (1989) Tamper evident packaging, Leatherhead: Pira International, ISBN 0 902799 24 X.
Paine, F.A. (ed.) (1987) Modern processing, packaging and distribution systems for food, Glasgow: Blackie, ISBN 0 216 92247 X.

Paine, F.A. (1987) in Herschdorfer, S.M., ed. *Quality control in the food industry*, **4**, London: Academic Press, ISBN 0123430046.
SEPTIC (1993) Guidelines for Life-Cycle Assessment. Brussels: Society of Environmental Toxicology and Chemistry (SEPTIC).
Styles, M.E.K., *et al*. (1971) The machine/material interface in flexible packaging production, *Packaging Technology*, May.
Ulrich, K.T. and Eppinger, S.D. (1995) Product design and development, McGraw-Hill Inc., ISBN 0–07–113742–4.

Other sources of information

American Society for Testing and Materials (ASTM), 100 Barr Harbor Drive, West Conshohocken, PA 19428-2959, USA. Website: www.astm.org.
British Standards Institute (BSI), 389 Chiswick High Road, London W4 4AL, UK. Website: www.bsi.co.uk.
Campden & Chorleywood Food Research Association (CCFRA), Chipping Campden, Gloucestershire GL55 6LD, UK. Website: www.campden.co.uk.
Deutsches Institut fur Normung e.V. (DIN), Burggrafenstrasse 6, 10787 Berlin, Germany. Website: www.din.de.
Food Science Australia – Commonwealth Scientific and Industrial Research Organisation (CSIRO), P.O. Box 52, North Ryde, NSW 1670, Australia. Website: www.dfst.csiro.au/.
Freight Transport Association (FTA), Hermes House, St John's Road, Tunbridge Wells, Kent N4 9UZ, UK. Website: www.fta.co.uk.
Industry Council for Packaging and the Environment (INCPEN) Suite 108, Sussex House, 6, The Forbury, Reading RG1 3EJ. Website: www.incpen.org.
Institute of Grocery Distribution (IGD), Letchmore Heath, Watford WD2 8DQ. Website: www.igd.com.
Institute of Logistics and Transport (IOLT), P.O. Box 5787 Corby, Northants NN17 4XQ. Website: www.iolt.org.uk.
Institute of Packaging (IoP), Sysonby Lodge Nottingham Road, Melton Mowbray, Leics., LE13 ONU, UK. Website: www.iop.co.uk.
Institute of Packaging Professionals (IOPP), 481 Carlisle Drive, Herndon, Virginia 22070, USA. Website: www.iopp.org.
International Air Transport Association (IATA), 800 Place Victoria, P.O. Box 113, Montreal, Quebec, H4Z 1M1, Canada. Website: www.iata.org.
International Maritime Organisation (IMO), 4 Albert Embankment, London SE1 7SR, UK. Website: www.imo.org.
International Organization for Standardization (ISO), 1, rue de Varembe, Case postale 56, CH-1211 Geneva 20, Switzerland. Website: www.iso.org.
International Trade Centre (ITC), Export Packaging Service Division of Trade Support Services, Palais des Nations, 1211 Geneva 10, Switzerland. Website: www.intracen.org/ep/.
Michigan State University, School of Packaging, East Lansing, MI 48824-1223, USA. Website: www.msu.edu/~sop/.
Netherlands Organisation for Applied Scientific Research (TNO), Nutrition and Food Research, Utrechtseweg 48, P.O. Box 360, 3700 AJ Zeist, The Netherlands. Website: www.tno.nl.
Organisation for Economic Co-operation and Development (OECD) Website: www.oecd.org.
Packaging Industry Research Association (PIRA), Randalls Road, Leatherhead, Surrey KT22 TRU. Website: www.pira.co.uk.
Processing & Packaging Machinery Association (PPMA), New Progress House, 34 Stafford Road, Wallington, Surrey SM6 9AA, UK. Website: www.ppma.co.uk.
Technical Association for the Worldwide Pulp, Paper and Converting Industry (TAPPI), P.O. Box 105113, Atlanta, GA 30348, USA. Website: www.tappi.org.

2 Food biodeterioration and methods of preservation
Gary S. Tucker

2.1 Introduction

Biodeterioration can be defined as the breakdown of food by agents of microbiological origin, either directly or indirectly from products of their metabolism. This chapter discusses how the agents of food biodeterioration operate and the commercial methods available to counteract these agents in order to produce safe and wholesome packaged foods. Contamination of packaged foods can arise from microbiological, chemical and physical sources. Microbiological sources can be present in foods prior to packaging or on surfaces of packaging materials; therefore, the shelf life will depend on their types and numbers, in addition to the hurdles to growth offered by the preservation techniques. Chemical sources often arise from enzymes released by microorganisms, in order to catalyse the rate at which food substrates decompose into smaller compounds that can move through cellular walls of a microorganism. Physical contamination may carry and introduce microorganisms which could cause food biodeterioration, but does not in itself play a role in biodeterioration; thus, physical contamination is not considered in this chapter.

The first section introduces the types of enzymes and microorganisms that are responsible for biodeterioration of foods. Growth conditions for microorganisms are presented, in addition to the factors that can be used to effect a reduction in their numbers. Bacteria and fungi warrant separate sub-sections because of the differing implications for introducing the organisms to the food and in methods for their destruction.

The second section outlines the principal methods of preserving packaged foods. Food preservation originated with traditional methods such as curing, salting and sugaring, which were developed before refrigeration was commonplace. Some of the major commercial developments in preservation were the introduction of canning and freezing processes for extending the useable life of fruits and vegetables. These methods still form a substantial part of the food preservation business, with thermal processing being part of many combination processes that are designed to present *hurdles* to microbial growth.

Food processors aim to produce foods that are safe and palatable. To achieve this aim pathogenic organisms must be eliminated or reduced to a safe level, spoilage organisms must be reduced and maintained at a low level and the eating quality of the food up to the time of consumption must be optimised. Some foods are processed in order to achieve commercial sterility of the product and its packaging, such that pathogenic bacteria are effectively eliminated. The definition of commercial sterility arose from the canning industry, where the primary aim is to eliminate pathogens and reduce the number of spoilage organisms capable of growth to an acceptably small number under intended storage conditions.

2.2 Agents of food biodeterioration

2.2.1 *Enzymes*

Enzymes are complex globular proteins found in living organisms and act as catalysts for speeding up the rate of biochemical reactions. Enzymes are naturally present in foods and can therefore potentially catalyse reactions which could lead to food biodeterioration. The action of enzymes can be used to beneficial effect by the food industry to produce food products and is, for example, commonly used in the manufacture of hard cheese. However, it is usually necessary to inactivate enzymes (i.e. denature the protein) present in food and on packaging surfaces using heat or chemical means in order to preserve and extend the shelf life of foods.

Fruit and vegetables being major sources of enzymes provide many examples of the nature and action of these agents of food spoilage. Enzymes associated with the deterioration of fruits and vegetables include peroxidase, lipoxygenase, chlorophyllase and catalase. Peroxidase is relatively heat-resistant and comprises a mixture of several enzymes of varying heat resistance. Some of these can be inactivated by mild heat treatments, whereas others require several minutes at sterilisation temperatures to effect a complete peroxidase inactivation. During the ripening of fruit, the activity of some enzymes (e.g. pectinesterase and polygalacturanase) increases and consequently causes a softening of the tissue. In potatoes, enzyme inhibitors play an important role in balancing the rate of biochemical reactions in relation to sugar accumulation. This has commercial importance for the storage and conditioning of potatoes prior to processing, where the presence of reducing sugars is undesirable because they can lead to enhanced browning reactions that cause discoloration. The active enzyme is invertase, which produces reducing sugars from sucrose, especially at low temperatures. At higher temperatures the inhibitor is active and reducing sugars are prevented from forming.

Another problem with fruits and vegetables is enzymic browning that results from damage or cutting of the surface and exposure to the air. This is due to the action of polyphenoloxidase, which in the presence of air oxidises phenolic constituents to indole quinone polymers. These reactions are particularly undesirable and several methods are used to prevent this type of browning.

These include the use of citric acid, malic acid or phosphoric acid to inactivate the enzyme, or alternatively preventing oxygen from coming into contact with the food by immersion in brine. Packing the fruit in an atmosphere that excludes air will reduce the extent of the browning reactions but can lead to quality problems and does not provide a viable solution.

Enzymes are also produced during microbial spoilage of foods and are often involved in the breakdown of texture. Many of the microorganisms that secrete enzymes are moulds; however, there are bacterial species (e.g. *Bacillus subtilis*, *Bacillus amyloliquefaciens* and *Bacillus licheniformis*) that produce heat-stable amylases. Amylase enzymes degrade starches, particularly naturally occurring starches, with the effect that the viscosity of the food is reduced as the macromolecular starch granules are broken down into their constituent sugars. Moulds are of particular concern for packaged foods because it is common for many species to produce spores as part of their reproductive cycle. Spores are easily carried in the air and can contaminate inside surfaces of exposed packaging. Other complex reactions can occur, for example *Rhizopus* species causes softening of canned fruits by producing heat-stable pectolytic enzymes that attack the pectins in the fruit. *Mucor piriformis* and *Rhizopus* species also cause the breakdown of texture in sulphite-treated strawberries as a result of similar production of enzymes. *Byssochlamys* species have been considered responsible for the breakdown in texture of canned foods, particularly strawberries, in which it is commonly found. This is a heat-resistant mould that requires temperatures in excess of 90°C for several minutes to adequately destroy it.

Failure to inactivate enzymes completely often shortens the storage life of packaged foods. This is rarely an issue with canned foods but is a factor to consider with frozen fruits and vegetables that receive only a blanching process prior to freezing. Blanching is intended to inactivate the majority of the enzymes, without imposing excessive thermal damage to the food, and hence it uses relatively mild temperatures (90–100°C) and short heating times (1–10 min). The renewed activity that often seems to be present in the thawed food after frozen storage is attributed to enzyme regeneration.

2.2.2 Microorganisms

The term microorganism includes all small living organisms that are not visible to the naked eye. They are found everywhere in the atmosphere, water, soil, plants and animals. Microorganisms can play a very important role in breaking down organic material. This action of degrading organic material is what food preservation techniques aim to counteract.

Temperature regulation is the most commonly used method to kill or control the number of microorganisms present within foods and on packaging material surfaces. Five categories of temperature-sensitive microorganisms are used to define the preferred temperature range for their growth.

- *Psychrotrophic* (cold tolerant), which can reproduce in chilled storage conditions, sometimes as low as 4°C. Having evolved to survive in extremes of cold, these are the easiest to destroy by heat.
- *Psychrophilic* (cold loving), which have an optimum growth temperature of 20°C.
- *Mesophilic* (medium range), which have an optimum growth temperature between 20 and 44°C. These are of greatest concern with packaged foods.
- *Thermophilic* (heat loving), which have an optimum growth temperature between 45 and 60°C. In general, these organisms are only of concern if packaged foods are produced or stored in temperate climates.
- *Thermoduric* (heat enduring), which can survive above 70°C, but cannot reproduce at these temperatures.

2.2.2.1 Bacteria

Bacteria are single-celled microorganisms that normally multiply by binary fission, that is they divide into two cells following a period of growth. If conditions are favourable for reproduction, one bacterium can divide into two by fission, so that after 11 hrs there can be more than 10 million cells ($>1 \times 10^{-7}$). This is a level where organoleptic spoilage of the food is apparent due to the production of off-flavours, unpleasant odours and slime or it can result in toxin release. There are four stages in bacterial growth (Fig. 2.1):

- lag phase, during which the bacteria are acclimatising to their environment, which can be several hours long;
- log phase, during which reproduction occurs logarithmically for the first few hours. Conditions for growth are ideal during this period;
- stationary phase, during which the bacteria's reproduction rate is cancelled by the death rate;
- mortality or decline phase, during which exhausted nutrient levels or the levels of toxic metabolites in the environment prevent reproduction, with the result that the bacteria gradually die off.

The simplest method of identifying bacteria is according to their appearance, which approximates to spherical (cocci), rod (bacilli) and spiral shapes. Cocci occur in different formulations, for example diplococci in pairs, staphylococci in clusters and streptococci in chains. Bacilli, mostly form chains. Cocci can vary in size between 0.4 and 1.5 μm, whereas bacilli can vary between 2 and 10 μm in length. Some cocci and many bacilli can move within their liquid environment by using flagella, which are similar in appearance to hairs. The most widely used method of identifying bacteria was introduced by the Danish bacteriologist Gram and is called Gram dyeing. Bacteria are divided into two main groups according to their Gram stain characteristics: red, Gram negative and blue, Gram positive.

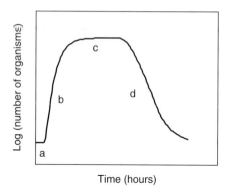

Figure 2.1 Typical growth curve for a bacterium (Phases: a – lag, b – log, c – stationary, d – mortality).

Bacteria require water, proteins, carbohydrates and lipids for growth. In addition, small quantities of vitamins and trace elements are needed to support and catalyse metabolism. Nutrients must be available in soluble form to aid transport through the cytoplasmic cell membrane. Large and complex organic molecules cannot pass through the membrane without first being broken down into smaller molecules. This is achieved by the bacteria releasing enzymes into the surrounding food that catalyse the break down of complex molecules into a form which can be used by the organism. An example of this is the release of amylase by various *Bacillus* species such as *subtilis*, *amyloliquefaciens* and *licheniformis*, which breaks down complex carbohydrates into simple sugars that can be absorbed through the cell membrane.

Water is essential for bacterial growth because it facilitates the transport of small molecules through the outer cytoplasmic membrane of the bacterial cell via osmotic pressure gradients. Bacteria require higher levels of available water than yeasts or moulds. At 20% available water, their growth is good, but it is limited when reduced to 10%, and at 5% there is no bacterial growth. The available water or water activity (a_w) is the amount of free water in a food and excludes moisture that is bound and unavailable for the microorganisms to use. Table 2.1 shows typical a_w values for a range of foods. Most bacteria cannot grow below an a_w of 0.91–0.94 The equilibrium relative humidity (ERH), defined as the ratio of the vapour pressure of food over that of water. ERH is calculated by $a_w \times 100$ and is expressed as a percentage. Therefore a food with an a_w of 0.90 has an ERH of 90%. Section 2.3.3 discusses methods of drying and control of water activity.

All bacteria need a supply of oxygen to oxidise their food in order to produce energy and for growth. Some bacteria obtain their oxygen directly from the air (aerobic bacteria), whereas others obtain oxygen from their food (anaerobic bacteria). The latter are usually killed when exposed to oxygen from the air. Some bacteria can be facultatively anaerobic, which means that they can consume oxygen from the air if present, but can also grow in its absence. Adjusting the atmosphere above a packaged food is used as a means to prevent bacterial growth, typically in combination with chilled storage as a further hurdle to growth. This allows foods to be manufactured with minimal or no heat processing, yet delivers an extended shelf life.

Light is not an essential requirement for bacterial growth because the cells do not synthesise food using light energy. Instead, light has a destructive effect on bacteria because of the ultraviolet (UV) component that causes chemical changes in the cell proteins. Bacteria prefer to grow in conditions where light is excluded. This effect is utilised by using UV light to sterilise bottled water where the limitation of transparency is not a restriction.

If the environmental conditions become adverse to the growth of certain bacterial species they have the ability to form a protective spore. Examples of conditions that can initiate spore formation are extremes of temperature, presence of adverse chemical environments (e.g. disinfectants) and low quantities of available moisture and low concentrations of nutrients. During spore formation, the vegetative part of the cell dies and will only form again if the environmental conditions become favourable. Spores do not metabolise and hence can lie dormant for years, in conditions that could not support the growth of the bacteria. This presents a challenge to the food processor, in that it is often necessary to kill both bacterial cells and spores that collect on the surface of exposed packaging. Destruction of spores requires more severe conditions of heat or disinfectant than that required to destroy vegetative bacterial cells. In some situations, the bacteria can develop resistance against the adverse conditions as a result of their rapid evolution time.

Clostridium botulinum is a spore former that is of concern to the heat processed foods sector because of the resistant nature of the resulting spore and the deadly toxin produced by the bacteria. *C. botulinum* spores will germinate if the pH is above 4.5 and there is an absence of oxygen (i.e. anaerobic condition). Since it is the most heat-resistant and lethal pathogen all sterilisation processes used to kill it will also destroy the other less heat-resistant pathogens.

Several bacteria need consideration when designing a packaging and processing line. Of primary concern from a public health perspective are those that produce toxins such as *C. botulinum*, *Listeria monocytogenes*, *Salmonella*, *Escherichia coli*, *Staphylococcus aureus*, *Bacillus cereus* and *Camplylobacter*. These can be controlled by the use of sterilising solutions and/or heat, with the aim to achieve the condition of commercial sterility for the packaged food. The most lethal of these is *C. botulinum*, which produces a toxin that attacks the nervous system. The industry is fully aware of the risks associated with this organism and has taken measures to prevent it from being present in food.

C. botulinum spores only germinate in anaerobic conditions where there is available moisture, nutrient and where the pH is over 4.5. Products with pH over 4.5 are often referred to as *low acid* foods, whereas products with pH of 4.5 and below are referred to as *high acid* foods. This critical pH limit is an important determinant as to whether heat-preserved foods receive a pasteurisation or sterilisation treatment. Sterilisation processes (typically in the range 115.5–135°C) using heat have greater cooking effects on product quality than the relatively mild heat treatment of pasteurisation processes (typically 75–105°C). It is important to ensure that spoilage organisms in a high-acid food do not cause a shift in pH to the low acid level, thereby allowing the potential development of *C. botulinum*. Pasteurisation often requires foods to be acidified prior to thermal treatment (e.g. pickled vegetables). Refer to the work of Adam and Dickinson (1959) for pH values of various products.

Growth of most strains of *C. botulinum* is inhibited at refrigeration temperatures; however, there are psychrotrophic strains that can grow at low temperatures and are increasingly giving rise for concern in foods. This is of concern with sous-vide and vacuum packed foods that only receive a mild heat treatment and rely heavily on precise control of chill temperatures to prevent out-growth. *Listeria* is another bacterium that can survive and grow at low temperatures, but fortunately it is killed by mild temperatures. The process used to achieve a 6-log reduction in *Listeria* is 70°C for 2 min, a process also applicable to *Salmonella* and *E. coli* (the group of bacteria referred to as aerobic pathogens).

Not all bacteria are pathogenic or the cause of food spoilage. Bacteria have been used to beneficial effect in food fermentation and preservation processes to extend the shelf life of certain foods. One example that has been exploited for many years is the deliberate introduction of lactic acid bacteria for the fermentation of milk to produce yoghurts. Lactic acid bacteria can be either bacilli or cocci and are facultatively anaerobic. Their energy source during growth is the milk sugar, lactose, and during fermentation this is converted to lactic acid. The result is that the level of acidity increases until a predetermined pH value is reached whereby the yoghurt is ready to be packaged. Lactic acid bacteria do not form spores, and the cells can be killed by heating to 70°C once the fermentation is complete. However, many types of yoghurt are marketed as having additional health attributes because they contain live lactobacilli that may have beneficial actions within the human digestive system. Section 2.3.5 provides further detail on fermentation methods of preservation.

2.2.2.2 *Fungi*

Fungi are a group of microorganisms that are found in nature on plants, animals and human beings. Different species of fungi vary a great deal in their structure and method of reproduction. Fungi may be single-celled, round or oval organisms or threadlike multi-celled structures. The threads may form a network, visible to the naked eye, in the form of mould as seen, for example, on foods such as bread and cheese. Fungi are sub-divided into yeasts and moulds.

Yeasts are single-cell organisms of spherical, elliptical or cylindrical shape. The size of yeast cells varies considerably, for example, Brewer's yeast, *Saccharomyces cerevisiae*, has a diameter of the order of 2–8 µm and a length of 3–15 µm. Some yeast cells of other species may be as large as 100 µm. The yeast cells normally reproduce by budding, which is an asexual process, although other methods of reproduction can also occur. During budding, a small bud develops on the cell wall of the parent cell, with the cytoplasm shared by parent and offspring before the bud is sealed off from the parent cell by a double wall. The new cell does not always separate from its parent but may remain attached, while the latter continues to form new buds. The new cell can also form fresh buds of its own. This can result in large clusters of cells being attached to each other. Some yeasts can form spores, although these are part of the reproductive cycle and are quite different to bacterial spores where the spore is formed as means to survive adverse conditions. Hence yeast spores are relatively easy to kill within foods by mild heat or on packaging surfaces by mild heat or sterilising solutions.

Moulds belong to a large category of multi-celled threadlike fungi. Moulds attach themselves to their food, or substrate, using long threads called hyphae. These are the vegetative part of the mould and grow straight up from the substrate. Like yeasts, moulds can also multiply by sexual or asexual reproduction. Reproduction is normally by asexual methods and results in the production of a large number of spores. Sexual reproduction tends to be in response to changing environmental conditions, although this is not necessarily the case. Spores produced by sexual reproduction are more resistant to adverse conditions and, like bacterial spores, can lie dormant for some time. One of the key sources of contamination of exposed packaging materials is from mould spores. This is because the spores are very small and light, are produced in huge numbers and are designed to be carried by air to new environments. Moulds introduced to the package as mould spores are the usual source of post-process contamination of foods. For example, a gooseberry jam hot-filled into glass jars and capped without heat pre-treatment of closures resulted in fungal growth on the headspace surface of the jam. Steam treatment of closures, jar inversion immediately after sealing or a pasteurisation process would have prevented spoilage.

One of the most important medicines for treating bacterial infections, penicillin, is derived from the *Penicillium* mould. The appearance of the spore-forming hyphae is that of a miniature green bush and is responsible for much of the green mould found in nature. The *Penicillium* family of moulds produces powerful lipases and proteases (fat- and protein-degrading enzymes) that make them key agents in the ripening of blue cheeses (*Penicillium roqueforti*) and Camembert (*Penicillium camemberti*). Another fungus, *Oospora lactis*, which displays characteristics of both yeasts and moulds, occurs on the surfaces of cultured milk as a white velvety coating. This organism is used for ripening soft cheeses (Tetra Laval, 1977).

Conditions for the growth of yeasts and moulds are similar to those for bacteria. They can survive at lower available water levels, which is why bread is at risk of mould spoilage but not of spoilage by bacteria which are unable to grow. Fungi also have a greater resistance to osmotic pressure than bacteria and can grow in many commercial jams and marmalades. Fungi present on packaging surfaces and in food will be killed by the heat process applied to the packaged food, typically of the order 85°C for 5 min (CCFRA, 1992), but once the jar is opened, airborne contamination from mould spores can occur. The moulds that grow in high sugar conditions do not form toxins, and so the *furry* growth colonies are unsightly but do not represent a public health risk.

Both yeasts and moulds are more tolerant to high acidity levels, with yeasts being able to grow between pH 3.0 and 7.5, and moulds between pH 2.0 and 8.5. Optimum pH for the growth of fungi tends to be in the pH range 4.5–5.0. Few bacteria can survive these low pH conditions, although spores of *Alicyclobacillus* strains have been reported to exhibit extreme acid and heat tolerance. These bacteria are referred to as acido-thermophiles and have caused spoilage problems in fruit juices where the pH can be as low as 3.0.

Generally, fungi are less tolerant to high temperatures than bacteria, an exception being ascospores from moulds such as *Byssochlamys fulva*. To produce a commercially sterile food in which these ascospores are the target organism would require extended heating at temperature above 90°C. Strawberries are a common source of *B. fulva*. Typically, yeast or mould cells are killed after only 5–10 min heating at 60°C. Temperature for optimum growth of fungi are normally between 20 and 30°C, which is the main cause for increases in spoilage outbreaks in food production during summertime.

Yeast cells are facultatively anaerobic and moulds almost exclusively aerobic. In the absence of oxygen, yeast cells break down sugar to alcohol and water, while in the presence of oxygen, sugar is broken down to carbon dioxide and water. The former reaction is used in the fermentation of alcoholic drinks; however, within the fermenting liquid, conditions lie between anaerobic and aerobic and hence alcohol and carbon dioxide are produced.

2.2.3 Non-enzymic biodeterioration

One further category of biodeterioration worthy of mentioning is that of non-enzymic browning. An important reaction in foods takes place between the sugar constituents and amine-type compounds, which results in progressive browning and the development of off-flavours. An example of foods in which this type of quality deterioration takes place is dehydrated foods, especially dried potato and vegetables, fruit juices, both dried and concentrated, and wine. These complex chemical reactions are known as Maillard reactions, named after the French chemist who first investigated the interaction of sugars and amines in 1912. Essentially, the aldehyde groupings of the reducing sugars react with the free amino groups of amino acids to form furfuraldehyde, pyruvaldehyde, acetol, diacetyl, hydroxydiacetyl and other sugar-degradation compounds that in turn react with amines to produce melanoide-type macro-molecules (brown pigments). Despite intensive research into this subject, the only successful way of inhibiting these reactions is by using sulphurous acid and sulphites. The levels of sulphur dioxide allowed in food products are strictly controlled by legislation and also by the amount that can be tolerated before the taste becomes unacceptable. In the case of dried products, these are sulphited immediately after or during blanching. The use of sulphites for this purpose does not involve the antimicrobial protection for which these compounds are used in other applications. It is important to note that sulphite treatment of any fruit and vegetables intended for canning needs to be very tightly monitored in order to avoid the risk of severe accelerated internal de-tinning.

2.3 Food preservation methods

Food preservation is aimed at extending the shelf life of foods. In most cases, it is the growth of either spoilage or disease-causing microorganisms that limits the length of time that a food can be kept, and most preservation techniques are primarily based on reducing or preventing this growth. However, there are other factors that limit shelf life, such as the action of naturally occurring enzymes within the food, or natural chemical reactions between the constituents of the food, and these also have to be taken into consideration.

There are many methods that can be used to preserve foods, and it is common for these to be used in combination in order to reduce the severity of any individual method. This is referred to as *hurdle* technology. The sections that follow describe the most important preservation methods to the food industry. The importance of each method can be difficult to determine because of the synergistic effects of combining methods. For example, acidification of peach syrup with citric acid screens out the more heat-resistant bacterial flora by lowering the pH, thereby reducing the pasteurisation treatment required and consequently the cooking effect on peach texture.

2.3.1 High temperature

Microorganisms and enzymes are the major causes of undesirable changes in foodstuffs; both are susceptible to heat, and appropriate heating regimes can be used to reduce, inhibit or destroy microbial and enzymic activity. The degree of heat processing required to produce a product of acceptable stability will depend on the nature of the food, its associated enzymes, the number and types of microorganisms, the conditions under which the processed food is stored and other preservation techniques used.

Manufacture of a heat-preserved packaged food can be broken down into two basic processes: (a) heating the food to reduce the numbers to an acceptably small statistical probability of pathogenic and spoilage organisms capable of growing under the intended storage conditions, and (b) sealing the food within an hermetic package to prevent re-infection. Preservation methods, such as traditional canning, seal the food in its package before the application of heat to the packaged food product, whereas other operations such as aseptic, cook-chill and cook-freeze, heat the food prior to dispensing into its pack.

2.3.1.1 Blanching

Blanching is a process designed to inactivate enzymes and is usually applied immediately prior to other thermal preservation processes either using high temperatures (e.g. thermal processing) or low temperatures (e.g. freezing). Blanching is not designed to reduce the microbial population on the surface of foods, but it will nevertheless reduce the numbers of organisms of lower heat resistance, such as yeasts, moulds and certain bacteria (e.g. *Listeria, Salmonella, E. coli*). Without a blanching step, the shelf life of, for example, frozen vegetables would be substantially reduced as a result of chemical breakdown during storage. Freezing does not totally eliminate reactions at the storage temperatures used in commercial and domestic practice, but it does slow down those that rely on ionic transport. If enzymes were present in foods during their frozen storage life, the chemical reactions that cause food spoilage could occur, albeit at a slow rate. Inactivating the enzymes, will prevent such reactions from occurring and shelf life is extended. In thermal processing of fruits and vegetables, the blanching step is similar, but its objective is to prevent further enzymic breakdown of the foods if delays occur prior to processing the foods. One further advantage of this treatment is that a proportion of the air enclosed within cellular material (e.g. in strawberries) is removed and in doing so the tendency for the fruit or vegetable to float is diminished.

2.3.1.2 Thermal processing

Can retorting or processing is a term that is still widely used in the food industry to describe a wide range of thermal processes where the food is heated within the pack to achieve a commercially sterile packaged food. The heating takes place in retorts that are basically batch-type or continuous hot water and/or steam-heated pressure cookers. The principal concept of food canning is to heat a food in a hermetically sealed container so that it is commercially sterile at ambient temperatures, in other words, so that no microbial growth can occur in the food under normal storage conditions at ambient temperature until the package is opened (Department of Health, 1994). Once the package is opened, the effects of canning will be lost, the food will need to be regarded as perishable and its shelf life will depend on the nature of the food itself. Various packaging materials are used in the canning process and includes tin-plate, glass, plastic pots, trays, bottles and pouches, and aluminium cans. Most canned foods are sterilised, but there is a growing trend towards applying additional hurdles to microbial growth that allow the processor to use a milder heat treatment referred to as pasteurisation (covered in more detail in Section 2.3.1.4).

The most heat-resistant pathogen that might survive the canning process of low-acid foods is *C. botulinum*. This bacterium can form heat-resistant spores under adverse conditions, which will germinate in the absence of oxygen and produce a highly potent toxin that causes a lethal condition known as *botulism* which can cause death within seven days. As the canning operation generates anaerobic conditions (i.e. no oxygen), all canning processes target this organism, if no other effective hurdle to its growth is present.

In practical terms, the thermal process applied must reduce the probability of a single spore surviving in a can of low-acid product to one in one million million (i.e. 1 in 10^{12}). This is called a *botulinum cook*, and the standard process is 3 min equivalent at 121.1°C, referred to as $F_0 3$. It is important to recognise that many factors inherent in the food may either allow a reduced level of processing to be applied (e.g. pasteurisation) or necessitate an increased degree of processing. For example, the latter may depend on factors such as desired product texture (e.g. softening of canned haricot beans in tomato sauce occurs beyond $F_0 3$), consistency of product formulation, retort operation control and, in cases of concern over thermophilic organisms, where the packaged food is intended for distribution and sale. The $F_0 3$ botulinum cook is not designed to reduce numbers of thermophilic organisms to a significant extent because these will not germinate from their spore forms during the shelf life. However, if the ambient conditions are likely to allow the growth of thermophilic organisms then a more severe process must be applied, of the order up to 15–20 min at 121.1°C. The thermophilic spoilage risk tends to be more of an issue with certain types of products with high initial counts in the raw ingredients (e.g. canned mushrooms in brine). $F_0 3$ is regarded as the absolute minimum, but most canned foods receive a much higher heat treatment ($F_0 6$ or more) to ensure a good level of safety against possible spoilage due to any uncertainties over variations in product and/or thermal process control.

In the traditional food canning process, the filled aluminium or tinplate cans are hermetically sealed, with can ends attached by a double seaming operation, and the cans heated in a batch retort. Heating is usually achieved by steam or water, and care must be taken to ensure that the heat penetrates to the slowest heating point in the can, so that no part of the food is left under-processed. At the same time, it is desirable not to overcook the food, as this will result in reduction in the quality of the food. A metal can is the ideal package from a processor's perspective because, relative to other packaging media, it offers the possibility of high production speeds, as well as good pack size flexibility, and the high compression strength of cans enables them to withstand physical abuse during processing and distribution. The metal can offers a high degree of reliability of pack integrity throughout the distribution chain as well as high space efficiency in warehouse storage due to the feasibility of high pallet stacking. In contrast to thicker plastic and glass packs that permit slower heat transfer, the thin-walled metal can is a good conductor of thermal energy from the heating and to the cooling media.

After heating, the food needs to be cooled, and it is vital that no post-process contamination occurs through the package seals or seams. Therefore, the seal integrity is vital, and there are strict regimes for container handling to minimise abuse to the seals. Cooling water must be of high quality microbiologically, and the containers must not be handled while wet, as this could potentially result in contamination, with the water acting as a conduit for any microorganisms present. There is evidence that cans do not create a hermetic (gas-tight) seal while they are hot, because of the expansion of the metal in the double seams. Good practice in canneries avoids manual handling of hot and wet cans to reduce the risk of post-process introduction of microbial contaminants into the container.

Thermal processing can be achieved using either batch retorts or continuous cooker-coolers. Batch retorts (see Fig. 2.2) operate with a variety of heating media, which include condensing steam, mixtures of steam and air, water

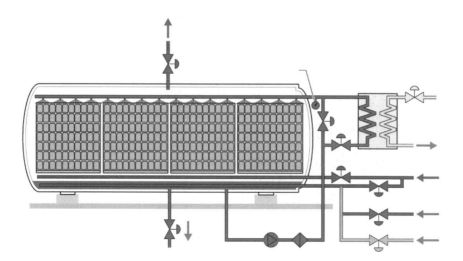

Figure 2.2 Example of a batch retort operating on the steam and air principle (courtesy of Lagarde).

immersion or water droplets that can be sprayed onto the packs. This offers considerable flexibility for changing the food and/or package. The retaining baskets or crates within the retort can be rotated in order to induce mixing inside the food by end-over-end agitation of the containers, and thus increase the rate of heat transfer to the thermal centre (i.e. slowest heating point) of the pack. Typical rotation speeds can vary between 2 and 30 rpm, depending on the strength of the pack and the convective nature of the food inside the pack. For example, a plastic pouch containing rice would be rotated slowly (e.g. 2–5 rpm) so that the delicate pack and its contents are not damaged. However, the rotation is sufficient to reduce the process times to an extent that economic gains are made and measurable quality benefits are achieved. Continuous cooker-coolers come in two types: reel and spiral, and hydrostatic. Both use the ability of the metal can to roll along a constrained pathway. A reel and spiral unit (see Fig. 2.3) forces the cans to rotate about their axis in the bottom third of a helical path where gravity maintains contact between the cans and the metal can guides. This is known as fast axial rotation (FAR) and delivers a very rapid rate of heat penetration to the can centre. Soups, sauces and foods that can move within the can are processed in reel and spiral cooker-coolers. Relative to stationary retort static heating, significant reductions in process times can be

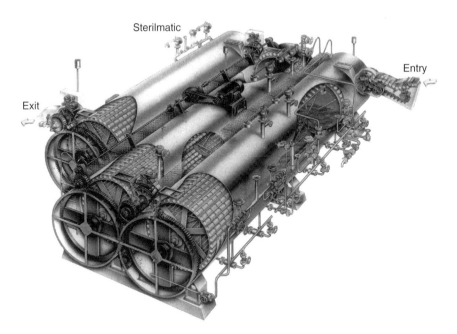

Figure 2.3 Reel and spiral cooker-cooler for processing cylindrical food cans (courtesy of FMC Foodtech).

achieved by utilising FAR. In general, a reel and spiral unit is the most efficient method of sterilising or pasteurising packaged foods with a liquid-to-solid ratio that allows heat from the circulated medium to penetrate more swiftly and evenly to the thermal centre. It is claimed that, in most cases, the quality of agitated products is by far superior to those processed by the static method. A hydrostatic cooker-cooler (see Fig. 2.4) does not invoke such dramatic rotation but instead carries the cans on carrier bars through various chambers, where they are pre-heated at 80–90°C, sterilised at 120–130°C, pre-cooled at 80–90°C and finally cooled below 40°C. The only rotation is a half-turn as the cans move between the chambers. Hydrostats are used for thicker foods where rotational forces cannot be utilised, for example solid petfoods and meat products.

Although metal, in the form of cans or trays, is the traditional packaging medium for thermally processed foods, other media including glass bottles/jars, plastic cans/bowls/pots/trays and flexible pouches can also be filled and processed in a similar way. These are increasing in popularity because of the consumer benefits they offer, e.g. the possibility of microwave and/or light transparency and handling characteristics. Each of these packs requires greater handling

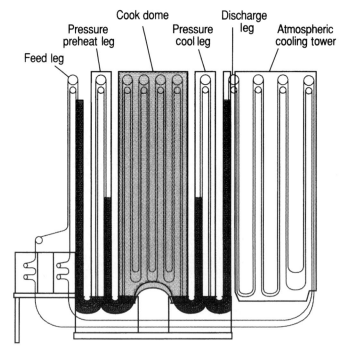

Figure 2.4 Hydrostatic cooker-cooler for processing food cans (courtesy of FMC Foodtech).

care during and after the process than metal containers because they are more easily damaged. Air overpressure, greater than that generated in the main by water vapour pressure within the pack, is needed during heating and cooling. For glass jars, this prevents the lids from being forced off, and it ensures that plastic packs retain their shape and size. Without air overpressure, flexible packs (constrained within perforated stainless steel racks) would expand, resulting in excessive stress on the seals and the possibility of packs bursting. Overpressure profile is one of the critical control points (CCPs) in a thermal process and must be established by viewing or measuring pack deflection inside the retort. The retort pressure is adjusted in order to maintain the original pack shape. If this has not been set up correctly, there will be chances of damage to the sealing area, which may not be visible to the naked eye but sufficient for a microorganism to enter the pack. A hole need only be a few microns in size to allow a bacterium to pass through. Since most packs are closed to form an internal vacuum, which helps eliminate oxidative deterioration of the food and reduce internal corrosion, the external pressure acts as a driving force moving water from outside of the sealing area to inside the pack. The first few minutes of cooling are also of particular concern, because of the substantial pressure swings in the retort that arise as steam condenses to leave a vacuum. This must be counteracted by the introduction of sterile air into the retort. A high proportion of pack damage is thought to occur during the onset of cooling, but modern computer-controlled retorts have reduced this risk.

2.3.1.3 *Continuous thermal processing (aseptic)*
The term UHT (ultra high temperature or ultra heat treatment) has been used to describe food preservation by in-line continuous thermal processing. In the aseptic packaging process UHT treatment is followed by packing in a sterilised container in a sterilised environment. The major difference here is that the package and food are sterilised separately and then the package is filled and sealed in a sterile environment, i.e. the aseptic form, fill and seal (FFS) process. The liquid food or beverage is sterilised or pasteurised in a continuous process in which it travels through a heat exchanger before being filled cold into the package. This technique is particularly suitable for liquid foods such as soups, fruit juices, milk and other liquid dairy products. Aseptic packaging has been carried out with metal cans, plastic pots and bottles, flexible packaging and foil-laminated paperboard cartons. Since air overpressure is not required in this process the range of suitable containers is greater than that with thermally processed foods.

Suitable heat exchangers for heating and cooling the foods are plate packs for thin liquids (see Fig. 2.5), tubular heat exchangers for medium viscosity foods (see Fig. 2.6) and scraped surface heat exchangers for high-viscosity foods that may contain particulates (see Fig. 2.7). It is assumed that the thermal process is delivered solely within a holding or residence section that usually

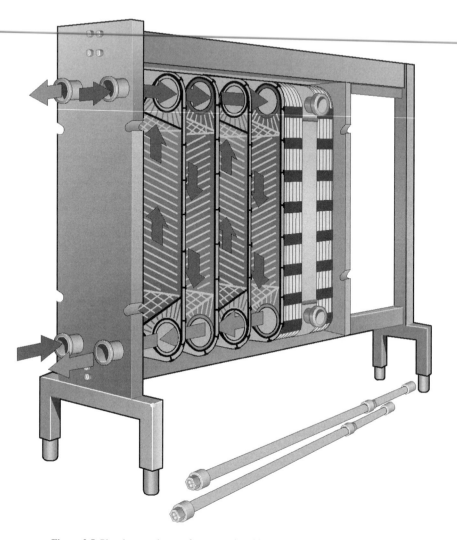

Figure 2.5 Plate heat exchanger for processing thin liquids (courtesy of Tetra Pak).

comprises a long length of tube. This does take into account the high temperature periods at the end of heating and at the start of cooling, therefore allows generous safety margins. Sterilisation values are calculated from the holding tube outlet temperature and the residence time taken from the measured flowrate. Control of these parameters must be to very high levels of accuracy because of the high temperatures and short times (HTST) used.

One potential advantage of UHT processing is that of enhanced food quality, as the problem of overcooking can be significantly reduced. Typical tempera-

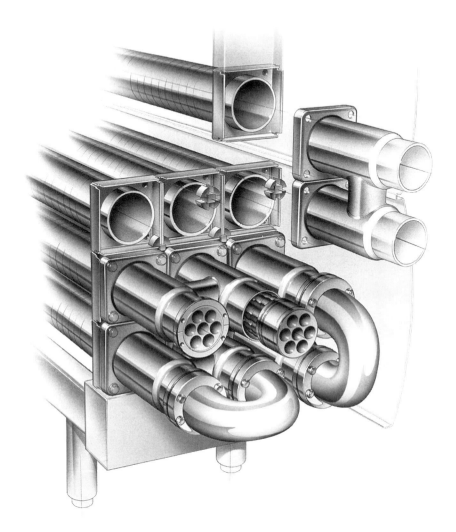

Figure 2.6 Tubular heat exchanger for processing medium viscosity liquids, including those with small particulates (courtesy of Tetra Pak).

tures and holding times in a UHT process are of the order 140°C for a few seconds. At these elevated temperatures, the rate of lethal kill to *C. botulinum* spores is substantial, whereas the rate of the cooking reactions is less significant. This is explained using the temperature sensitivity (z-value). For *C. botulinum* spores, this is taken as 10°C, which means that for every 10°C rise in the food temperature, the lethal effect increases by a factor of 10. Hence to achieve the minimum botulinum, cook of 3 min at 121.1°C or it would only require 0.03 min

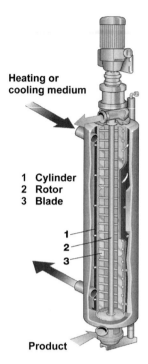

Figure 2.7 Scraped surface heat exchanger for processing medium and high viscosity liquids, including those with particulates (courtesy of Tetra Pak).

(1.8 s) at 141.1°C. Many cooking processes such as browning, on the other hand, are typified by higher z-values, in the range 25–33°C; therefore the step from 121.1 to 141.1°C would result in less than an one-log change in the cooking criterion. In effect, UHT allows extremely short process times with minimal detriment to quality. This benefit is used in the production of sterilised milk and cream that would end up too brown (caramelised) with associated off-flavours if it was processed in the pack, whereas UHT processing followed by aseptic filling can result in a more acceptable end product. In the case of fruit juice and fruit drinks, a benefit other than improved colour and flavour retention is increased residual content of vitamin C, a heat-sensitive substance.

Aseptic filling accounts for a high proportion of heat-processed foods, where high line speeds can be used. The packages are usually constructed within the sterile filling environment from reels of film. For example, most liquid foods and fruit juices/juice drinks, are packed in Tetra Brik Aseptic (TBA) foil-laminated paperboard cartons which are filled aseptically. Firstly, the liquid is pasteurised or sterilised using a plate heat or tube exchanger and the carton material sterilised using a combination of hydrogen peroxide and heat or UV light. The TBA system is a continuous process that involves the sterilised or pasteurised liquid being filled into a continuous tube of sterilised unreeled carton material and a heat seal made through the liquid prior to carton forming. The TBA system permits high production speeds to be achieved with a high assurance of seal integrity and normally there is no headspace, therefore it uses less packaging. A headspace can be induced if necessary – for example, by nitrogen bubble dispersion in chocolate milk prior to filling and sealing – such that the product can be thoroughly mixed by hand shaking the pack prior to consumption. The TBA system, however, imposes certain restrictions on the food types that can be filled, almost eliminating aseptic filling for foods containing discrete particulates that may not allow an effective seal to be formed. One possible solution is Rexam's Combi-bloc aseptic carton system that involves form-fill-seal of flat ready-made carton units and subsequent filling of the open-top carton. The Combi-bloc system has lower production speeds, but one of its main advantages is pack size flexibility. A further drawback to UHT processing of foods with particulates is that the particulates heat by conduction, which is slow relative to convection heating of a liquid component, and a typical UHT process takes only a few seconds. This is not sufficient time to sterilise the centre of the particulates, and so one solution is to lower the temperature in order to increase the residence time, which may negate the benefits of UHT processing. Ohmic heating, which relies on the electrical conductivity of particulates, offers another possible solution (refer to Section 2.3.7.2).

2.3.1.4 *Pasteurisation*

This is a heating regime (generally below 105°C) that primarily aims to achieve commercial sterility by virtue of additional factors that contribute towards preserving the food. The actual degree of heat process required for an effective pasteurisation will vary depending on the nature of the food and the types and numbers of microorganisms present. In certain cases, an extended pasteurisation treatment may be required to inactivate heat-resistant enzymes.

Milk is the most widely consumed pasteurised food in the UK, and the process was first introduced commercially in the UK during the 1930s, when a treatment of 63°C for 30 min was used. Modern milk pasteurisation uses an equivalent process of 72°C for 15 s.

Pasteurisation is used extensively in the production of many different types of food, including fruit products, pickled vegetables, jams and chilled ready meals. Food may be pasteurised in a sealed container (analogous to a canned food) or in a continuous process (analogous to an aseptic filling operation). It is important to note that pasteurised foods are not sterile and will usually rely on other preservative mechanisms to ensure their extended stability for the desired length of time. Chilled temperatures are often used, but some foods have a sufficiently high salt, sugar or acid content to render them stable at room temperature (e.g. pasteurised *canned* ham).

Whereas heat-preserved packaged foods for ambient storage (as described above) have shelf lives normally measured in years, cook-chill foods typically have shelf-lives measured in weeks or months, and in those cases where only chilling is the preservation factor they are significantly less than this.

Further detail on the hurdles to microbial growth that allow pasteurisation processes to be applied can be found in the following sections. Many of these hurdles are used in combination with heat to deliver a commercially sterile packaged food.

2.3.2 Low temperature

2.3.2.1 Freezing

Freezing of food does not render it sterile. Although the freezing process can reduce the levels of some susceptible microorganisms, this is not significant in the context of the overall microbial quality of the food. However, at commercial freezing temperatures (−18 to −24°C), all microbial activity is suspended and the length of time for which the food can be kept is dependent on other factors. It is important to note, however, that once a frozen food is defrosted, those viable microorganisms present will grow and multiply.

At freezing temperatures, enzymic activity may continue, albeit at a reduced rate, and over time could alter the organoleptic properties of the food. The potential problems with enzyme activity will depend on the particular food. For example, the sugar in peas is rapidly converted to starch once the pods are picked, and if not prevented, would result in an un-sweet product. For this and related reasons, vegetable products are frozen within hours of harvesting and are blanched before freezing to ensure that enzymes are inactivated. A typical blanching process involves heating to 90–95°C for a few minutes prior to rapid freezing. The rate of freezing is important to the quality of the food. Rapid freezing in blast freezers is desirable to prevent the formation of large ice crystals that will tend to adversely affect the texture of the food by disrupting cell integrity in fruits and vegetables or degrading the muscle proteins of meat, fish and poultry.

Apart from enzymic activity, there are many other chemical and physical changes which may limit the shelf life of frozen food; examples include fat oxidation and surface drying, both of which may occur over a period of months, depending on the food. The interaction of the food and its package is critical in reducing these undesirable changes.

Packaging of frozen foods uses a variety of materials and formats, including paper, plastic and metal. Unlike heat-processed foods, stored under ambient conditions, the requirement for packaging materials is less stringent. Migration of gases such as oxygen through the packaging material has less effect on the food because the chemical reactions do not occur at significant rates; therefore the need for gas barrier materials is less critical. Also, as the foods are frozen solid, which gives the pack greater rigidity, there is no need for the packages to be commercially sterile as in aseptic filling.

Storage life of frozen foods tends to be dictated more by consumer handling than the effectiveness of the freezing processes. Typical domestic freezers operate at temperatures much higher than those used for production and distribution. Repeated freeze–thaw cycles damage the food structure around the edges and promote chemical and physical breakdown. Ice cream is an example of a high-quality food that is manufactured to contain small (invisible) ice crystals within a complex matrix. However, the ice crystals grow in size as the food is abused, with the result that the smooth structure gradually breaks down and is replaced with a harder, coarse textured food with visible ice crystals.

2.3.2.2 Chilling and cooling

Chilling may be referred to as the process that lowers the food temperature to a safe storage temperature of between 0 and 5°C, whereas cooling is a more general term applied to the lowering of a food temperature. Chilled foods can potentially present a greater risk to public safety than frozen foods. Keeping products at a low temperature reduces the rate of microbiological and chemical deterioration of the food. In most processed chilled foods, it is the microbial growth that limits the shelf life; even the slow growth rates that occur under chilled conditions will eventually result in microbial levels that can affect the food or present a potential hazard (see Table 2.1). This microbial growth can result in the spoilage of the food (it may go putrid or cloudy or show the effects of fermentation), but pathogens, if present, may have the potential to grow and may show no noticeable signs of change in the food. Under the UK's Food Protection Act of 1990, a *use by* date must be declared on labels for packaged chilled dairy, meat, egg, seafood and poultry products which must be distributed, stored and retailed at 5°C or below.

To reduce microbial effects to a minimum, chilled prepared foods are usually given a pasteurisation heat process, sufficient to eliminate a variety of pathogens such as *Salmonella*, *Listeria* and *Escherichia coli* O157. A process equivalent

Table 2.1 Minimum growth temperatures for selected pathogens (taken from Betts, 1996)

Pathogenic microorganism	Minimum growth temperature (°C)
Bacillus cereus	4.0
Clostridium botulinum (psychrotrophic)	3.3
Escherichia coli O157 (and other VTEC)	7.0
Listeria monocytogenes	−0.4
Salmonella species	4.0
Staphylococcus aureus	6.7
Vibrio parahaemolyticus	5.0
Yersinia enterocolitica	−1.0

to 70°C for 2 min is generally considered to be sufficient, but the exact process given will depend on the nature of the food. Application of hurdles to growth such as pasteurisation and chilling can allow shelf lives of several days. If the pasteurisation regime is more severe, for example, 90°C for 10 min (CCFRA, 1992), it is possible to extend the shelf life to between 18–24 days or more. The exact length depends on the suitability of the food to support microbial growth, and it is common for a company to apply the same heat process to two different foods, yet the declared shelf life of one may be 14 days while the other may allow 20 days. The use of product ingredients with low microbial counts, ultra-clean handling and filling conditions in combination with sterilised packaging (i.e. near aseptic conditions) will serve to reduce the initial microbial loading of pasteurised product and thereby extend shelf life.

Chilling is also used to prolong the shelf life of many fresh fruits and vegetables. Here, low temperatures not only retard the growth of naturally occurring microorganisms (which might rot the food), but also slow down biochemical processes that continue after the food has been harvested. However, each individual fruit and vegetable has its own ideal storage temperature, and some are susceptible to chill injury. For example, storing bananas below 12°C will result in blackening of the skin. Many fresh fruits and vegetables are simply cooled to temperatures above 5°C.

Packaging for chilled foods shows a greater variety than for other preservation systems. This is because it is the microbiological growth within the food that limits the shelf life and not the interaction between the food and package. The package only has to survive a few days before the consumer uses it, which compares with frozen or canned foods where the pack has to offer protection to the food for up to 3 years. Thus, the barrier properties for short shelf life foods are less restrictive. A chilled food pack needs to be clean but not sterile, and this also opens up new packaging opportunities that cannot be realised with aseptic filling. Partial sterilisation of the open packs with sanitising solutions is sometimes used to reduce the microbial population, although it is more common for packs to receive just a water wash or air blast.

Exceptions to pack simplicity for chilled foods are found where hurdle approaches are used, for example, when using modified atmospheres or in-pack pasteurisation. These are discussed in detail in the relevant sections that follow.

2.3.3 Drying and water activity control

Microorganisms need water to grow. Reducing the amount of water in a food that is available to the microorganism is one way of slowing or preventing growth. Thus, dried foods and ingredients such as dried herbs and spices will not support microbial growth and provided they are stored under dry conditions, can have an expected shelf life of many months if not years. Many staple foods are available in dried forms (e.g. cereals, pulses and rice) and, provided they remain dry, will be edible for a long period of time. The shelf life of breakfast cereals is usually limited by texture changes caused by moisture ingress through the packaging, with the food losing its crispness and becoming soft. Selection of suitable packaging materials is therefore critical in extending the shelf life of dried foods. Laminated paperboard with a plastic moisture barrier, such as polyethylene, is a common pack format for dried foods such as pasta, fruit and breakfast cereals, although alternative pack formats include moisture barrier bags and pouches. Shelf life for dried foods can extend to several years.

Most dried foods achieve moisture levels that are low enough to prevent chemical reactions from occurring, and in doing so, chemical deterioration is removed as a factor that affects shelf life. The moisture content of foods is measured by ERH. It represents the ratio of the vapour pressure of food divided by that of pure water, and is given the symbol a_w water activity. As stated in an earlier section on microbial growth, most bacteria cannot grow below a_w levels of 0.91, yeasts cease growing at a_w levels of 0.85 and moulds at a_w levels of 0.81. The target a_w levels for dried foods are around 0.3, substantially below the value which supports microbial growth. Whether there is a killing effect at such low a_w levels is uncertain and probably depends on whether the microorganism can produce resistant spores in the time available when the moisture content is within growth limits. Rapid drying processes such as spray drying may not allow time for this to happen, but the traditional method of sun drying will involve a longer time as the food's moisture content is decreasing at a much slower rate, which may be sufficient for spore production.

Various drying methods can be used in the production of dried foods (see Table 2.2). Choice of the method and the packaging format are dependent on the food and its intended use. Each of the listed methods can reduce the a_w level close to 0.3 and thus eliminate microbial and enzymic breakdown reactions.

A traditional method for reducing a_w levels in foods is to use sugar to generate an osmotic pressure gradient. Some foods such as jams and marmalades may contain fairly high levels of water, but much of this is *tied up* or *bound* (i.e.

Table 2.2 Commercial drying methods with examples of food products

Drying method	Food products	Packaging format
Spray dryer	Powdered milk, coffee granules	Plastic bottles, glass jars, tinplate cans, multi-wall paper sacks
Freeze dryer	Granulated coffee	Glass jars
Perforated plate	Fruit, e.g. raisins, sultanas	Plastic film, laminated board
Fluidised bed	Peas	Cartonboard
Drum dryer	Breakfast cereals, flaked products	Plastic laminated cartonboard, cartonboard with plastic inner bag.
Sun	Tomatoes, meat	Glass jar, packed in oil to prevent contact with moisture

the food has a low water activity) by the sugar and pectin present in the jam and is not available for the microorganisms to use. As a result, traditional jams can be kept for many months without spoiling. Conversely, many low-sugar jams have to be refrigerated after opening, as the sugar levels are not sufficient to prevent microbial growth. Microorganisms present in the jam and on the packaging are destroyed during manufacture by applying a pasteurisation process, but once opened, airborne contamination from mould spores can be introduced. This is a good example of how the consumer-driven desire for foods with altered characteristics (i.e. jam with less sugar) has resulted in a food that has lost one of its major, original characteristics, a long-term stability at room temperature. Screw top glass jars and heat-sealed plastic pots are common packaging types for foods that use osmotic pressure to control the a_w levels. The key criterion for the chosen packaging is a high moisture barrier (and good heat resistance if hot filling) to prevent moisture ingress over an extended shelf life of up to 18 months, and in the case of glass jars, the top must also be re-sealable to provide an additional shelf life of several weeks after opening.

2.3.4 Chemical preservation

The addition of specific chemicals to foods to inhibit microbial growth and chemical reactions is a major method of preserving food. Antimicrobial additives (*preservatives*) probably receive the most attention, much of it being unjustly adverse. There are relatively few preservatives permitted for use in the UK and EU, and in many cases there are specific limits on how much can be used and in which foods. The use of some preservatives is limited to just a few types of food (e.g. nitrate and nitrite salts to specific meat, cheese and fish products). The two preservative types used most widely are sorbic and benzoic acids and their salts, and sulphur dioxide and its derivatives. There is currently a major consumer-led move to reduce the range of foods containing preservatives and the levels of preservatives actually used. This poses a significant problem for the food industry, as a reduction in preservative level necessitates either another preservation technique to be used (e.g. heating or freezing) or a significant reduction in shelf life. Both of these alternatives may result in a food of actual or perceived poorer quality.

In addition to preservatives, antioxidants are widely used to prevent chemical deterioration of foods; this includes the rancidity caused by oxidation of fats and the browning of cut vegetables caused by the formation of high molecular weight compounds due to the action of the enzyme polyphenol oxidase.

Introduction of the preservative to the packaged food can be achieved in one of two ways. By far the most common is to mix the preservative into the food prior to the package being filled and sealed. This is the method used in the production of soft drinks that require benzoate, metabisulphate and sorbate salts to inhibit microbial growth once the package (e.g. plastic bottles of juice concentrate) has been opened. Without the preservatives, yeasts and moulds would contaminate the juice, grow in the ambient storage conditions and very quickly cause it to spoil. That is why many packs, such as single-shot drinks bottles, do not possess a re-close facility. Some meat products, for example canned ham and tongue, have nitrite salts and/or nisin added before the thermal process is applied. Their function is to allow a reduced thermal process to be given (a sub-botulinum cook, e.g. $F_0 0.5$–1.5) but still ensuring product safety and avoiding excessive thermal degradation of the meat. The alternative method of introducing the preservative to the food is to integrate it into the packaging or introduce it as a component of the packaging; these are examples of *active packaging*. This is discussed in detail in Chapter 9. Costs for active packaging materials are considerably greater than that for conventional materials; therefore, these packs are almost exclusive to foods that can command higher retail prices. Preservatives included in packaging are called bacteriocins. The antimicrobial agents are slowly released to the food or the atmosphere above the food and will prevent microbial growth over the short shelf life. Extensions to shelf life of several days can be achieved by using active packaging, which has developed into a growth sector of considerable commercial value.

Unlike the relatively novel active packaging techniques, some forms of chemical preservation are well-established, traditional techniques, as outlined below.

2.3.4.1 Curing

Strictly speaking, curing actually means saving or preserving, and processes include sun drying, smoking and dry salting. However, curing now generally refers to the traditional process that relies on the combination of salt (sodium chloride), nitrate and nitrite to effect chemical preservation of the food, usually meat, but also to a lesser extent fish and cheese. The method of preservation relies on the available water for microbial growth being chemically or physically bound to the curing agent, and thus, not available to the microorganism. For example, salt achieves this by creating ionic bonds between the polarised hydrogen and hydroxide ions in water with the sodium and chloride ions from the salt.

In cured meat products, the salt has preservative and flavour effects, while nitrite also has preservative effects and contributes to the characteristic colour of these foods. Bacon, ham and gammon are cured pork products, and there are a number of variations on the curing technique. For example, the traditional Wiltshire curing of bacon (Ranken *et al.*, 1997) involves injection of brine into the pig carcass and immersion in a curing brine is which contains 24–25% salt, 0.5% nitrate and 0.1% nitrite. The curing brine is used from one batch to another is *topped up* between batches, and is a characteristic deep red colour due to the high concentration of accumulated protein.

Typical cured food products such as bacon, ham and fish are shrink-wrapped in plastic film and usually stored under chilled conditions. Vacuum packaging is common because this increases the shelf life by reducing the rate of oxidative damage.

2.3.4.2 Pickling
This commonly refers to the preservation of foods in acid or vinegar, although the term can occasionally be applied to salt preservation. Most food-poisoning bacteria (e.g. *C. botulinum*) stop growing at acidity levels below pH 4.5, the minimum level attained during the pickling process, although yeasts and moulds require a much higher degree of acidity (pH 1.5–2.3) to prevent their growth.

A number of vegetables are pickled in vinegar in the UK, such as beetroot, gherkins and cucumbers, onions, cabbage, walnuts and eggs. In some foods, the raw or cooked material is simply immersed in vinegar to effect preservation, but in others, additional processes such as pasteurisation are required to produce a palatable and safe end-product. For example, a typical process for pickled shredded beetroot would involve peeling of the beetroot, size grading and steam blanching (to inactivate the betanase enzyme which degrades the red pigment, betanin, in the presence of oxygen), shredding, filling the container with shredded beetroot and hot pickle brine, exhausting (to remove entrapped air), brine top-up to correct headspace level, steam flow closing and pasteurisation in a retort with water at 100°C.

Glass jars with tinplate regular twist-on/off (RTO) lug caps and fully internally lacquered tinplate cans are typical packs used for pickled and low pH foods. The package seal must not leak as pickle brines have a corrosive effect on the external surface of the tinplate cans and, in the case of glass jars, rusting of tinplate cap lugs onto the glass thread may prevent cap removal. Also, the absence of a vacuum or poor vacuum levels in bottled pickles may serve to promote corrosion beneath the internal lacquer coating (i.e. under-film attack) and compound lining of caps, thereby effecting leakage. Ambient shelf-stable low pH foods have emerged over the last 10 years as a major area of growth. Naturally occurring acids, such as lactic and citric acid, are added to a variety of foods in order to lower the pH to a level where the highly heat-resistant and toxic *C. botulinum* organism cannot grow. While spores of this organism may be present in the foods, it is not a risk, because the spores cannot germinate at pH levels below 4.5. Production of pasteurised foods with high acid/low pH is often achieved using glass as the packaging format. By applying hurdle technologies of low pH and mild heat, it is possible to produce high-quality foods, and glass offers excellent visual characteristics combined with consumer appeal.

2.3.4.3 Smoking
This is another traditional preservation method that relies on chemicals to effect preservation of the food. Meat smoking derives from the practice of hanging

meats in a chimney or fireplace to dry out. This had a variety of effects: the meat was partially dried, which itself assisted with preservation, but polyphenol chemicals in the smoke has direct preservative and antioxidant effects, and imparts a characteristic flavour on the food. In modern smoking techniques, the degree of drying and of smoke deposition and cooking are usually controlled separately.

Smoked salmon is an example of a food in which brining is used in combination with smoking to give an end product with an extended shelf life under chilled conditions. Three preservation technologies (brining, smoking and low temperature) provide the hurdle for microbial growth. Packaging for smoked foods is usually in transparent shrink-wrapped plastic film that excludes air from the package and provides an odour barrier to prevent loss of flavour from the product.

Many modern manufacturers now use synthetic smoke solutions into which the food is dipped in order to achieve higher rates of production and exercise better control over penetration of flavours as well as preservative chemicals into the product. It also serves to limit the presence of undesirable poly-aromatic hydrocarbons (PAH). Liquid smoke does not dry out the food as does the traditional smoking method, and combined with the different chemical profile deposited on the food it is likely to result in a different microbial population on the surface of the food. Liquid-smoked products may spoil in a different way than its traditionally smoked equivalent.

2.3.5 Fermentation

Fermentation represents one of the most important preservation methods in terms of the calorific proportion of food consumed by an individual, which can be as high as 30%. In fermented foods, preferred microorganisms are permitted or encouraged to grow in order to produce a palatable, safe and relatively stable product. Those microorganisms present prevent or retard the growth of other undesirable spoilage or pathogenic organisms, and may also inhibit other undesirable chemical changes. There are three main types of fermentation in the food industry: bacterial fermentation of carbohydrates (as in yoghurt manufacture); the bacterial fermentation of ethanol to acetic acid (as in vinegar production); and yeast fermentation of carbohydrates to ethanol (as in beers, wines and spirits). Many fermented milk products involve lactose (milk sugar) fermentation to lactic acid. In many cases the fermentation results in the production of chemicals (e.g. acetic acid or ethanol) that act as chemical preservatives.

In yoghurt production, the milk is first pasteurised to reduce the natural microbial population and destroy pathogens before the bacterial cultures are added. As fermentation progresses, lactic acid is produced and the pH drops to around 4.0–4.3. At these pH levels, few pathogenic bacteria can grow and the yoghurt is ready to be cold-filled into heat-sealed plastic pots. Many of the filling systems for yoghurts operate on the form-fill-seal principle, in which the packaging is presented to the filler in two reels, and the pots are formed within the filler environment. Although a yoghurt operation does not need to be aseptic (the yoghurt has a low pH and short chilled shelf life), this type of packaging machine could easily be converted to operate in aseptic conditions. Long, ambient shelf life yoghurts are filled in this way.

2.3.6 Modifying the atmosphere

This is a technique that is being increasingly used to extend the shelf life of fresh foods such as meat, fish and cut fruit, as well as of various bakery products, snack foods and other dried foods. Air in a package is replaced with a gas composition that will retard microbial growth and the deterioration of the food. For example, grated cheddar cheese for retail sale is packed in an atmosphere of 30% carbon dioxide and 70% nitrogen. In most cases, it will be the microbial growth that is inhibited (carbon dioxide dissolves into the surface moisture of the product to form a weak acid, carbonic acid, and the absence of oxygen prevents the growth of aerobic spoilage bacteria and moulds), but in dried foods, the onset of rancidity and other chemical changes can be delayed. The exact composition of the gas used will depend entirely on the type of food being packaged and the biological process being controlled (Day, 1992; Air Products, 1995). Modified atmosphere packaging (MAP refer to Chapter 10) is generally used in combination with refrigeration to extend the shelf life of fresh, perishable foods (see Table 2.3 for typical gas mixtures used for selected foods). Most MAP foods are packaged in transparent film to allow the retail customer to view the food.

There are several techniques related to MAP that are worthy of mention. Unprocessed fruits and vegetables continue to respire after being packed, consuming oxygen and producing carbon dioxide. Using packaging with specific

Table 2.3 Typical gas mixtures use in MAP of retail products (from Air Products, 1995)

Food product	Gas mixture
Raw red meat	70% O_2; 30% CO_2
Raw offal	80% O_2; 20% CO_2
Raw, white fish and other seafood	30% O_2; 40% CO_2; 30% N_2
Raw poultry and game	30% CO_2; 70% N_2
Cooked, cured and processed meat products	30% CO_2; 70% N_2
Cooked, cured and processed fish and seafood products	30% CO_2; 70% N_2
Cooked, cured and processed poultry products	30% CO_2; 70% N_2
Ready meals and other cook-chill products	30% CO_2; 70% N_2
Fresh pasta products	50% CO_2; 50% N_2
Bakery products	50% CO_2; 50% N_2
Dairy products	100% CO_2
Dried foods	100% N_2
Liquid foods and drinks	100% N_2

permeability characteristics, the levels of these two gases can be controlled during the shelf life of the food. Alternatively, *active* packaging can be used, in which chemical adsorbents are incorporated, e.g. to remove gases or water vapour from the package. Active packaging is discussed in Chapter 9.

An alternative to controlling or modifying the atmosphere is vacuum packaging, where all of the gas in the package is removed. This can be a very effective way of retarding chemical changes such as oxidative rancidity development, but care needs to be taken to prevent the growth of the pathogen, *C. botulinum*, which grows under anaerobic conditions. A specific pasteurisation process, referred to as the *psychrotrophic botulinum* process, is applied to the packaged food to reduce its numbers to commercially acceptable levels. By using vacuum packaging, mild heat and chilled storage, greatly extended shelf-lives have been achieved. This was the basis of *sous-vide* cooking, which originated in France as a method of manufacturing high-quality meals for restaurant use with up to 42 days shelf life when stored below 3°C. The thermal process has evolved since the original sous-vide concept of pasteurising at 70°C for 40 min, and the target process is now 90°C for 10 min. Cleanliness of the packaging materials is a key requirement to achieve the extended shelf life.

2.3.7 Other techniques and developments

Food manufacturers are continually looking for new ways to produce food with enhanced flavour and nutritional characteristics. Traditional thermal processes tend to reduce the vitamin content of food and can affect its texture, flavour and appearance. Processes which are as effective as traditional thermal systems in reducing or eliminating microorganisms, but do not adversely affect the constituents of the food, are being actively developed. In addition to those mentioned below, ultrasound, pulsed light, and electric field and magnetic field systems are all being actively investigated. In the UK, before any completely novel food, ingredient or process can be marketed, it has to be considered by the Advisory Committee on Novel Foods and Processes (ACNFP). The primary function of the ACNFP is to investigate the safety of novel food or process and to advise the government of their findings. The European Union has also formulated *Novel Foods* legislation.

2.3.7.1 High pressure processing
High pressure processing (HPP) was originally considered in the 1890s, but it was not until the 1970s that Japanese food companies started to develop its commercial potential. Pressures of several thousand atmospheres (500–600 Mpa) are used to kill microorganisms, but there is little evidence that high pressure is effective on spores or enzymes. Thus, chilled storage or high acidity is an essential hurdle to microbial growth. Jams were the first products to be produced in this way in Japan, and the process is now being investigated in Europe and

the USA. Sterilisation of the package is not possible using high pressure, and without aseptic filling, this may restrict its widespread use.

2.3.7.2 Ohmic heating
Ohmic heating achieves its preservation action via thermal effects, but instead of applying external heat to a food as with in-pack or heat exchangers, an electric current is applied directly to the food. The electrical resistance of the food to the current causes it to heat it up in a way similar to a light bulb filament. The advantage is that much shorter heating times can be applied than would otherwise be possible, and so the food will maintain more of its nutritional and flavour characteristics. The limitation is that ohmic cooling, or some other means of effecting rapid cooling, cannot be applied and so cooling relies on traditional methods that are slow in comparison with ohmic heating. Foods containing large particulates are suited to ohmic heating, because the electrical properties of the particulate and carrier liquid can be designed so that the particulate heats preferentially and instantaneously. The only commercial ohmic heater in operation in the UK (at the time of writing) is used to pasteurise fruit preparations, in which good particle definition is a key requirement. The packaging takes the form of stainless steel tanks that are transported to yoghurt manufacturers for inclusion in yoghurts, or occasionally plastic Pergall bags are used for food service use. Filling is aseptic in both examples.

2.3.7.3 Irradiation
Irradiation has seen much wider applications in the USA than in the UK, where public opinion has effectively sidelined it. In the UK there is a requirement to label food that has been irradiated or contains irradiated ingredients. In addition to killing bacterial pathogens, such as *Salmonella* on poultry, it is especially effective at destroying the microorganisms present on fresh fruits such as strawberries, thus markedly extending their shelf life. It can also be used to prevent sprouting in potatoes. Its biggest advantage is that it has so little effect on the food itself that it is very difficult to tell if the food has been irradiated. It also has some technical limitations, in that it is not suitable for foods that are high in fat, as it can lead to the generation of off-flavours. The only commercial foods that are currently licensed for irradiation in the UK are dried herbs and spices, which are notoriously difficult to decontaminate by other techniques, without markedly reducing flavour. A major application for irradiation is in decontaminating packaging. The Pergall bags used for filling ohmically heated fruit preparations are irradiated to destroy microorganisms.

2.3.7.4 Membrane processing
Membrane processing has been used for many years in the food industry for filtration and separation processes, usually taking the form of porous tubes, hollow fibres or spiral windings, and ceramics. The membrane retains the

dissolved and suspended solids, which are referred to as the concentrate or retentate, and the fraction that passes through the membrane is referred to as the permeate or filtrate. The product can be either the permeate (e.g. clarification of fruit juices or wastewater purification) or the concentrate (e.g. concentration of antibiotics, whey proteins), or in occasional circumstances both. By selection of the membrane pore size it is possible to remove bacteria from water or liquid foods, and in doing so, cold pasteurise the food. Commercial examples are the cold pasteurisation of beer, fruit juices and milk. Since the membrane is used to remove the microorganisms in the food, those present on the packaging surfaces must be killed using sterilisation solutions and/or heat.

2.3.7.5 Microwave processing

Microwave processing, like ohmic heating, destroys microorganisms via thermal effects. Frequencies of 950 and 2450 Hz are used to excite polar molecules, which produces thermal energy and increases temperature. In Europe, a small number of microwave pasteurisation units are in operation, primarily manufacturing pasta products in transparent plastic trays. Benefits of rapid heating can result in improved quality for foods that are sensitive to thermal degradation. The technology has not received widespread adoption because of the high capital costs of the equipment and the wide distribution in temperatures across a package. Heat generated by the microwaves pasteurises the food and the package together, and the products are sold under chilled storage to achieve extended shelf-lives. Microwave sterilisation has not developed much because of the need for air overpressure to maintain the shape of the flexible packages during processing. This creates complications with continuous systems, in that transfer valves are required between the chambers.

Domestic microwave use has a far greater impact on the food industry, with a very wide range of foods available in microwave re-heatable packaging. Package design can be complex, utilising susceptor technology to shadow regions that lead to more uniform re-heating performance. Research into methods to enhance the desirable browning and crisping of certain foods is beginning to find its way into commercial packages.

References

Adam, W.B. and Dickinson, D. (1959) Scientific Bulletin No. 3, Campden Food & Drink Research Association.
Air Products (1995) The freshline guide to modified atmosphere packaging.
Betts, G.D. (1996) A code of practice for the manufacture of vacuum and modified atmosphere packaged chilled foods with particular regard to the risks of botulism. Guideline No. 11, Campden & Chorleywood Food Research Association.
CCFRA (1992) Pasteuristation treatments. CCFRA Technical Manual No. 27, CCFRA, chipping Campden, Glos., GL55 6LD.

Day, B.P.F. (1992) Guidelines for modified atmosphere packaging. Technical Manual No. 34, Campden & Chorleywood Food Research Association.
Department of Health (1994) Guidelines for the safe production of heat-preserved foods.
Ranken, M.D., Kill, R.C. and Baker, C.G.J. (1997) Food Industries Manual, 24th edn, Blackie Academic and Professional.
Tetra Laval (1977) Dairy Handbook, Tetra pak, Lund, Sweden.

3 Packaged product quality and shelf life
Helen Brown and James Williams

3.1 Introduction

The intention of this chapter is to illustrate how the product quality and shelf life of packaged foods can be affected by the appropriate selection of packaging materials. Factors that affect product quality and shelf life are considered with examples of how packaging has been used to influence them to extend shelf life. Packaging can become a shelf life limiting factor in its own right. For example, this may be as a result of migration of tainting compounds from the packaging into the food or the migration of food components into the packaging. This chapter, therefore, also discusses the adverse effects that inappropriate packaging materials can have on product quality and shelf life.

Achieving a consensus on the definition of shelf life is never easy. Different groups within the food chain, i.e. consumers, retailers, distributors, manufacturers and growers, proffer subtly different perspectives of shelf life, reflecting the aspect of greatest importance and significance to them. For consumers, it is imperative that products are safe and the quality meets their expectations. Consumers will often actively seek the product on the shelf with the longest remaining shelf life as this is considered to be indicative of freshness. Consumer handling of products in terms of storage and use impacts on shelf life and is perhaps the biggest *unknown* for manufacturers when designing shelf life trials. For retailers, product quality must meet or exceed consumer expectations for repeat purchases. Product shelf life must be set to ensure that this is the case over the entire product life, allowing sufficient product life for the distribution chain and retail turnover of product and some life for the consumer. Manufacturers, who are responsible for setting product shelf life must be able to justify the validity of the shelf life assigned. They are also under considerable pressure to produce products to meet the shelf life requirements of retailers and often this will dictate whether or not a product is stocked. Achieving the desired product shelf life is a powerful driver for product and packaging innovation to extend product life. Many new packaging materials such as modified atmosphere packaging (MAP), have been developed to complement developments in new preservation techniques. The role of packaging in the maintenance and extension of shelf life cannot be over emphasised.

The Institute of Food Science and Technology (IFST) Guidelines (1993) provide a definition of shelf life: 'shelf life is the period of time during which the food product will remain safe; be certain to retain desired sensory, chemical, physical and microbiological characteristics; and comply with any label declaration of nutritional data.' This definition encapsulates most perspectives and leaves some flexibility, i.e. 'desired...characteristics', in assigning product shelf life. This is essential because once considerations of safety have been met, quality is generally a commercial consideration dependent upon the marketing strategy of the companies.

The shelf life of a product is best determined as a part of the product development cycle. As the packaging may be one of the means by which shelf life limiting processes are controlled, or the packaging *per se* may limit the product shelf life, it is also important that the packaging requirements for the product are considered early during product development. Shelf life testing is carried out by holding representative samples of the final product under conditions likely to mimic those that the product will encounter from manufacture to consumption. Once the microbiological safety of the product has been determined, quality issues can be considered. This may be based on microbial numbers, chemical specifications or sensory assessment. In most cases it is likely to be a combination of these. For products with long shelf lives it is desirable to have indirect or predictive methods for determining shelf life. Increasing temperature is the most common means of accelerating shelf life, but other parameters such as humidity, shaking or exposure to light, known to affect product stability can be used. Such tests are often product specific and based on a considerable knowledge of the product and its response under the accelerating conditions. The danger with this approach is that chemical reactions or microbiological growth is initiated, that would not take place under normal storage conditions.

Another approach to accelerated shelf life testing is predictive modelling, where mathematical models are used to predict either the shelf life or the level of a shelf life limiting attribute as a function of the composition of the product. A food's equilibrium relative humidity (ERH) is the atmospheric humidity condition under which it will neither gain nor lose moisture to the air (value often expressed as a_w, the water activity). The relationship between the mould-free shelf life and the ERH has been established for a large number of manufactured bakery products. The ERH of a product can be calculated by using conversion factors to obtain the *sucrose equivalents* that its ingredients contribute and this can be used to estimate the mould-free shelf life. Predictive food microbiology is based on mathematical models that describe the growth of microorganisms under specified conditions such as temperature, pH or level of preservative. There are a number of predictive food microbiology models available for use, for example the Campden & Chorleywood Research Association FORECAST Service and the USDA Pathogen Modelling Program. These models can be used to predict the likely growth of organisms of interest and to quickly assess the effects that changes in product formulation will have on microbial growth and therefore shelf life, where microbial number is the shelf life limiting criterion.

In the EU, shelf life is not defined in law, nor is there legislation about how shelf life should be determined. According to Directive 92/59/EEC on general product safety, the manufacturer is responsible only for putting safe products on the market. The EU Directive on food labelling (79/112/EEC) requires pre-packaged foods to bear *a date of minimum durability* (*best before* or *best before end*) or, for highly perishable foods from a microbiological point of view, a *use by date*. The date of minimum durability is defined as 'the date until which the foodstuff retains its specific properties when properly stored'. This date (and therefore the shelf life) is fixed under the producer's responsibility. The decision as to whether a food requires a *use by* or *best before* indication rests with those responsible for labelling the food since they are in the best position to assess its properties. It is an offence to sell any food after the stated *use by* date and only authorised persons may alter or remove the date.

Other specific pieces of legislation impact indirectly on shelf life: for example, if additives are used to achieve the desired shelf life, then legislation relating to the use of additives (those permitted for use and the permitted levels) is relevant. Legislation that prescribes the maximum temperatures for the storage and distribution of chilled and frozen products has a significant impact, especially on those foods that use temperature as a key factor to control shelf life limiting processes.

Legislation regarding food contact materials is intended to ensure that no components of food contact materials likely to endanger health or food quality are transferred into foods. In the United Kingdom, specific requirements for food contact materials are set out in two food regulations under the Food Safety Act. 'The Materials & Articles in Contact with Food Regulations 1987 Statutory Instrument No. 1523, as amended by Statutory Instrument 1994 No. 979', originated from EC Directive 76/893/EEC, includes the requirements that:

> Materials and articles must be manufactured in compliance with good manufacturing practice so that, under their normal or foreseeable conditions of use, they do not transfer their constituents to foodstuffs in quantities that could:
> - endanger human health
> - bring about an unacceptable change in the composition of the foodstuffs or a deterioration in the organoleptic characteristics thereof.

Limits are prescribed for the quantity of vinyl chloride monomer, that may be transferred to food, in materials and articles manufactured with polyvinyl chloride. 'The Plastic Materials & Articles in Contact with Food Regulations 1998 Statutory Instrument No. 1376, as amended by Statutory Instrument 2000

No. 3162' prescribes limits in relation to the content of materials and articles intended for food contact, and the migration of constituents into food. They also define methods required for testing migration into foods. The Regulations implement three European Council and Commission Directives: Council Directive 82/711/EEC, which lays down the basic rules necessary for measuring the migration of constituents of plastic materials and articles intended to come into contact with foodstuffs; Council Directive 85/572/EEC, which specifies the list of food simulants to be used for testing migration from plastic food contact materials and articles; and Commission Directive 90/128/EEC, a key directive for plastic materials intended for food contact. The Directive relates to the composition of plastic materials, defined broadly as organic polymers, but does not cover many auxiliary packaging components such as regenerated cellulose film, elastomers and rubber, paper and board, surface coating containing paraffin or microcrystalline waxes and ion-exchange resins, which have their own Directives. A general overall migration limit of $10\,mg\,dm^{-2}$ contact area sets a limit on the maximum quantity of constituents allowed to transfer out of plastic materials and articles into food. The Directive, and its five amendments, establishes a *positive list* of approved monomers and starting substances, which are the only such substances permitted for use in food contact plastics. Specific migration limits are included for some monomers and starting substances, which restrict residual levels in the finished material.

3.2 Factors affecting product quality and shelf life

For many foods, the product shelf life is limited by specific or *key* attributes that can be predicted at the time of product development. This is either on the basis of experience with similar products or observations of them, or from a consideration of

- the make-up of the product (intrinsic factors)
- the environment that it will encounter during its life (extrinsic factors)
- and the *shelf life limiting processes* that this combination of intrinsic and extrinsic factors is likely to result in.

Intrinsic factors are the properties resulting from the make-up of the final product and include the following:

- water activity (a_w) (available water)
- pH/total acidity; type of acid
- natural microflora and surviving microbiological counts in final product
- availablility of oxygen
- redox potential (E_h)
- natural biochemistry/chemistry of the product
- added preservatives (e.g. salt, spices, antioxidants)

- product formulation
- packaging interactions (e.g. tin pickup, migration).

Selection of raw materials is important for controlling intrinsic factors, since subsequent processing can rarely compensate for poor-quality raw materials.

Extrinsic factors are a result of the environment that the product encounters during life and include the following:

- time–temperature profile during processing
- temperature control during storage and distribution
- relative humidity (RH) during storage and distribution
- exposure to light (UV and IR) during storage and distribution
- composition of gas atmosphere within packaging
- consumer handling.

Product packaging can have significant effects on many of these extrinsic factors, and many developments in packaging materials have been driven by the need to reduce the impact of these environmental factors and extend shelf life. In some instances the packaging alone may be effective in extending shelf life, e.g. a complete light and oxygen barrier, whereas in most instances it acts as one of a number of *hurdles* each of which is ineffective alone, but which together effect a shelf life limiting factor.

The interactions of intrinsic and extrinsic factors affect the likelihood of the occurrence of reactions or processes that affect shelf life. For ease of discussion these shelf life limiting reactions or processes can be classified as: chemical/biochemical, microbiological and physical. The effects of these processes are rarely mutually exclusive but these categories provide a convenient framework for discussion. The effects of these factors are not always detrimental and in some instances they are essential for the development of the desired characteristics of a product.

3.3 Chemical/biochemical processes

Many important deteriorative changes can occur as a result of reactions between components within the food, or between components of the food and the environment. Chemical reactions will proceed if reactants are available and if the activation energy threshold of the reaction is exceeded. The rate of reaction is dependent on the concentration of reactants and on the temperature and/or other energy, e.g. light induced reactions. A general assumption is that for every 10°C rise in temperature, the rate of reaction doubles.

Specialised proteins called *enzymes* catalyse biochemical reactions. They can be highly specific catalysts, lowering the activation threshold so that the rate of reaction (of thermodynamically possible reactions) is dramatically increased. The specificity of enzymes for a particular substrate is indicated in the name, usually by attachment of the suffix *-ase* to the name of the substrate on which it acts; for example, lipase acts on lipids, and protease on proteins. In this chapter some examples of chemical and biochemical reactions which affect shelf life, and how they can be affected in turn by packaging will be discussed.

3.3.1 Oxidation

A number of chemical components of food react with oxygen affecting the colour, flavour, nutritional status and occasionally the physical characteristics of foods. In some cases, the effects are deleterious and limit shelf life, in others they are essential to achieve the desired product characteristics. Packaging is used to both exclude, control or contain oxygen at the level most suited for a particular product. Foods differ in their avidity for oxygen, i.e. the amount that they take up, and their sensitivity to oxygen, i.e. the amount that results in quality changes. Estimates of the maximum oxygen tolerance of foods (Table 3.1; Salame, 1986) are useful to determine the oxygen permeability of packaging materials required to meet a desired shelf life.

Foods containing a high percentage of fats, particularly unsaturated fats, are susceptible to oxidative rancidity and changes in flavour. Saturated fatty acids oxidise slowly compared with unsaturated fatty acids. Antioxidants that occur naturally or are added, either slow the rate of, or increase the lag time to, the onset of rancidity. Three different chemical routes can initiate the oxidation of fatty acids: the formation of free radicals in the presence of metal ion catalysts such as iron, or heat, or light – termed the classical free radical route; photo-oxidation in which photo-sensitisers such as chlorophyll or myoglobin affect the energetic state of oxygen; or an enzymic route catalysed by lipoxygenase. Once oxygen has been introduced into the unsaturated fatty acids to form

Table 3.1 The estimated maximum oxygen tolerance of various foods (from Salame, 1986)

Food/beverage	Maximum oxygen tolerance (ppm)
Beer (pasteurised)	1–2
Typical autoclaved low acid foods	1–3
Canned milk	
Canned meat and vegetables	
Canned soups	
Baby foods	
Coffee (freshly ground)	2–5
Tomato based products	3–8
High acid fruit juices	8–20
Carbonated soft drinks	10–40
Oils and shortenings	20–50
Salad dressings, peanut butter	30–100
Liquor, jams, jellies	50–200+

hydroperoxides by any of these routes, the subsequent breakdown of these colourless, odourless intermediates, proceeds along similar routes regardless of how oxidation was initiated. It is the breakdown products of the hydroperoxides – the aldehydes, alcohols and ketones that are responsible for the characteristic *stale*, *rancid* and *cardboard* odours associated with lipid oxidation.

Lowering the storage temperature does not stop oxidative rancidity because both the first and second steps in the reaction have low activation energies. Reduction of the concentration of oxygen (both dissolved and in the headspace) to below 1%, removal of factors that initiate oxidation and the use of antioxidants are strategies employed to extend shelf life where rancidity is a shelf life limiting factor.

In milk chocolate, the presence of tocopherol (vitamin E), a natural antioxidant in cocoa liquor provides a high degree of protection against rancidity. However, white chocolate does not have the antioxidant protection of cocoa liquor and so is prone to oxidative rancidity, particularly light induced. Even with light barrier packaging, its shelf life is shorter than that of milk or plain chocolate. However, the cost of eliminating oxygen from the pack would be prohibitive and not worth the additional cost for the relatively small increase in shelf life that this change would result in.

In snack products and particularly nuts the onset of rancidity is the shelf life limiting factor. Such sensitive products are often packed gas flushed to remove oxygen and packed with 100% nitrogen to protect against oxidation and provide a cushion to protect against physical damage. The packaging material generally used for commodity products with short shelf life is PVdC-coated OPP/LDPE laminates, whilst higher added value products with longer shelf life requirements are often packed in metallised polyester/LDPE laminates.

Investigations of rancidity in potato chips in relation to the light barrier properties of various films showed that improved light barrier properties of packaging films gave extended shelf life with respect to rancidity. Prolonged storage under fluorescent lights at ambient humidity caused the shelf life of crisps to become limited by rancidity rather than texture changes due to moisture uptake (BCL, 1985).

Vacuum packaging extends the shelf life of chilled fatty fish. Trout stored on ice packaged in polyethylene (high oxygen permeability) can develop a markedly rancid taste after 8 days. For trout vacuum packed in a plastic material with low oxygen permeability the shelf life at 0°C is increased to 20 days. For frozen fish, low storage temperatures combined with good packaging slows down degradation. The packaging must be tight fitting and must have a low water vapour transmission rate (WVTR) as the surface of fish easily suffers freezer burn. For fatty fish it is especially necessary to use a packaging material with low oxygen permeability, and vacuum packaging is preferred.

Oxidation of lycopene, a red/orange carotenoid pigment in tomatoes, causes an adverse colour change from red to brown and affects flavour. In canned tomato products this can be minimised by using plain unlacquered cans. The purpose of the tin coating is to provide protection of the underlying steel, but it also provides a chemically reducing environment within the can. Residual oxygen is consumed by tin dissolution into the product, minimising product oxidation that would otherwise lead to quality loss. However, the extent of dissolution of tin into the product needs to be taken into account in the assigned shelf life of the product as the maximum permitted level of tin in canned foods in the UK is $200\,\text{mg}\,\text{kg}^{-1}$ (Tin in Food Regulations 1992, SI 1992 No. 496). Tin dissolution can be avoided by using fully lacquered cans but oxygen-induced quality loss is more likely to occur.

Tomato ketchup used to suffer from *black neck* – the top of the ketchup in contact with oxygen in the headspace turned black. To disguise this, a label was placed around the neck of the bottle, hiding the discolouration. It has since been shown that oxidation depends on the level of iron in the ketchup and blackening has now been prevented.

Oxygenated myoglobin, oxymyoglobin, is the pigment in raw meat that is responsible for the bright attractive red colour, which consumers associate with freshness and good eating quality. In conventionally packed fresh beef, meat on a plastic tray is overwrapped with a highly gas permeable plastic material which allows an almost unrestricted supply of oxygen to the myoglobin, favouring the red colour. By vacuum packaging fresh beef, the reduced oxygen level in the pack results in a considerable increase in shelf life, however, the meat colour becomes purple, due to myoglobin being converted to the reduced form, which most consumers are not familiar with. By using MAP, where the meat is placed on a tray with a volume about 2–3 times that of the meat and the air is drawn out and replaced with a gas mixture of about 80% oxygen to maintain the bright red colour and 20% carbon dioxide to reduce bacterial growth, life can be extended to approximately 2–3 times that of conventionally packed fresh beef.

Nitrosohemochrome is the pigment responsible for the pink colouration of cooked cured meats. This pigment fades rapidly on exposure to air (oxygen) and light. Therefore vacuum packaging or MAP is generally used to achieve the desired shelf life of cooked cured meats. The oxygen content must be less than 0.3%, and in MAP a commonly used mixture is 60% nitrogen and 40% carbon dioxide. Oxygen scavengers are one means of reducing the levels of oxygen to a minimum. Light will often be excluded from such packs by the use of large attractive labels, protecting the product from retail lighting.

For freezing, whole chickens are generally packed after the chilling process in thin polyethylene bags. Turkeys and ducks are generally vacuum packed or shrink packed in more expensive plastic materials with low WVTR and low oxygen permeability. Turkeys especially demand good packaging because of the tendency for turkey meat to become rancid more quickly than other poultry.

3.3.2 Enzyme activity

Fruits and vegetables are living commodities and their rate of respiration affects shelf life – generally the greater the rate of respiration, the shorter the shelf life. Immature products such as peas and beans have much higher respiration rates and shorter shelf life than products that are mature storage organs such as potatoes and onions. Respiration is the metabolic process whereby sugars and oxygen are converted to more usable sources of energy for living cells. Highly organised and controlled biochemical pathways promote this metabolic process. Depletion and exhaustion of reserves used for respiration leads to metabolic collapse and an appearance associated with senescence. Disruption of tissues that occurs during the preparation of fruits and vegetables for the fresh-cut market leads to leakage of cell contents and encourages invasion by microorganisms. It also leads to an increase in respiration rate (Zhu *et al.*, 2001) that depletes reserves and results in quality loss. In non-storage tissues where there are few reserves, such as lettuce and spinach, or immature flower crops such as broccoli, this effect is even greater. Use of temperature control reduces the respiration rate (Table 3.2), extending the life of the product. Temperature control combined with MAP further suppresses the growth of yeasts, moulds and bacteria, extending shelf life further.

Ethylene (C_2H_4) is a plant growth regulator that accelerates senescence and the ripening process. It is a colourless gas with a sweet ether-like odour. All plants produce ethylene to differing degrees and some parts of plants produce more than others. The effect of ethylene is commodity dependent but also dependent on temperature, exposure time and concentration. Many commodities if exposed over lengthy periods are sensitive to ethylene concentrations as low as 0.1 ppm. Climacteric fruits such as apples, avocados, melons and tomatoes are particularly sensitive to ethylene. Physical (cutting) or chill injury induces the production of ethylene particularly in fruiting tissue due to its effect on the rate limiting enzyme (1-aminocyclopropane-1-carboxylic acid synthase) in the biochemical pathway leading to ethylene formation and increases tissue sensitivity to ethylene (Kato *et al.*, 2002). As ethylene induces senescence it has a significant impact on quality loss (Table 3.3). Removal of ethylene by absorption using activated carbon or potassium permangante has been shown to be effective in

Table 3.2 Respiration of intact and fresh-cut honeydew melon stored at chill and ambient temperature (Watada *et al.*, 1996)

	Respiration rate (arbitary units)				
	Temperature (°C)			Q_{10}	
	0	10	20	0–10	10–20
Intact	1.4	5.2	10.0	3.7	1.9
Cut	2.3	8.3	62.0	3.6	7.5
% increase	64.3	59.6	520.0		

Table 3.3 Effect of ethylene of the quality of fruit and vegetables

Commodity group	Affects of ethylene
Leafy vegetables	Turn yellow due to chlorophyll loss
	Russet spotting on leaves
	Accelerated leaf abscission leading to loss of leaves
	Phenolic synthesis leading to browning and bitter flavours
Cucumbers	Turn yellow and become soft
Unripe fruits	Accelerates ripening – tissue softening
Asparagus	Toughening and thickening of fibres

extending the shelf life of a number of whole and cut fruits and vegetables by extending the time of ripening. The storage life of MAP packaged mangoes was extended from 16 to 21 days by the inclusion of activated carbon and potassium permanganate (Illeperuma & Jayasuriya, 2002).

3.4 Microbiological processes

Under suitable conditions, most microorganisms will grow or multiply. Bacteria multiply by dividing to produce two organisms from one, their numbers increasing exponentially. Under ideal conditions some bacteria may grow and divide every 20 min, so one bacterial cell may increase to 16 million cells in 8 hrs. Under adverse conditions this doubling or generation time is prevented or extended – a feature that is exploited when developing food products and processes to achieve the desired shelf life.

During growth in foods, microorganisms will consume nutrients from the food and produce metabolic by-products such as gases or acids. They may release extra-cellular enzymes (e.g. amylases, lipases, proteases) that affect the texture, flavour, odour and appearance of the product. Some of these enzymes will continue to exist after the death of the microorganisms that produced them, continuing to cause product spoilage.

When only a few organisms are present, the consequences of growth may not be evident, but when numbers increase the presence of many yeasts, bacteria and moulds is evident from the formation of visible colonies, the production of slime or an increase in the turbidity of liquids, and from the effects that gas production, acidification and the off-odours caused by secondary metabolites have on the acceptability of the food product. The relationship between microbial numbers and food spoilage is not always clear. Whether or not high numbers of microorganisms result in spoilage is dependent upon the numbers present, the make-up of this number in terms of the type of microorganism and its stage of growth or activity, and the intrinsic and extrinsic factors of the food in which it is present. The key to achieving the desired shelf life of a product is to understand which microorganisms are likely to be the ones that will give rise to product spoilage, and what conditions can be used to either kill or reduce the rate of growth and multiplication. This requires careful selection or manipulation of the intrinsic and extrinsic factors of a food.

The presence of food poisoning organisms (pathogens) is not necessarily evident from changes in the food, and may only be apparent from the effects they produce, ranging from mild sickness to death. With many human pathogens, the greater the number of cells consumed, the greater the chance of infection and the shorter the incubation period before the onset of disease. Therefore, destruction, inhibition or at least control of growth is essential. For some invasive pathogens e.g. viruses, *Campylobacter*, the infectious dose is low and growth in the food may not be necessary. From the point of view of product shelf life, the first question must always be 'is the product safe?' once this safety is assured quality and commercial aspects can be considered.

Using knowledge of the initial levels of microorganisms and the conditions that destroy them or reduce their growth rate, food products are developed and designed by use of the best combination of intrinsic and extrinsic factors. The development and implementation of predictive microbiological growth models, particularly for chilled foods, has assisted in targeted product formulation and definition of packaging requirements to achieve a desired shelf life in terms of microbiological numbers.

3.4.1 Examples where packaging is key to maintaining microbiological shelf life

Heat processing which kills microorganisms is a widely used means of achieving safe products and extending shelf life. The amount of heat treatment required depends on the characteristics of the most harmful microorganism present, the nature of the food in terms of its viscosity, the pH of the food, the shape of the pack and the shelf life required. However, the heat process also changes the texture, taste and appearance of the product. This has prompted the move to minimally processed foods where a number of factors are combined to achieve the desired shelf life, e.g. a mild heat treatment, antioxidant action and controlled atmosphere packaging each restricting microbial growth, such that their combined effect allows the product to retain its sensory and nutritional properties.

In canning, low acid foods are filled into containers that are hermetically sealed and sterilised, typically at 115.5–121°C or above, to ensure all pathogens, especially *Clostridium botulinum*, are destroyed. The critical factor is the thermal treatment, which is the integration of product temperature with time throughout the heating cycle at a given process temperature and initial product temperature, required for the coldest part of the product to receive the required minimum *botulinum cook* of $F_0 3$. The size and shape of the container is important. Retort pouches are flat in shape, hence processing time can be reduced compared to a conventional cylindrical can and the reduced processing time generally results in improved taste and texture. Similarly, for sous-vide processing, foods are cooked in vacuum in sealed evacuated heat stable pouches or thermoformed trays. In aseptic processing, the barrier that the packaging poses to heat transfer is removed completely – the product and packaging being sterilised separately and then brought together under clean (aseptic) conditions. Where heat processing has been used to achieve sterility, the use of packaging to maintain sterility throughout the subsequent life becomes a key factor to achieving the desired product shelf life – both the packaging and the pack seals must provide a barrier to ingress of microorganisms.

Low temperatures might inhibit the growth of an organism and affects its rate of growth. Some microorganisms are adapted to grow at chill temperatures, hence the composition of organisms in the natural microflora will change. For example, in fresh milk the dominant microflora is Gram-positive cocci and bacilli that may spoil the product by souring if stored at warm temperatures. At chill temperatures, the microflora becomes dominated by psychrotrophic Gram-negative bacilli (most commonly *Pseudomonas* spp.). When temperature is used as the key limiting factor to control the rate at which shelf life limiting processes proceed, from a microbiological point of view, the role of packaging is less significant to shelf life because regardless of the packaging, if the temperature is not maintained, spoilage will proceed. Storage at frozen temperatures will stop microbial growth and can kill some microorganisms, but is not necessarily a lethal process.

Where vacuum packaging or modified atmospheres are the key shelf life limiting factors controlling microbiological growth, packaging is a critical factor in achieving the desired shelf life. *Pseudomonas* species, the major spoilage group in chilled proteinaceous foods, require the presence of oxygen to grow. The use of vacuum packaging or modified atmospheres excluding oxygen will prevent the growth of this type of bacteria. Whilst other organisms can grow in the absence of oxygen, they generally grow more slowly and so the time to microbial spoilage is increased. In MAP, the gas mixture must be chosen to meet the needs of the specific product; this is usually some combination of oxygen, nitrogen and carbon dioxide. Carbon dioxide at 20–60% has bacteriostatic and fungistatic properties and will retard the growth of mould and aerobic bacteria by increasing the lag phase and generation time of susceptible microorganisms. Several factors influence the antimicrobial effect of carbon dioxide, especially the microbial load, gas concentration, temperature and permeability of the packaging film. The antimicrobial effect is enhanced at lower temperatures because carbon dioxide is more soluble in water at lower temperatures, forming carbonic acid, so good temperature control is essential to obtain the maximum potential benefits of MAP and vacuum packaging. However, the effect of carbon dioxide is not universal – it has little effect on yeast cells and the growth of lactic acid bacteria is improved in the presence of carbon dioxide and lower oxygen levels. Nitrogen is an inert gas that has no antimicrobial effect *per se*.

It is generally used to prevent package collapse in products that absorb carbon dioxide and is used to replace oxygen in products that are susceptible to the growth of aerobic microorganisms.

Packaging materials designed to have antimicrobial activity provide a hurdle for microbial growth but seldom act alone as the key shelf life limiting factor. Antimicrobial activity can be obtained in two ways. Preservative-releasing or migrating systems contain a preservative intended for migration into the food (Luck & Jager, 1997). Non-migrating systems contain or produce a compound that has antimicrobial activity when the target organism comes into contact (Kourai *et al.*, 1994). For both systems, the antimicrobial substance can be incorporated into the packaging material or applied to the surface. Maximum contact is required between food and packaging to ensure adequate protection. Therefore, it is particularly suitable for vacuum packed foods. A number of antimicrobial packaging materials are commercially available and their activities and effectiveness have been reviewed (Vermeiren *et al.*, 1999, 2002). One example of this technology is Microban by Microban Products Co., UK, which incorporates the biocide triclosan into almost any type of plastic in a way that it is still free to migrate to the surface to act against developing bacteria.

3.5 Physical and physico-chemical processes

Physical changes affecting shelf life can be brought about directly by physical damage or by physico-chemical processes resulting from the underlying food chemistry. Many packaging functions such as protection of the product from environmental factors and contamination such as dust and dirt, dehydration and rehydration, insect and rodent infestation, containment of the product to avoid leakage and spillage, and physical protection action against hazards during storage and distribution are taken for granted by the consumer. However, careful consideration of the extent of protection required for the product in the context of the rigours of the storage and distribution chain through which it is to pass, is required if the product is to achieve its shelf life. Packaging is very often the key factor to limiting the effects of physical damage on product shelf life. Whilst the threat of careless or deliberate tampering cannot be accounted for when assigning product shelf life, the use of tamper-evident packaging provides a means of signalling whether packaging and potentially a preservative system has been breached.

3.5.1 Physical damage

During product life, particularly in storage, distribution and consumer handling, products are subjected to vibration on vehicles, compressive loads during stacking in warehouses and sudden jolts and knocks. The formulation of the product must be sufficient to tolerate such shocks or extended periods of vibration (for example, emulsions must be stable enough to withstand vibration), and the packaging must be able to withstand and protect against such forces. Vulnerable areas on packs are heat seals and screw caps, where damage resulting in leakage may result in loss of the preservation effect provided by the packaging. For fragile products, that are susceptible to crushing, such as soft cheeses, breakfast cereals and biscuits, the outer carton provides protection from physical damage and from potential tampering. Fruit and vegetables that are susceptible to bruising require protection from rough handling and the outer packaging used for distribution purposes needs to withstand stacking to considerable heights and high and variable humidity. The design of packaging for this purpose should be based on the properties of the commodity in terms of the humidity level it can withstand, the airflow allowed, the rate of respiration of the product and its susceptibility to bruising.

3.5.2 Insect damage

Infestation of foods with insects can be extremely unpleasant for the consumer because it is often not detected until the opening of the product. Insect infestation can occur at any point after manufacture, but is most likely during extended storage periods or during shipment. Although the problem may not be the fault of the food manufacturer, loss of materials can be expensive and legal cases can severely damage the reputation of brands. Package pests are classified in two groups – penetrators and invaders (Highland, 1984, 1991). Penetrators are capable of boring through one or more layers of flexible packaging materials. It is possible to reduce infestation with penetrators by preventing the escape of odours from the package through the use of barrier materials (Mullen, 1997). A rapid method to evaluate the usefulness of odour barriers has been developed (Mullen, 1994). Invaders are more common and enter packages through existing openings, usually created from poor seals, openings made by other insects or mechanical damage. It is therefore important that seals are not vulnerable to attack from insects. The corners of square packages can also be potential points of entry for insects. Various new packaging systems have been devised to minimise potential infestation (Hennlich, 2000).

3.5.3 Moisture changes

Moisture changes leading to loss or gain of moisture is a significant physical cause of the loss of shelf life of foods. Hygroscopic foods require protection from moisture take up which in dry products such as breakfast cereals and biscuits causes loss of texture, particularly crispness. For breakfast cereals the inner liner provides most protection to the food. Its main purpose is to protect from moisture transfer so as to preserve the product characteristics. The most effective type of liner will be determined during shelf life testing or by combining information from break point testing (holding at increasing humidities) and knowledge about the characteristics of the moisture permeability of the packaging material.

Protection or prevention of moisture loss is best achieved by maintaining the correct temperature and humidity in storage. In chilled and frozen foods, water loss (desiccation, dehydration or evaporation) can result in quality loss. However, it is often the resulting weight loss that is of greater importance due to the high monetary value of the products sold on a weight basis. The impact that packaging can have is illustrated by the losses that occur in chilled foods sold unwrapped from delicatessen counters, particularly cooked fresh meat, fish, pâtés and cheese. The shelf life of such products differs markedly from the wrapped equivalent – six hours versus a few days to weeks. Evaporative losses result in a change in appearance (Table 3.4), to such an extent that the consumer will select products that have been loaded into the cabinet most recently in preference to those which have been held in the display cabinet. The direct cost of evaporative loss from unwrapped foods in chilled display cabinets was estimated to be in excess of £5 million per annum in 1986 (Swaine & James, 1986). In stores where the rate of turnover of products is high, the average weight loss will be greater because of the continual exposure of freshly wetted surfaces to the air stream.

Weight loss during storage of fruit and vegetables is mainly due to transpiration. Most have an equilibrium humidity of 97–98% and will lose water if kept at humidities less than this. For practical reasons, the recommended range for storage humidity is 80–100% (Sharp, 1986). The rate of water loss is dependent on the difference between the water vapour pressure exerted by the produce and the water vapour pressure in the air, and air speed over the product. Loss of as little as 5% moisture by weight causes fruit and vegetables to shrivel or wilt. Films used in MAP packaging should have low water vapour transmission rates (WVTR) to minimise changes in moisture content inside the pack. As the temperature of air increases, the amount of water required to saturate it increases (approximately doubling for each 10°C rise in temperature).

Table 3.4 Evaporative weight loss from, and the corresponding appearance of, sliced beef topside after six hours' display (James, 1985)

Evaporative loss ($g\,cm^{-2}$)	Change in appearance
Up to 0.01	Red, attractive and still wet; may lose some brightness
0.015–0.025	Surface becoming drier; still attractive but darker
0.025–0.035	Distinct obvious darkening; becomes dry and leathery
0.05	Dry, blackening
0.05–0.10	Black

If placed in a sealed container, foods will lose or gain water until the humidity inside the container reaches a value characteristic of that food at that temperature i.e. the ERH. If the temperature is increased and the water vapour in the atmosphere remains constant, then the humidity of the air will fall. Minimizing temperature fluctuations is crucial for the prevention of moisture loss in this situation.

Severe dehydration in frozen foods leads to freezer burn – the formation of greyish zones at the surface due to cavities forming in the surface layer of the food. Freezer burn causes the lean surfaces of meat to become rancid, discoloured and physically changed. Packaging in low water vapour permeability materials protects against water loss during storage and distribution. However, as temperatures tend to fluctuate rather than be constant, dehydration will still occur if the package used does not fit tightly round the product. As water is removed from the food it will remain inside the package as frost. Frost in packages can amount to 20% or more of product weight and the concomitant desiccation of the product results in an increased surface area and thus greater access to oxygen, increasing the rate of quality loss at the food surface. Removing as much air from the pack as possible can reduce the problem. However, this is difficult to achieve for retail packed frozen foods such as vegetables, and such products are very susceptible to internal frost formation, particularly if they are allowed to spend a long time in the outer layers of display cabinets. By using laminates that include a layer of aluminium foil, internal frost formation can be reduced considerably.

One of the most widely experienced quality changes involving the transfer of moisture is sogginess in sandwiches. Moisture transfer from the filling to the bread can be reduced by the use of fat-based spreads to provide a moisture barrier at the interface (McCarthy & Kauten, 1990). In pastry-based and crust-based products such as pies and pizzas, movement of moisture from fillings and toppings to the pastry and crust causes similar problems. The movement of moisture or oils may be accompanied by soluble colours; for example, in pizza toppings where cheese and salami come into contact, red streaking of the cheese is seen, and in multi-layered trifles migration of colour between layers can detract from the visual appearance unless an appropriate strategy for colouring is used. Transfer of enzymes from one component to another, for example when sliced unblanched vegetables are placed in contact with dairy products, can lead to flavour, colour or texture problems depending on the enzymes and substrates available (Labuza, 1985). Packaging sensitive components in separate compartments for the consumer to mix them at the point of consumption has been one means of using packaging, to overcome such problems. Edible films, some as obvious as enrobing a wafer biscuit with chocolate to prevent moisture uptake, or as simple as an oil layer or a gelatin film over pâté are alternative solutions to such problems.

3.5.4 Barrier to odour pick-up

In practice, several commodities are sometimes stored or distributed in the same container or trailer. Dairy products, eggs and fresh meat are highly susceptible to picking up strong odours. Chocolate products have a high fat content and sometimes bland flavour. Unacceptable flavour pick-up can result if they are inadequately wrapped and stored next to strong smelling chemicals, such as cleaning fluids, or in shops close to strongly flavoured sweets, such as poorly wrapped mints. Packaging reduces the problem but most plastic materials allow quite a significant volatile penetration – the plastic materials used for vacuum packs and MAP have low permeability, and this reduces, but does not prevent, the uptake of foreign odours i.e. taints.

3.5.5 Flavour scalping

If a chemical compound present in the food has a high affinity for the packaging material, it will tend to be absorbed into or adsorbed onto the packaging until equilibrium concentrations have been established in food and packaging. This loss of food constituents to packaging is known as scalping. Scalping does not result in a direct risk to the safety of the food, or in the introduction of unpleasant odours or flavour. However, the loss of volatile compounds which contribute to a food's characteristic flavour affect sensory quality. It is common for unpleasant flavours naturally present in a food to be masked by other flavours, *the high notes*. If the *high notes* are scalped by the packaging material, the product is either bland or unpleasant flavours are more perceptible. The degree of scalping is dependent partly on the nature of the polymer and partly on the size, polarity and solubility properties of the aroma compound. A comparison of the loss of apple aroma compounds through scalping by low density polyethylene (LDPE), linear low density polyethylene (LLDPE), polypropylene (PP), nylon 6 (a polyamide, PA) and polyethylene terephthalate (PET) showed that polypropylene absorbed the highest quantities of volatiles (Nielsen *et al.*, 1992).

Where particular food-packaging combinations are susceptible to scalping, introducing an effective barrier layer may reduce the problem. Such systems must be carefully designed and tested to high levels of sensitivity to ensure that permeation through the barrier is minimised (Franz, 1993).

3.6 Migration from packaging to foods

The direct contact between food and packaging materials provides the potential for migration. Additive migration describes the physico-chemical migration of molecular species and ions from the packaging into food. Such interactions can be used to the advantage of the manufacturer and consumer in active and intelligent packaging, but they also have the potential to reduce the safety and quality of the product, thereby limiting product shelf life.

Much work has been carried out into the migration of substances from packaging into food. This has included the development of methods for the identification and diagnosis of problems using chemical and sensory assessment. The kinetics of migration have been modelled to allow prediction of the extent of migration over the shelf life of the product. In the UK, a number of surveillance exercises have been undertaken into the levels of particular migrants in foods (Table 3.5). The history of these surveillance exercises provides a useful summary of the concerns surrounding food contact materials over recent years.

Table 3.5 A summary of Food Surveillance Information Sheets (FSIS) on food contact materials published by MAFF/FSA (FSA, 2002)

FSIS	Date	Title
1	Jul 93	Metallic compounds in plastics
15	Oct 93	Hydrocarbons in chocolate
25	Feb 94	Composition of films used to wrap food
26	May 94	Formaldehyde in tea-bag tissue
35	Sep 94	Survey of benzene in food contact plastics
38	Oct 94	Survey of styrene in food
47	Jan 95	Fluorescent whitening agents
59	Apr 95	Dioxins in PVC food packaging
60	May 95	Phthalates in paper and board packaging
66	Jun 95	Grease-proofing agents in paper and board
72	Jul 95	Curing agents in carton board food packaging
90	May 96	Survey of paper and board food-contact materials for residual amine monomers from wet strength agents
125	Oct 97	Survey of BADGE epoxy monomer in canned foods
139	Dec 97	Survey of pentachlorophenol in paper and board packaging used for retail foods
157	Sep 98	Survey of chemical migration from can coatings into food and beverages – 1. Formaldehyde
169	Jan 99	Diisopropylnaphthalenes in food packaging made from recycled paper and board
170	Jan 99	Survey of chemical migration from can coatings into food and beverages – 2. Epichlorohydrin
174	Apr 99	Survey of retail paper and board food-packaging materials for polychlorinated biphenyls (PCBs)
186	Sept 99	Epoxidised soya bean oil migration from plasticised gaskets
189	Nov 99	Total diet study: styrene
6/00	Oct 00	Benzophenone from cartonboard
7/00	Oct 00	Chemical migration from can coatings into food – terephthalic and isophthalic acids
9/00	Nov 00	BADGE and related substances in canned foods
10/00	Jan 01	2-mercaptobenzothiazole and benzothiazole from rubber
13/01	Apr 01	Survey of bisphenols in canned foods
27/02	July 02	Paper and board packaging: not likely to be a source of acrylamide in food

3.6.1 Migration from plastic packaging

The popularisation of polymeric packaging materials has resulted in increased concerns over the migration of undesirable components into foods. This has the potential to affect product quality (Tice, 1996) as well as safety (Carter, 1977). These concerns are generally focused on the levels of residual monomers and plastics additives, such as plasticisers and solvents, present in polymers intended for direct or close contact with food. It is therefore important that the formulation of plastic packaging materials is designed so that the polymerisation process is as complete as possible.

A typical modern plastic packaging material may comprise many different constituents, which all have the potential to result in safety and/or quality problems if the material is poorly designed or there are errors in the manufacturing procedure.

The material itself is a polymer or copolymer manufactured from one or more types of monomers such as styrene, vinyl acetate, ethylene, propylene or acrylonitrile. All polymers contain small quantities of residual monomers left unreacted from the polymerisation reaction. These constituents are potentially available to migrate into foods. A food surveillance exercise undertaken in the UK measured the levels of styrene monomer in a total of 248 samples of food from a wide variety of manufacturers and in a wide variety of pack types and sizes (MAFF, 1994). Samples of milk and cream products, sold as individual portions (~10 g), contained the highest levels of styrene, ranging from 23 to 223 $\mu g\,kg^{-1}$, with a mean value of 134 $\mu g\,kg^{-1}$. Two samples of low fat table spread product contained styrene at an average concentration of 9 $\mu g\,kg^{-1}$. Styrene was detected at levels between 1 $\mu g\,kg^{-1}$ and 60 $\mu g\,kg^{-1}$ in the majority of remaining samples, with an average of less than 3 $\mu g\,kg^{-1}$. A total diet study estimated that the average daily intake of styrene is between 0.03 and 0.05 $\mu g\,kg^{-1}$ body weight for a 60 kg person (MAFF, 1999b), with part of this coming from styrene naturally formed in food. This compares favourably with a provisional maximum tolerable daily intake of 40 $\mu g\,kg^{-1}$ body weight set by the Joint FAO/WHO Committee on Food Additives (WHO, 1984). The presence of styrene in food, however, also has the potential to affect flavour. Taste recognition threshold concentrations for styrene in a number of different food types are below the level detected in foods (Table 3.6) and customer complaints due to tainting with styrene do still occur.

Additives are used to aid the production of polymers and to modify the physical properties of the finished material. For instance, plasticisers, added to give a plastic the desired flexibility, have been identified as a potential threat to health. The World Health Organisation has published opinions on a number of commonly used plasticisers with comments on toxicity.

The use of plasticised PVC as cling film has been targeted as a potential problem in terms of migration. Studies have been conducted on the migration

Table 3.6 Taste recognition threshold concentrations for styrene in different food types (Linssen *et al.*, 1995)

Food type	Fat content (%)	Taste recognition threshold concentration ($\mu g\,kg^{-1}$)
Water	0	22
Emulsions	3	196
	10	654
	15	1181
	20	1396
	25	1559
	30	2078
Yoghurts	0.1	36
	1.5	99
	3	171
Yoghurt drinks:		
Natural	0.1	82
Strawberry	0.1	92
Peach	0.1	94

of the plasticiser di-(2-ethylhexyl)adipate from PVC films into food during home-use and microwave cooking (Startin *et al.*, 1987) and in retail food packaging (Castle *et al.*, 1987). The level of migration increased with both the length of contact time and temperature of exposure, with the highest levels observed where there was a direct contact between the film and food, and where the latter had a high fat content on the contact surface. Use of a thinner PVC film was suggested as means of reducing the migration of this plasticiser (Castle *et al.*, 1988).

Mineral hydrocarbons, including liquid paraffin, white oil, petroleum jelly, hard paraffin and microcrystalline wax, may be used in certain polymers as processing aids. They have been detected in polystyrene containers at levels between 0.3 and 3% (Castle *et al.*, 1991), in polystyrene and ABS pots and tubs at between 0.3 and 5.5% (Castle *et al.*, 1993a), and cheese coatings at up to 150 mg kg^{-1} (Castle *et al.*, 1993b). The same studies also detected migration of these compounds into various foods, with the extent of migration broadly dependent on temperature and fat content.

The high performance of plastic packaging materials means that often only one layer of packaging is necessary to protect the product throughout shelf life. This means that the material used to protect the product is also required to promote it as a brand and provide the consumer with information on ingredients and nutritional information. Consequently, the food must be stored in contact with or in close proximity to printing inks, which can often pose a greater threat to product safety and quality than the base packaging material itself.

One of the most common types of ink and varnish formulations to be used is UV curable ink. This is made up of monomers, initiators and pigments.

Following application to the packaging, the ink is exposed to a source of ultraviolet radiation, converting the photoinitiator to a free radical that reacts with a monomer, thus initiating the polymerisation. During polymerisation, or curing, polymers are formed that irreversibly bind to the base packaging and trap the colourant into the polymer matrix, leaving a high quality, fast and safe printed surface. Other ink and varnish systems comprise pigment resin and a vehicle that may be either an organic solvent or water. Adequate drying with these inks requires removal of this solvent or water.

However, the quality of the print can be highly dependent on a number of factors. In the case of UV cured inks, excessive residual quantities of monomer or photo-initiator can result if the ink formulation is unbalanced, or through the incorrect function of the UV source. The migration of these constituents into a food may pose a risk to health and affect sensory quality. In addition to the inherent odour of these constituents, it is known that interactions between migrants and food components can lead to the formation of more potent tainting compounds. For example, the migration of benzophenone, a commonly used odourless photoinitiator, has been found to result in the generation of alkylbenzoates, which contribute undesirable flavours (Holman & Oldring, 1988).

A survey of printing inks on a selection of packaging materials for products including confectionery, snacks, chips, potatoes, chocolate bars and biscuits, in England and Spain found plasticisers above detectable levels (Nerin *et al.*, 1993). Plasticisers are commonly used in printing inks to contribute to adhesion to packaging, imparting improved flexibility and wrinkle resistance. Phthalates were identified as the major plasticisers in printing inks, although some samples contained N-ethyl- and N-methyl-toluenesulphonamide and tris(2-ethylhexyl) trimellitate. Although the printing inks were generally applied to the outer surface of the packaging, and consequently were not in direct contact with the food, it has been shown that plasticisers can migrate through the plastic layer to the food (Castle *et al.*, 1988).

Adhesives used to seal packaging can also be a source of migrating constituents. Common types of adhesive for use in food packaging are hot melt, pressure sensitive, coldseal, water-based, solvent-based, solvent-free, acrylics and polyurethanes. An adhesive should be chosen for an application on the basis of the nature of the product (e.g. it would be inappropriate to use a hot melt adhesive on a milk chocolate bar pack), the type of packaging used and special requirements, such as the inclusion of desirable odour volatiles into a cold seal to enhance product perception on opening.

A list of substances used in the manufacture of adhesives for food was compiled from a survey of adhesive manufacturers (Bonell & Lawson, 1999). More than 360 substances were listed as being used in one or more of the commonly used adhesives. A subsequent study into the chemical composition and migration levels focused on polyurethane adhesives (Lawson & Barkby, 2000). Major migrants (10–$100\,\mu g\,dm^{-2}$) were identified to be residual polyether polyols, cyclic reaction products formed during the preparation of polyester polyols, and low levels of additives used in the preparation of the adhesive. It was also suggested that constituents in the printed surface of packaging could migrate into the adhesive layer, when stored in direct contact, and then migrate into food after the packaging was formed.

The potential for interactions between different components of a packaging material is increased in multi-layer laminate systems. These complex packaging materials are manufactured by combining different polymeric and non-polymeric (e.g. metals) constituents to fulfill property requirements. The presence of numerous components, and adhesives used to bind them, can increase the probability of problems occurring and the difficulty associated with identifying the cause.

3.6.2 Migration from other packaging materials

Although the majority of research into interactions between food and packaging is concentrated on plastics, more traditional materials such as paper, board and cans also present problems.

Paper and board has been used to package food products for many years. Their composition is generally less complex than plastics, presenting less scope for migration. However, a number of taint problems in foods have been attributed to paper and board packaging.

Chlorophenols can be responsible for *antiseptic* taints. An investigation into a particularly disagreeable odour and taste in a shipment of cocoa powder found that the paper sacks used to package the product contained a number of chlorophenols in the paper itself at levels up to $520\,\mu g\,kg^{-1}$, and in the glued side seams at up to $40\,000\,\mu g\,kg^{-1}$ (Whitfield & Last, 1985). It was concluded that 2,4-dichlorophenol, 2,4,6-trichlorophenol and 2,3,4,6-tetrachlorophenol had been formed during the bleaching of wood pulp for paper manufacture, while pentachlorophenol had been used as a biocide in adhesives. It is important that the wood used in pallet construction is also free from such biocide treatment.

Chloranisoles can be formed from chlorophenols through fungal methylation (Crosby, 1981). These compounds have lower sensory thresholds than the equivalent chlorophenol, and can be responsible for strong musty taints. It has been known for chloranisoles to be formed from chlorophenols in damp paper and board materials (Whitfield & Last, 1985).

Health concerns have been raised over the use of recycled paper and board for food contact because of the possible migration of diisopropylnaphthalenes (DIPNs). These constituents, usually present in a number of isomeric forms, are commonly used in the preparation of special papers, such as carbonless and thermal copy paper. Research has shown that DIPNs are not eliminated during the recycling of paper (Sturaro *et al.*, 1995) and have the potential to migrate into dry foods, such as husked rice, wheat semolina pasta, egg pasta and cornflour (Mariani *et al.*, 1999). A UK survey confirmed that the use of recycled paper and board packaging materials can result in the migration of DIPNs into food (MAFF, 1999a). It was proposed that migration may be reduced if a film or laminate was placed between the food and the board. However, there was no simple relationship between the presence of an intervening film wrap and reduced levels of DIPNs in foods. Ironically, there may also be an argument for the use of DIPNs in functional packaging, as they have been identified as potential suppressants of sprouting in potatoes (Lewis *et al.*, 1997).

Migration from can lacquers into canned foods has been another area of concern over recent years. The migrants considered to pose the greatest health concern are the monomers bisphenol-A and bisphenol-F, and their diglycidyl ethers, known as BADGE and BFDGE, respectively. A study into the mechanisms involved in the migration of bisphenol-A from cans into drinks found that it was necessary to heat the can to a temperature above the glass transition temperature of the epoxy resin (105°C) in order for the compound to be mobilised (Kawamura *et al.*, 2001).

It has been proposed that BADGE can react with food components following migration (Richard *et al.*, 1999). After addition to homogenates of tuna in water, 97% of BADGE could not be detected either as BADGE or as its hydrolysate or HCl derivative. Following addition to some foods, a small percentage of added BADGE was converted to methylthiol derivatives but no other products could be identified, possibly because BADGE reacted with such a large number of components that the resulting chromatographic peaks were not detectable. A consequence of these results is the possibility that migration of BADGE into foods may have been grossly underestimated.

Another example of interactions between food and migrants from packaging material is *catty* taint, studied extensively in the 1960s. An initial study (Neely, 1960) reported on the formation of a substance having an objectionable odour similar to that of cat urine. Subsequent work suggested that the cause was a sulphur derivative of a compound related to acetone.

It was later found that the *catty* odour was formed from mesityl oxide, which is readily formed from acetone (migrating from the packaging) in the presence of dehydrating reagents (Pearce *et al.*, 1967). The mesityl oxide formed reacts rapidly with hydrogen disulphide at room temperature to generate the taint, but the *catty* odour is only evident at extreme dilutions. Higher concentrations of the tainting compound are perceived as having a mercaptan-like odour.

A study into the structure–odour relationship for *catty* smelling mercapto compounds confirmed 4-methyl-4-mercaptopentan-2-one to be the volatile principally responsible for *catty* taint (Polak *et al.*, 1988) and that it has a sensory threshold as low as $0.1\,\mu g\,kg^{-1}$ (ppb) in bland foods (Reineccius, 1991).

The most recent published study on an occurrence of a *catty* food taint, concerned separate complaints of an obtrusive off-odour in *cook-in-the-bag* ham products produced by two different manufacturers (Franz et al., 1990). The laminate packaging used in both products contained acetone, a precursor of mesityl oxide, although at concentrations considered too low to induce the formation of *catty* taint (0.4 mg m^{-2} and 1 mg m^{-2}, respectively). Subsequent work found the source to be diacetone alcohol (DAA), present at a concentration of 3 mg m^{-2} and 9 mg m^{-2}, respectively. The DAA was converted to mesityl oxide within the laminate packaging material through dehydration by an ethylene ionomer. The authors concluded that avoiding the use of mesityl oxide precursors, such as DAA, in packaging materials, could prevent the occurrence of *catty* taint in sulphur-rich foods.

3.6.3 *Factors affecting migration from food contact materials*

The extent of migration from food contact materials into food is dependent on a number of factors. Most obviously, the quantity of available potential migrants in the packaging material itself is of paramount importance. These levels must be minimised by careful design and production of the packaging.

The degree of contact between food and packaging also has a direct influence on migration, and in cases where particular problems have been encountered, it may be necessary to protect a food from direct contact with, for example, a printed surface.

As migration is a process that usually occurs gradually, the period of time for which the food and packaging are in contact should also be considered when trying to anticipate potential migration issues. Thus, there may be less concern over migration for a chilled dairy product with a short shelf life than over a box of shortbread biscuits with a shelf life of six months.

The intrinsic factors of a food are of great significance to the degree of migration likely to occur. A potential migrating constituent of the packaging is gradually transferred to the food causing the concentration of that constituent to gradually decrease in the packaging and increase in the food. Eventually, a point of equilibrium is reached when the concentration of the constituent stays constant in food and packaging. The quantity of the constituent in the food at the point of equilibrium is dependent on the physical affinity of the constituent for the packaging and food; for example, the degree of migration of a hydrophobic monomer such as styrene is partially dependent on the lipid content of the food. This was demonstrated in a study of migration from polystyrene into cocoa powder and chocolate flakes (Linssen et al., 1991). Residual styrene in the packaging material was quantified at around 320 mg kg^{-1}, and the amounts of styrene ranged from 7 to 132 µg kg^{-1} in cocoa drinks and from 414 to 1447 µg kg^{-1} in the higher fat chocolate flakes. Sensory assessment detected a styrene taint only in the chocolate flakes. Therefore, by considering the composition of the foodstuff, it may be possible to minimise the level of migration into foods from food contact materials. The same process of partitioning constituents between food and packaging to a point of equilibrium occurs in flavour scalping, discussed earlier in this chapter.

An investigation into the migration of paramagnetic additives from rigid PVC into a food simulant of pure or mixed fatty esters found that there was a tendency for the fatty esters to migrate into and plasticise the PVC matrix (Riquet *et al.*, 1994). This had the effect of increasing the mobility of the additive and accelerated its migration into the food simulant.

3.6.4 *Packaging selection to avoid migration and packaging taints*

For general selection of packaging, it is of paramount importance to ensure that the material complies with relevant legislation. This may entail specific and global migration measurements to check that the packaging is safe. When selecting a packaging material for a defined purpose, it is important to consider all components of the end product, how they are likely to interact, and the effect that the interaction will have on the food. The potential for taints can be evaluated by considering three questions: firstly, the composition of the packaging material – is it optimised to minimise the quantity of potential migrating components that are available to migrate into food? Secondly, what is the probability that any available migrating components might migrate into the food – this will depend on the composition of the food, which determines the affinity of migrants for the food matrix. The majority of migrating constituents likely to result in taints are hydrophobic and so are more likely to present problems in high fat foods. Thirdly, what impact is the migrating compound likely to have on the product? This is influenced by how strongly flavoured the product is. For example, similar levels of migration into a white chocolate product and a meat pie may make the chocolate unpalatable, but may not be detectable in the pie. Thus, the levels of migration that can be tolerated (within legislative limits) are dependent on the flavour characteristics of the food.

3.6.5 *Methods for monitoring migration*

There are two commonly used approaches to monitoring the occurrence and impact of migration from packaging into food. The first, and most standardised approach, is sensory assessment, in which a panel assesses a product to determine whether contact with the packaging has affected sensory properties. The advantage of this approach is the relative ease with which a panel can be assembled and trained. Data obtained from a properly designed assessment is of great relevance to a packaging or food manufacturer because the panel are using the same means of analysis that a consumer will use upon opening a product, namely the olfactory senses. Therefore, the panel should detect any problems that would be apparent to a consumer.

The disadvantage of a sensory approach is the inability to diagnose identified problems. Where problems have been identified, either through a quality assurance sensory panel or a consumer complaint, or are anticipated, for example, when developing a new packaging material or when starting to use a new supplier for a constituent of the packaging, it is appropriate to use instrumental chemical analysis. Chromatographic methods can be used to identify differences between a suspect packaging sample and a reference sample. This can provide detailed information on problems such as residual solvents or monomers, high levels of plasticisers and the presence of impurities.

A typical sensory method, based on British Standard BS3755:1964 (BSI, 1971) 'The Assessment of Odour from Packaging Material used for Foodstuffs', can be applied either to food simulants or specific food products. If performing a potential taint test on a packaging to be used on a range of foods, it is typical to use fairly bland foods, such as cream or butter (high fat), icing sugar or plain biscuits (high carbohydrate and dry), cooked chicken breasts (high protein) and melon pieces (high water).

The samples of food are stored in two glass tanks, one containing the material under test, the other the food alone. Usually the food is not placed in direct contact with the material under investigation. This is to restrict the types of constituents that can migrate into the food, and is for the safety of the sensory panel. The volatile nature of odour compounds allows odours to transfer indirectly through the air inside the chamber to the food. In one of the tanks, a known area of the packaging material is placed so that it surrounds the bowl into which the food is placed. The tank is sealed with lid and held at ambient temperature for a set time, usually 24 h. The other tank is set up and stored in exactly the same way omitting the packaging material.

The most suitable method to test for differences between control and test food samples is usually a triangle test, based on British Standard BS5929: 3:1984 (BSI, 1984) 'Sensory Evaluation on Sensory Analysis of Food, Part 3, Triangular Test'. Aliquots of food samples from the tanks are placed into identical containers for presentation to assessors, who are given three samples, two of which are identical. The assessor is then requested to identify which of the samples is different to the other two, and should give a response even if he/she believes there is no perceivable difference. It is important that sufficient assessors are used to allow distinction between significantly perceived differences and random responses. Care should also be taken to avoid distractions such as external odours or noise, and to ensure that the presentation and appearance of all samples is equivalent, using controlled lighting if necessary.

Instrumental chemical analysis can be used both to identify the cause of observed tainting problems and to routinely screen for targeted analytes known to be potential migrants. Following the report of a packaging related taint, a typical chemical investigation would begin with headspace analysis of the suspect packaging material and food contained within, using a gas chromatogram/mass spectrometer (GC/MS), and similar, untainted reference samples for comparison. A comparison of the GC/MS chromatograms of the two packaging samples can often reveal clear differences that may explain the cause of the taint, such as a large peak for benzophenone indicating that a UV curable ink may have been insufficiently cured. Reliable sensory data can be helpful in identifying analytes to be targeted in the analysis. Once differences have been identified between the two packaging samples, the chromatograms of the food samples can be compared with respect to those analytes to confirm whether the suggested compound could have been responsible for the taint.

In cases where this approach is unsuccessful or reference samples are not available, the technique of GC-sniffing can be invaluable. For this technique, the GC outlet is split so that some of the flow is directed to a detector (e.g. an MS) and the remainder passes through a sniffer port where the odour is assessed. It is common for tainting compounds to have very low sensory thresholds, meaning that a very low concentration may generate an easily detected odour but only give a very small chromatographic peak. Using GC-sniffing, the chromatographic retention time of the tainting compound can be determined by the response of the sensory assessor, and the identity of that compound confirmed using mass spectrometry.

If specific problems have caused a taint, it may be considered necessary to regularly screen samples of packaging material for indicators that the problem may be returning. This is possible by establishing a routine chromatographic method or *electronic nose* sensors may be used in-line to continually monitor for potential problems (Squibb, 2001).

3.7 Conclusion

The objective of this chapter has been to illustrate by way of examples, factors that affect the quality and safety of packaged food products. The desire to extend product shelf life will continue to stimulate the development of new processing and packaging innovations. Selection of the most appropriate product packaging requires a knowledge and understanding of the food chemistry and microbiology of the product, the environmental conditions that it will encounter from production to consumption and how this affects interactions between the packaging and the food.

References

British Cellophane Limited (BCL) (1985) Crisp rancidity – the effect of a clear window in metallised film on crisp shelf life. Coutalds Films Polypropylene, UK.
Bonell, A.E. and Lawson, G. (1999) Compilation of list of substances used in the manufacture of adhesives for food packaging in the United Kingdom. *MAFF R&D and Food Surveillance Report 2223.* London, Ministry of Agriculture, Fisheries and Food.

BSI (1971) Method of expose based on the assessment of odour from packaging material used for foodstuffs. *BS3755:1964*.
BSI (1984) Sensory evaluation on sensory analysis of food, *Part 3, Triangular Test. BS5929:3:1984*.
Carter, S.A. (1977) The potential health hazard of substances leached from plastic packaging. *Journal of Environmental Health*, **40**, 73–76.
Castle, L., Kelly, M. and Gilbert, J. (1991) Migration of mineral hydrocarbons into foods. I. Polystyrene containers for hot and cold beverages. *Food Additives and Contaminants*, **8**, 693–700.
Castle, L., Kelly, M. and Gilbert, J. (1993a) Migration of mineral hydrocarbons into foods. II. Polystyrene, ABS, and waxed paperboard containers for dairy products. *Food Additives and Contaminants*, **10**, 167–174.
Castle, L., Kelly, M. and Gilbert, J. (1993b) Migration of mineral hydrocarbons into foods. III. Cheese coatings and temporary casings for skinless sausages. *Food Additives and Contaminants*, **10**, 175–184.
Castle, L., Mercer, A.J. and Gilbert, J. (1988) Migration from plasticized films into foods. 4. Use of polymeric plasticizers and lower levels of di-(2-ethylhexyl)adipate plasticizer in PVC films to reduce migration into foods. *Food Additives and Contaminants*, **5**, 277–282.
Castle, L., Mercer, A.J., Startin, J.R. and Gilbert, J. (1987) Migration from plasticized films into foods. II. Migration of di-(2-ethylhexyl)adipate from PVC films used for retail food packaging. *Food Additives and Contaminants*, **4**, 399–406.
Crosby, D.G. (1981) Environmental chemistry of pentachlorophenol. *Pure and Applied Chemistry*, **53**, 1052–1080.
Franz, R. (1993) Permeation of volatile organic compounds across polymer films – Part 1: Development of a sensitive test method suitable for high barrier packaging films at very low permeant vapour pressures. *Packaging Technology & Science*, **6**, 91–102.
Franz, R., Kluge, S., Lindner, A. and Piringer, O. (1990) Cause of catty odour formation in packaged food. *Packaging Technology & Science*, **3**, 89–95.
FSA (2002) *Food Contact Materials and Articles: Explanatory Note on the Surveillance Programme*, London, Food Standards Agency.
Hennlich, W. (2000) A new approach to insect-proof packaging. *The European Food and Drink Review*, 83–86.
Highland, H.A. (1984) Insect infestation of packages, in *Insect Management for Food Storage and Processing* (ed. F.J. Baur), American Association of Cereal Chemists, St. Paul, pp. 309–320.
Highland, H.A. (1991) Protecting packages against insects, in *Ecology and Management of Food-Industry Pests* (ed. J.R. Gorham), Association of Analytical Chemists, pp. 345–350.
Holman, R. and Oldring, P. (1988) *UV and EB Curing Formulations for Printing Inks, Coatings and Paints*, London, SITA-Technology.
James, S.J. (1985) Display conditions from the products points of view. Institute of Food Research, Bristol Laboratory, Teach-in on retail display cabinets.
Illeperuma, C.K. and Jayasuriya (2002) Prolonged storage of 'Karuthacolomban' mango by modified atmosphere packaging at low temperature. *Journal of Horticultural Science and Biotechnology*, **77**, 153–157.
Kato, M., Kamo, T., Ran-Wang, Nishikawa, F., Hyodo, H., Ikoma, Y., Sugiura, M. and Yano, M. (2002) Wounds induced ethylene synthesis in stem tissue of harvested broccoli and its effect on senescence and ethylene synthesis in broccoli florets. *Postharvest Biology and Technology*, **24**, 69–78.
Kawamura, Y., Inoue, K., Nakazawa, H., Yamada, T. and Maitani, T. (2001) Cause of bisphenol A migration from cans for drinks and assessment of improved cans. *Journal of the Food Hygienic Society of Japan*, **42**, 13–17.
Kourai, H., Manabe, Y. and Yamada, Y. (1994) Mode of bactericidal action of zirconium phosphate ceramics containing silver ions in the crystal structure. *Journal of Antibacterial and Antifungal Agents*, **22**, 595–601.
Labuza, T.P. (1985) An integrated approach to food chemistry: illustrative cases. In: *Food Chemistry* (ed., O.R. Fennema) Marcel Dekker, New York, pp. 913–938.

Lawson, G. and Barkby, C.T. (2000) Chemical composition and migration levels of packaging adhesives. *MAFF R&D and Surveillance Report 557*, London, Ministry of Agriculture, Fisheries and Food.

Lewis, M.D., Kleinkopf, G.E. and Shetty, K.K. (1997) Dimethylnaphthalene and diisopropylnaphthalene for potato sprout control in storage. I. Application methodology and efficacy. *American Potato Journal*, **74**, 183–197.

Linssen, J.P.H., Janssens, J., Reitsma, J.C.E. and Roozen, J.P. (1991) Sensory analysis of polystyrene packaging material taint in cocoa powder for drinks and chocolate flakes. *Food Additives and Contaminants*, **8**, 1–7.

Linssen, J.P.H., Janssens, J.L.G.M., Reitsma, J.C.E. and Roozen, J.P. (1995) Taste recognition threshold concentrations of styrene in foods and food models. in *Foods and Packaging Materials – Chemical Interactions* (ed. T. Ohlsson), Royal Society of Chemistry, Cambridge, pp. 74–83.

Luck, E. and Jager, M. (eds) (1997) *Antimicrobial Food Additives: Characteristics, Uses, Effects*, Springer, Berlin.

MAFF (1994) Survey of Styrene in food, *Food Surveillance Information Sheet Number 38*. London, Ministry of Agriculture, Fisheries and Food.

MAFF (1999a) Diisopropylnaphthalenes in food packaging made from recycled paper and board, *Food Surveillance Information Sheet Number 169*, London, Minstry of Agriculture, Fisheries and Food.

MAFF (1999b) *Total Diet Study: Styrene. Food Surveillance Information Sheet Number 189*. London, Ministry of Agriculture, Fisheries and Food.

Mariani, M.B., Chiacchierini, E. and Gesumundo, C. (1999) Potential migration of diisopropyl naphthalenes from recycled paperboard packaging into dry foods. *Food Additives and Contaminants*, **16**, 207–213.

McCarthy, M.J. and Kauten, R.J. (1990) Magnetic resonance imaging applications in food research. *Trends in Food Science and Technology*, 134–139.

Mullen, M.A. (1994) Rapid determination of the effectiveness of insect resistant packaging. *Journal of Stored Product Research*, **30**, 95–97.

Mullen, M.A. (1997) Keeping bugs at bay. *Feed Management*, **48**, 29–33.

Neely, J.W. (1960) Elimination by chlorination of noxious efflents. *Petroleum Times*, 419–420.

Nerin, C., Cacho, J. and Gancedo, P. (1993) Plasticizers from printing inks in a selection of food packagings and their migration to food. *Food Additives and Contaminants*, **10**, 453–460.

Nielsen, T.J., Jagerstad, I.M., Oste, R.E. and Wesslen, B.O. (1992) Comparative absorption of low molecular aroma compounds into commonly used food packaging polymer films. *Journal of Food Science*, **57**, 490–492.

Pearce, T.J.P., Peacock, J.M., Aylward, F. and Haisman, D.R. (1967) Catty odours in food: reactions between hydrogen sulphide and unsaturated ketones. *Chemistry and Industry*, 1562–1563.

Polak, E., Fetison, G., Fombon, A.-M. and Skalli, A. (1988) Structure–odour relationships for 'Catty'-smelling mercapto compounds in humans. *Journal of Agricultural and Food Chemistry*, **36**, 355–359.

Reineccius, G. (1991) Off-flavours in foods. *CRC Critical Reviews in Food Science and Nutrition*, 381.

Richard, N., Biedermann, M. and Grob, K. (1999) Reaction of bisphenol-A-diglycidyl ether (BADGE) from can coatings with food components. *Mitteilungen aus Lebensmitteluntersuchung und Hygiene*, **90**, 532–545.

Riquet, A.M., Wolff, N. and Feigenbaum, A. (1994) ESR study of the migration of paramagnetic additives from rigid PVC to pure or mixed fatty esters used as food simulators. *Packaging Technology & Science*, **7**, 175–185.

Salame, M. (1986) The use of barrier polymers in food and beverage packaging. *Journal of Plastic Film and Sheeting*, **2**, 321–334.

Sharp, A.K. (1986) Humidity: measurement and control during the storage and transport of fruits and vegetables. *CSIRO Food Res. Quart.*, **46**, 79–85.

Squibb, A. (2001) Don't sniff at electronic noses. *Food Processing*, UK, **70**, 22.

Startin, J.R., Sharman, M., Rose, M.D., Parker, I., Mercer, A.J., Castle, L. and Gilbert, J. (1987) Migration from plasticized films into foods. I. Migration of di-(2-ethylhexyl)adipate from PVC films during home-use and microwave cooking. *Food Additives and Contaminants*, **4**, 385–398.

Sturaro, A., Parvoli, G., Rella, R., Bardati, S. and Doretti, L. (1995) Food contamination by diisopropylnaphthalenes from cardboard packages. *International Journal of Food Science & Technology*, **29**, 593–603.

Swaine, M.V.L. and James, S.J. (1986) Evaporative weight loss from unwrapped meat and food products in chilled display cabinet. in *Meat Chilling '86* International Institute of Refrigeration Commission C2, Bristol.

Tice, T. (1996) Packaging material as a source of taints, in *Food Taints and Off-flavours – Second Edition* (ed. M.J. Saxby), Blackie Academic & Professional, Glasgow, pp. 226–263.

Vermeiren, L., Devlieghere, F. and Debevere, J. (2002) Effectiveness of some recent antimicrobial packaging concepts. *Food Additives and Contaminants*, **10**, 77–86.

Vermeiren, L., Devlieghere, F., van Beest, M., de Kruijf, N. and Debevere, J. (1999) Developments in the active packaging of foods. *Trends in Food Science and Technology*, **19**, 163–171.

Watada, A.E., Ko, N.P. and Minott, D.A. (1996) Factors affecting quality of fresh-cut horticultural products. *Postharvest Biology and Technology*, **9**, 115–125.

Whitfield, F.R. and Last, J.H. (1985) Off-flavours encountered in packaged foods, in *The Shelf Life of Foods and Beverages* (ed. G. Charambalous), Elsevier, Amsterdam, pp. 483–499.

WHO (1984) Toxicological evaluation of certain food additives and food contaminants. *28th Report of the Joint FAO/WHO Expert Committee on Food Additives. WHO Food Additives Series Number 19*. Geneva, World Health Organisation.

Zhu, M., Chu, C.L., Wang, S.L. and Lencki, R.W. (2001) Influence of oxygen, carbon dioxide, and degree of cutting on the respiration rate of rutabaga. *Journal of Food Science*, **66**, 30–37.

4 Logistical packaging for food marketing systems
Diana Twede and Bruce Harte

4.1 Introduction

This chapter discusses logistical packaging and its role in food marketing systems. Logistical packaging is called by a number of different names such as *distribution packaging*, *transit packaging*, *industrial packaging*, *intermediate packaging* and *shipping containers*. The term *logistical packaging* is used throughout this chapter as it stresses the importance of integrating packaging with all of the activities in a supply chain.

Logistical networks deliver food via complex distribution channels that encircle the earth. Supply chains range from hand delivery of a neighbor's garden vegetables to the importation of exotic and rare processed foods using specialized trans-global distribution systems. Farm markets, conventional grocery stores, restaurants, fast food take-outs, food service institutions and direct marketing systems are supplied by a myriad of operational variations. They are also supplied by a wide range of package types, sizes and formats.

Logistical activities in a typical processed food supply chain begin at the farm. Commodities are transported to factories in bulk or semi-bulk packages, where the food is processed and packaged to add value. Unit loads are transported to wholesalers or retail distribution centers (RDC) where orders are picked into mixed loads, delivered to retail stores and broken down for retail display. There, consumers buy an assortment of packages and transport them home, where all of the packages are emptied, discarded and either shipped to a recycling facility or collected and transported to a landfill. This process is shown in Figure 4.1.

Packaging affects the cost of every logistical activity in a supply chain, and has a significant impact on productivity. Transport and storage costs are directly related to the size and density of packages. Handling cost depends on unit loads. Inventory control depends on the accuracy of identification systems. Customer service depends on how well the packages protect products and how easy the package is to open, display and sell. The environmental impact depends on the materials, method of manufacture, reuse and disposal of the packaging.

The first section of this chapter discusses the functional requirements for logistical packaging to protect, add value and communicate. The second section discusses how packaging is related to the physical activities in factory operations, in transit, in warehouses and in retail stores. The third section

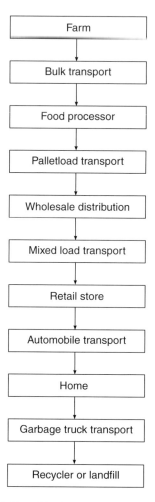

Figure 4.1 Flow of food package from farm to landfill.

discusses testing and evaluating packages for shock, vibration and compression performance. The final section describes the most common logistical packaging forms: corrugated fiberboard boxes, shrink-film bundles, plastic totes, stretch wrapping and pallet platforms.

4.2 Functions of logistical packaging

There are three inter-related functions of logistical packaging: protection, utility and communication. There is an increasing trend to view packaging in terms of the functions and value that it provides, rather than just in terms of traditional materials. (Materials are discussed in the third section of this chapter.) Packaging is part of a total system, with responsibility to minimize the cost of delivery as well as to maximize sales.

4.2.1 Protection

The first function is to protect the food and the consumers. Protection is an important packaging function because spoilage and distribution damage, wastes production and logistics resources. Replacement orders add further costs, and delays can result in lost customers. A loss of integrity in certain food packages can lead to product quality and safety issues. The type and amount of protection that a package is expected to provide depends on the characteristics of the product and the distribution environment with its associated hazards. A key aim of packaging is to provide the required protection using the lowest cost materials. The relationship can be conceptualized thus:

Product characteristics + Logistical hazards = Package protection

The relevant product characteristics are those that deteriorate or can be damaged over time. Food products are particularly vulnerable to biological and chemical changes that can affect quality and food safety. The hazards of the distribution environment range from exposure to extreme temperatures, dynamic forces, and insect infestation to ambient foes such as oxygen, moisture and time. The preservation ability of food packages and their characteristics and properties of packaging importance are discussed elsewhere in this book. It is important to note here that the required length of a food package's shelf life is directly related to how long it is in storage, transit and on the supermarket shelf. Short temperature-controlled channels for fresh food require less shelf life from their packages.

Protection from dynamic forces, handling impacts, in-transit vibration and warehouse stacking, is usually provided by the shipping container. Testing can determine how much abuse a product can withstand, and can be used to predict how well its package will prevent physical damage such as bruising, breaking, denting and smashing. Some standard dynamic tests are described later in this chapter.

The dynamic hazards of a logistical system and hence the most appropriate tests conducted depend on handling, and, the types of transportation and storage used. Firms that use a number of different types of distribution channel may need to package for a variety of conditions.

Damage is a symptom of an underlying problem that can be solved by changing the packaging or by changing distribution practices. In many cases, it costs less to reduce the hazards than to *improve* the packaging. For example, it costs less to reduce the force exerted by a clamp truck than to switch to stiffer boxes. Alternative methods of transportation (e.g. special equipment, refrigeration and/or dedicated carriers) and storage racks in warehouses can reduce damage; appropriate hygiene and pest control practices during distribution can reduce the need for packaging to protect against pest infestation.

4.2.2 Utility/productivity

The second packaging function, utility, is defined as value to a user. In the case of logistical packaging, the user is the logistical system and the value is productivity. Productivity in logistics is a very important concern because distribution is labor and capital intensive. Productivity is measured simply as the ratio of real output to real input:

Productivity = Number of packages output/logistics input

Logistical productivity is the ratio of the output of an activity, e.g. the number of packages loaded into a truck, to the input activity, e.g. the labor and forklift time required. Most logistical productivity studies center around better utilization of *inputs*, particularly labor, work harder. On the other hand, packaging initiatives like unitization and size reduction can easily increase the *output* of logistical activities. A good example is palletization, which dramatically improves the productivity of most material handling operations compared to break-bulk handling. Unitization enables a single person and a forklift to handle thousands of kilograms in an hour.

Almost all logistical productivity measures are described in terms of number of packages. Some examples include the number of cartons loaded per hour into a trailer, the number of packages picked per hour at a distribution center, the number of packages that fit into a cubic metre (*cube utilization*) of vehicle or warehouse space, the time to stock retail shelves and the cost of waste disposal. Packaging configuration directly affects the number that can be handled per hour or the number that fit into a vehicle.

Ergonomics is also a utility issue because healthy workers are more productive than employees engaged in personal injury lawsuits. Most injuries in physical distribution activities involve shipping containers. There are two types: accidents, usually involving an unstable package falling on a person, and chronic stress injuries due to manual handling of goods. Routine manual handling of packages has always been taken for granted, but it has a reputation for causing chronic back injuries. Many retail and warehouse workers are hurt by packages that are heavy, bulky, or must be lifted to a top shelf. In order to protect workers, the US Occupational Safety and Health Administration (OSHA) has issued guidelines for maximum weight of manually handled packages and appropriate handholds, and the EU has set ergonomic standards in Directive 90/269/EEC. The recommended package weight is related to how far and how often a package is lifted, how far the worker's hands are extended, how far he/she must twist and the adequacy of the hand grip. For most routine material handling jobs, the recommended weight limit is between 20 and 30 lb (9–14 kg).

4.2.3 Communication

The third packaging function, communication, is becoming more important as logistical information systems become more comprehensive. Electronic data interchange (EDI) and control has been key to the development of effective and integrated management of material flow, inventory, transportation and warehousing. For EDI to succeed, accurate timely information on the status of the packaged product is required. For all practical purposes, the package symbolizes the product throughout the distribution. Every time that a product changes status, for example when it is picked for a warehouse order, information about the status change is registered in various logistics records.

The information systems that record a status change include inventory records, shipping records, bills-of-lading, order picking lists, order receiving verification, accounting payables and receivables, manufacturing and logistics system tracking, and retail pricing. Packaging codes are also sometimes used for sorting products to various destinations in a factory, warehouse or transport terminal. International shipments additionally require the language of shipping origin, destination and intermediate stops, as well as international markings for handling instructions.

Correct identification of stock-keeping units (SKU) including SKU number, name, brand, size, color, lot, code dates, weight and number in the package are critical for good information management. Every logistical activity entails reading the package and recording/changing its status in an information system. Accuracy is essential. SKU information must be clearly legible. Workers must be able to quickly recognize a package from its label.

The most popular trend for reducing errors and increasing the efficiency of the information movement is to use automatic identification. Barcodes and radio frequency identification (RFID) enable a systems approach to managing information where every input is standardized, thus reducing errors. Bar codes require a line of sight to be read by a scanner. RFID enables packages to *call home* from a distance when prompted via radio frequency. Furthermore, new information can be added to RFID tags as they move through the supply chain. RFID promises to revolutionize package identification, since in theory the packages could be linked directly to a supply chain's information management system. The readability of these automatic identification forms depends on technological and symbolic compatibility of the package's label with every reader in the system. If automatic identification is intended to be used throughout a logistical system, it is necessary to use a common symbology. A number of standards-setting organizations exist for this purpose.

4.3 Logistics activity-specific and integration issues

Although they have been discussed separately, the three packaging functions, protection, utility and communication directly influence each other. The importance of each function varies between logistical operations. Sometimes there are trade-offs between these functions. For example, unitization increases handling protection and productivity, but may decrease in-transit cube utilization and require different markings from packages shipped break-bulk.

From the packages' point of view, there are many variations in logistical systems. In typical food and consumer product supply chains, unit loads are shipped by full truckload (TL) to the wholesale level. Here orders are picked, in one of several common methods, and then distributed in mixed loads on various types of pallets, carts or platforms to a variety of retail store types. On the other hand, some food products are floor loaded in trailers and delivered directly to a retail store without the use of pallets. Food service supply chains likewise include both direct and wholesale channels. And in e-commerce, each package may travel alone, or at least with *strangers*, using a small parcel transport system.

In each type of system, packaging must not only provide protection, utility and communication for specific operations, but it must also facilitate transitions between operations. In logistical systems, products change ownership and location, and packaging must be designed to satisfy a number of functional needs and users. For example, a slipsheet packaging system should not be implemented to improve trailer cube utilization without considering how it will increase the consignee's unloading time if he/she does not have the appropriate handling equipment. The more complex the system, the greater the need to study its handling methods, transport modes, facility dimensions, damage sources and communication needs, in order to design packages that will optimize the system. The next sections review the general functional requirements for packaging during each link in the supply chain: food processor, transport, warehousing and customer service issues.

4.3.1 Packaging issues in food processing and retailing

Food processors are concerned primarily with the packing operation. Their emphasis is on the cost and quality of incoming packaging materials, controlling the quality and productivity of the packing/filling process, hygiene and mechanization. In instances where food safety is threatened, packaging can be a critical control point in a product manufacturer's hazard analysis critical control (HACCP) plan. These are such important packaging issues that many food processors never find the time to look beyond their receiving and shipping docks. But there are other opportunities for packaging to improve both inbound and outbound logistics. Outbound opportunities focus on the customer's needs, and are described in subsequent sections of this chapter.

Inbound raw materials and ingredients being delivered to a factory need to be protected, and made easy to handle and identify. Their packages need to be easily emptied and inexpensively discarded. The high cost of land filling has been an incentive for producers to seek less expensive disposal means. There is a financial incentive for manufacturers to recycle or reuse the shipping containers that they receive. Reuse has been an especially popular strategy when hygiene can be assured, because a food processor generally has a close relationship to its suppliers with a quick turn-around cycle. Both of these factors are necessary for an efficient reusable packaging system. The quick cycle is necessary in order to minimize the number of containers in the system. A close relationship is needed in order to negotiate responsibilities for return or rental.

4.3.2 Transport issues

Packaging protection during transportation is so important that shock and vibration testing is often called *transit testing*. Damage that occurs during transit is particularly noticeable and well documented because of the relationship of carriers to the logistical system. Under common law, transportation carriers do not own the goods that they carry, but are *bailees* (contractors) entrusted to carry goods that are in their temporary possession without causing damage. When carriers damage products, they are liable for the items' full or partial value as stipulated in a contract or the bill of lading. The carrier is liable unless it can plead a common law defense and blame it on the shipper (e.g. inadequate labeling or packaging) or the perishability of the product (so-called *inherent vice*).

As a result of the importance of packaging issues to transport carriers, there is an history of carriers providing packaging guidance to shippers. Carriers reserve the right to refuse freight that they think is improperly packaged, and tests may be required before packaging can be approved. Some shippers develop partnerships with their carriers with the aim of reducing costs and preventing damage.

Dynamic forces vary by mode of transport. The technical principle of each mode affects service, economics, and dynamic forces as well as the packaging requirements for protection, utility and communication. The technical principle of truck transport, pneumatic wheels on pavement, is the major contributor to in-transit vibration, generally below 25 Hz. This is the force which vertical vibration testing aims to simulate. In addition, potholes and other surface discontinuities cause transient impacts that can disorganize a load. This is the reason why packages should be loaded to a level height in the trailer. Stopping, starting and turning forces are generally moderate, but do necessitate minimal restraints or void fillers when there are voids in the TL. The fact that roads are ubiquitous, and that every truck has its own driver, makes truck transport relatively fast compared to rail transport. Door-to-store service for a full TL minimizes the need for packaging to add much shelf life. Perishable products are usually shipped in refrigerated trailers. In full TL shipments, the load remains intact for the whole journey. The trailer can be loaded by the shipper in such a way as to minimize damage. Products shipped in full TLs can be minimally packaged to protect from impacts because they are handled under relatively controllable conditions by the shipper and consignee.

In less-than-truckload (LTL) shipments such as those used for e-commerce, a package is relayed from the origin through a number of the carrier's terminals until it is placed on the local delivery truck. The relay process results in handling under temperatures and dynamic conditions that cannot be controlled by the shipper, and an occasional package is dropped, frozen or thawed. Therefore LTL packaging needs more protection. In each successive trailerload, packages are consolidated jigsaw style, which results in an assortment of product and package stacking patterns, and so LTL packaging needs more protection. Packaging can improve the productivity of an LTL truck terminal if it is designed to be unitized or has ergonomic features. Small parcel carriers expect packaging to be conveyable and very protective, since packages can be repeatedly dropped, kicked or tossed. Packages may need to fit within specified dimensions or pass performance tests. Products shipped LTL are always marked with the shipper's and consignee's addresses, and for security reasons they should not identify the contents.

Truck transport is expensive, compared to rail and ocean transport. There is a decided advantage to optimizing the cubic and/or weight capacity of the trailer, and there is a trend towards making packages as small and lightweight as possible. Improving cube utilization is packaging's greatest opportunity to provide logistics value. If package size can be reduced by 50%, transportation efficiency doubles. There are many ways to reduce package size, such as concentrating products like orange juice, or by eliminating space inside packages by shipping items nested, with minimal head space and dunnage.

Cube minimization is most important for lightweight products like breakfast cereal that *cube out* a trailer below its weight limit. Trailer sizes vary; a typical size in the US is 48″ long × 102″ × 102″ ($14.4 \times 2.5 \times 2.5$ m), and payload weight limits are generally about 40 000 lb (18 000 kg). On the other hand, heavy products (like liquid in glass bottles) can *weigh out* a trailer before it is filled. For such products, the value-adding strategy is to reduce the package weight; for instance substituting plastic bottles for glass significantly increases the number of bottles that can be legally transported in a trailer.

The technical principle of rail transport – steel wheels on steel rails – restricts rail movement to fixed routes. The inflexibility of the wheels and rails restrict the vertical vibration, but staggered rail joints can exacerbate a tendency for low frequency (below 5 Hz) sidesway. As trains are assembled and disassembled in their relay across country, railcars are often slammed together to engage couplers. To avoid damage cargo must be securely blocked and braced, or voids must be filled within the railcar. Various national railroad organisations have researched rail damage and have developed tests and restraint recommendations. The fixed routes, route switching process, and train schedules make rail transportation slower than by truck. Packaging may need to provide a longer shelf life, more protection from climate changes, or products may need to be shipped in insulated or refrigerated cars. Railcars are generally larger than trailers and have no weight limit restrictions. Rail transport is less expensive than truck transport, and so there is less value in package cube or weight reduction when shipping by rail. Because of the long time, damaging conditions, and lack of weight restrictions, US railroads tend to carry more of the bulk and durable cargo, having lost the higher value consumer product cargo to trucking.

Air transportation's technical challenge is to overcome gravity. This makes it the most expensive form of transportation. The size and density of packages dramatically influence the cost of air transport, and every weight and volume reduction adds value. Air shipment is fast, and many products, including perishables, can be shipped with minimal shelf life protection. Air transport, itself, has very little dynamic input, except for some vertical vibration during take-off and touchdown, and a little high frequency vibration in transit. Decompression is a potential problem since some cargo holds are not pressurized. Temperature abuse is common.

Packages shipped by air are always picked up and delivered by truck, and so they have to withstand truck dynamics also. They are handled repeatedly, sometimes outdoors in the rain or other extreme weather conditions, and therefore, packaging needs to protect from impacts, moisture, temperature extremes and being stacked with other cargo. Many small parcel carriers ship by air, and the packaging requirements are similar to those for LTL shipments. When quantities shipped are sufficient, unitization can reduce damage and handling costs. Packages shipped by air need to be marked with addresses, but in order to deter theft, should not be marked to identify the contents.

The technical principle of ocean or river carrier transport – gliding through the water – determines the dynamic forces that ships encounter. Waves, swells and storms cause a ship to move in every direction. Cargo must be well-secured inside its packages, and within the vessel or intermodal container. Since waterborne transport usually interfaces with a land-based transport mode, packages also need to be suitable for rail or truck transport. The air aboard ship is high in humidity. Ordinary day/night cycles and climate changes in temperature can cause condensation, rust and rot. Packaging solutions to resist moisture damage include using desiccant sachets or treated films, ensuring efficiently sealed packs which prevent moisture ingress, and controlling the moisture content during packing.

Ocean transport is the slowest of all modes, and cargo can be held in port areas awaiting further shipment. As a result, packaging is often required to add shelf life to food products. Some intermodal containers are capable of refrigeration and/or controlled atmosphere. Cube optimization can yield savings, but the most significant ocean transport cost affected by packaging is the cost of ship-loading. An idle ship in port costs almost as much per day as it costs when it is underway. Furthermore, break-bulk loading operations are costly and long-shoremen's handling practices can be a great source of product damage and injury. Therefore, there is a great incentive to unitize or containerize cargo.

Intermodal containers combine the best and worst of ocean, rail, air and truck modes, since the container can travel on any or all of these modes in succession. Trailer-on-flatcar shipping combines railroad economies with door-to-door service, and has packaging requirements similar to those for truck, but with the added dynamics of a double set of springs while the trailer is on the flatcar. Container-on-flatcar shipping generally involves ocean transport, reducing the time to load ships, but with the same dynamics and shelf life problems as rail and ocean transport. Other combinations, such as air-surface containers, road-railers, roll-on-roll-off ships, and lash barges, have their own unique operating characteristics and packaging opportunities.

All intermodal transportations have a common advantage; individual packages are not handled. This reduces handling costs, impacts and the trip time. It should be remembered that intermodal containers are often opened overseas, and individual packages are shipped further in a break-bulk situation, and so the demands of this journey also need to be considered. International transportation ranges from flatbeds to bikecarts. Whether transportation is by one mode, multi-modes, or is intermodal, the rule for packaging protection is to package for the roughest leg of the journey.

4.3.3 Warehousing issues

Traditional warehouse order picking can be considered as a packaging operation. In a typical warehouse for consumer goods, large packages, e.g. full TLs or full palletloads are received. These are dismantled into smaller units, e.g. shipping containers or consumer packages as orders are picked. The orders are then repacked into mixed loads for delivery to logistical customers such as retailers. Sometimes the orders are so small that individual items are picked from shipping containers and re-boxed into a delivery package. Sometimes the delivery package is simply an original box or pallet repacked with a mixed load. In other cases, returnable pallets or totes are used to deliver to the retailer.

Modern warehouses are called *distribution centres*, to emphasize that they only profit by moving goods. To a distribution centre, storage represents unproductive assets. The productivity of warehousing is very important because order picking and materials handling are generally very labor-intensive. There is a great deal of emphasis on tracking and managing productivity. Cases or orders picked per hour, trucks loaded or unloaded per hour, and pallets received and put away per hour are examples of warehousing productivity measures.

Packaging can add productivity to order picking operations when products are packed in order quantities and cases that do not require additional labor to open or split. For example, if five is the standard order quantity, then products should be bundled into fives, rather than the traditional case quantities of 12, 24 or 48. The persistence of packing old-fashioned dozens is curious, given more rational reasons to specify other counts.

The trend in distribution is to speed up the order filling process, and many warehouses are now no more than cross-dock operations. They quickly assemble the mixed load orders as shipments are received, or simply transfer already mixed loads from a single manufacturer to delivery trailers. In cross-dock situations, there is an increased demand for packaging to be modular and automatically identifiable to facilitate mechanized sorting.

On the other hand, some warehouses still serve a significant storage function. Packaging can improve storage efficiency when packages are dense and maximize the use of cube space. Most warehouses store in, and pick from, racks. Shipping containers and palletloads should be sized to fit the racks (and/or visa versa). The stacking height, and therefore the required compression strength, is determined by the warehouse configuration. The stack height is only one palletload if racks are used, or it may be 3–4 palletloads high on the floor. Compression strength is affected by the strength of the shipping container and whether the contents help to support the load and the stacking pattern. Corrugated fiberboard box walls alone can rarely provide sufficient compression strength for long term storage, because the corrugated fiberboard weakens in the presence of relative humidity and over time. Stack compression failure damages product and is also a safety hazard because a stack can fall over and injure or kill workers.

Two other protective factors are best controlled by the warehouse: pests and temperature. Although packages can be made to resist insects and mice, it is better to keep the warehouse clean to eliminate their presence. Effective hygiene and pest control procedures must be implemented. Spillage should be promptly cleaned up, and traps should be set in key pathways.

Most food warehouses have at least three temperature zones: ambient storage for shelf-stable foods and general merchandise, a refrigerated area for fresh food, and a frozen storage area. With more varieties of fresh produce, freshly prepared meals, dairy products and bakery items, the demand for refrigerated storage has grown. Some refrigerated warehouses, especially those dealing with a wide variety of fresh produce, have several different-temperature rooms where each fruit or vegetable can be stored in its most advantageous climate. Most warehouse loading docks, however, are not refrigerated, and this can be a severe breach of the cold chain, especially if loads are staged on the dock for a long period of time. Although insulation can be added to packaging, it is usually more efficient to control the temperature of the facilities, especially in conventional supply chains for food.

Since packages are handled many times in warehouses, they must provide protection from impacts. And since they are often packed into mixed loads for shipment, they need to provide protection from stacking and puncture by other types of packages. They should be compatible, if not modular, to facilitate stacking and cube minimization.

Packaging also needs to make it easy to find the right items when picking orders. Easy to read stock keeping unit (SKU) identification is essential, regardless of whether it is read automatically or in the old-fashioned way. The markings should be concise and legible, on all four sides if necessary. The name of the manufacturer, brand, size and count should not be obscured by advertising messages. Packages have to be *read* when they are received, put away in the correct location, picked, repacked, and shipped. Good packaging communication can prevent shipping mistakes.

In many cases coordination is lacking among the members of a supply chain to implement a common automatic identification symbology. One packaging solution for warehouses is the use of slave pallets or in-house stickers which have a license plate bar code, magnetic strip, or RFID tag. The license plate can be used to track and record the status of palletloads throughout the single facility.

4.3.4 Retail customer service issues

Once a supply chain customer, such as a retailer, manufacturing plant, or contractor, receives a shipment, logistical packaging has to perform a new set of functions. The package needs to be opened easily without damaging the contents. Handling and the unpackaging operation should be quick and efficient, and reclosure may be desired. The product should be easy-to-identify, and the package may be required to display, provide special instructions, for installing and using the product. The package should minimize the customer's cost of disposal. Traditionally, retailers did not have much control over the packaging that they received. Most large manufacturers planned packaging to best suit their own operations, and small retail customers had little power to request improvements.

Increasingly, however, food marketing channels are dominated by large retailers who demand specific types of packaging. For example, grocery chains increasingly specify the durability of shipping containers and the information needed, prohibit some box styles, and help their suppliers to redesign packages to better maximize cube utilization in transit. Some retailers even control the consumer packaging and sell their own brands. Discount chains often specify that fast-moving products are to be packaged on display-ready small pallets that generate minimum waste. Some retailers demand that produce suppliers use standardized reusable totes, described in Section 4.5.3.

Increasingly, manufacturers find that designing packages to add retail value can be a profitable strategy to increase sales. Packaging affects the retailer's *direct product profitability* for every product, because a retailer's profit is directly related to the operational costs for opening packages, displaying and selling products. For example, bar codes that permit automatic scanning at a retail check-out counter reduce the cost of price marking, check-out, inventory control and reordering. Since packaging affects a retailer's productivity, cooperation between manufacturers and retailers can improve system-wide profitability.

The most important customer service trends are more display-ready cases, intelligent labelling and easy-open features like film wraps or reinforced strips that rip through corrugated fiberboard. Easy-open features also can help reduce customer-caused damage from razor blades. The graphics on most point-of-purchase display cases are of high quality and designed to attract attention. Some retailers are even requesting electronic security devices and RFID tags which could allow all of the cases on a mixed load to be scanned simultaneously.

Another retail trend is the *efficient consumer response* (ECR) strategy, in which products are packaged by suppliers in mixed cases to bypass the traditional warehouse order-picking operation. Packages may contain a number of different SKU's packed to order. They are shipped through a cross-dock operation controlled by the retailer, where packages are sorted using bar codes into trailers to be delivered directly to retail stores. In such cases, the markings and package size need to facilitate efficient sorting and control.

4.3.5 Waste issues

The cost for discarding shipping containers is generally paid by logistical customers like retailers. Besides the environmental impact, disposal is costly and can severely reduce a customer's productivity. There is a clear incentive for firms to reduce, reuse and recycle logistical packaging waste in order to avoid or reduce disposal costs.

This economic incentive has caused logistical packaging to be less of a target for legislation than has been consumer packaging waste, where disposal is an *external* cost. In Europe, where there is packaging waste legislation, the provisions for logistical packaging are different from those for consumer packaging. They usually result in the packer-filler paying a tax that funds a recycling infrastructure and reduces the retailer's disposal costs.

Each strategy – reducing, reusing and recycling – has an economic impact beyond disposal costs. Reduction of packaging materials also reduces package purchase costs. Packaging reuse generally adds some costs for sorting and return transportation but may reduce package purchase costs in the long run. The growth of recycling is reducing collection and processing costs and improving the market for recycled materials. The trend towards recycling and reusable packaging also has environmental benefits.

Recycling is an efficient disposal method for most logistical-packaging waste, since it naturally collects in large homogeneous piles at the facilities of manufacturers, warehouses and retailers. There are a limited number of materials: wooden pallets, corrugated fiberboard, polyethylene film, plastic foam and strapping. Recyclers welcome such concentrated and relatively clean sources (compared to sorting and cleaning curbside and food service wastes). As a result, logistical packaging has a very high recycling rate. Likewise, purchasing packages made from recycled material encourages the growth of a recycled-products market and infrastructure.

Recycling is sometimes erroneously called *reverse logistics*. Actually, the packaging materials are not taken back to the company that filled the packages, but rather the logistical system moves forward, through waste management companies, to reprocessors. In many cases, associations of packaging material manufacturers have set up their own networks for collecting and reprocessing.

4.3.6 Supply chain integration issues

Food can be prepared from commodities at any point in the channel, from the grower's facility to the consumer's kitchen. In this way, the supply chain concepts of *postponement* and *speculation* have always been applied to food production. Speculation strategies require the most packaging, since the food is prepared early in the channel and the package is expected to add more shelf life. On the other hand, postponement – waiting until the last minute to prepare the food – may not require the package to extend the shelf life. Likewise, the packages that supply these operations may also differ greatly in size because of economies of their large scale. Consumers buy flour in a very different package than does a food processing or food service business.

In more integrated supply chains, where the activities are coordinated and centrally planned, packaging can, likewise, be better planned to reduce system-wide costs. When supply chains are controlled by a powerful customer, one is more likely to find store brands, reusable totes and systemic automatic identification systems. Integrated supply chains are more likely to consider postponement versus speculation as a way to reduce costs.

When there is less integration, there is a tendency to suboptimize the system in favor of improving one operation. For example, the fresh produce industry has traditionally used a wide variety of box sizes, based on standard counts and/or the size of the fruit or vegetable. This works well for the grower because the fruit is sold by the standard count, and the box can be highly decorated to promote the identity of the grower. However, for an RDC which picks a mixed order, the variation in box sizes causes a number of problems because they cannot be automatically sorted and palletized and they do not fit securely in the mixed palletload. When they reach the retailer, the boxes may not fit together well in a display.

The recent trend to become larger and more powerful has led the retailers to better integrate their supply chain. It has also led many of them to demand that their produce suppliers use reusable modular totes, in order to minimize their sorting, stacking and stocking costs. Likewise, food service packaging for vertically integrated chain restaurants is often planned to improve operations and productivity in distribution centers and kitchens, in addition to fitting into the suppliers' operations.

For many food products, it is important to maintain a constant temperature throughout the supply chain. Any breach of the *cold chain* could exacerbate ripening, decay or the growth of bacteria. Temperature control is a responsibility of transport carriers, distribution centers and retailers, and many of these have a quality control process to ensure that the temperature is monitored and controlled. When food safety is threatened, the cold chain provides many critical control points in a good HACCP plan.

There are some ways that packaging can help to identify or bridge warm periods. Inexpensive temperature monitor tags can identify whether a package has exceeded a threshold temperature. Insulated packaging and/or frozen gel packs can be used when a gap in temperature control is expected. These play an important role for food packages in direct delivery and e-commerce, where the transport and delivery conditions cannot be assured.

Packages for extreme logistical systems, like those for the military and disaster relief food aid, are especially critical. Compared to commercial operations, these packages need to be stronger because distribution hazards are greater, shelf life requirements are longer, and the possibility of loss can be catastrophic. Optimization of handling and transport productivity are vital. Accurate package identification can be a matter of life and death. The more complex the system, the greater the need to understand its packaging requirements.

4.4 Distribution performance testing

The protection afforded by alternative packages can be evaluated and compared in laboratory and field tests. Performance testing is used to assess filled containers in situations that simulate distribution hazards and reproduce damage. It is important to distinguish distribution performance tests, which are used to aid packaging design, from material performance tests generally used for quality control. Quality control testing is not the subject of this chapter. Neither are shelf life testing and permeability, which are covered in other chapters.

The most common types of distribution performance tests are those that test the ability of packages to withstand the mechanical forces of shock, vibration and compression and those that test for the effects of temperature and humidity changes. There are a number of standardized tests available. The American Society of Testing and Materials (ASTM, 1999) has developed tests for packaging materials and containers since 1914. The International Standards Organization (ISO) has developed similar tests for international users. The International Safe Transit Association (ISTA, 2001) has developed a more limited set of standards, dealing solely with preshipment testing.

One of the most common standards is the test cycle, determined by distribution hazards and product vulnerability. ASTM D4169, ISO 4180, and ISTA's Projects include impact, vibration and stacking tests. There are similar French and Japanese standards. The Comite European de Normalisation (CEN) committee on packaging, TC/261, is developing a set of European standards. The test levels are determined by the product weight and the conditions expected in distribution. For example, a small air freight package is tested more severely than are packages shipped on pallets in full TLs; packages to be shipped overseas are tested more severely than those in domestic shipments.

Another type of standardized test is for specific package types. Some examples include ISO 10531, a stability test for unit loads, tests for intermediate bulk containers and tests for wooden boxes for fresh produce.

Some industries have developed performance tests that are specific to their product and logistical system. Such tests are often a variation of a standard test. For example, the makers of paperboard beverage bottle carriers have developed specific *jerk* tests for the handles to ensure that they will not tear when they are wet.

4.4.1 Shock and vibration testing

Dynamic testing of filled packages can be performed on a variety of testing equipment. The purpose of the test generally guides the choice of equipment and test methods. Impact tests can be performed on free-fall drop equipment (e.g. ASTM D775, ISO 2248 and ISTA 1/1A, 2/2A), or shock machines. A shock machine, generally used for fragility testing (e.g. ASTM D3332), has a higher velocity change for the same drop height because of its rebound, but can be used to produce repeatable impacts. The velocity change produced by a shock machine is generally two to three times greater than that of a free fall drop for a given drop height, depending on the machine and the distance of its rebound.

Some vibration tests, e.g. those specified by ISTA and ISO 2247, are very basic, performed on synchronous equipment with a fixed low frequency (about 4 Hz) and high displacement, generally 25 mm. Synchronous vibration tests are quite severe, and are sometimes called *vibratory impact* or *repeated impact* tests.

The ASTM standards specify electrohydraulic equipment, which is capable of varying the frequency and displacement. Some package/product tests (e.g. ASTM D3580) involve sweeping through a frequency range – usually 3–100 Hz; if an item or stack resonates in the range to be expected during transport, additional testing is done at that frequency to determine the likelihood of damage. Such testing is linked to the natural frequency of the product, the stack of packages and the transportation vehicle, and is generally below 25 Hz.

More sophisticated controls are capable of generating random vibration inputs, generating a spectrum of frequencies which have been statistically determined from measurements of transport conditions. ASTM standard D4728 gives examples of representative test profiles for various transport modes. As a result, product/package systems can be tested for their vulnerability to various truck suspension systems. Special fixtures may be used, for example, to simulate the vibration experienced by the bottom container in a stack.

Less used dynamic test methods are horizontal impact and revolving drum tests. Inclined plane tests (ASTM D880 and ISTA 1/1A, 2/2A) were originally developed to reproduce damage to appliances sliding down an inclined plane from the back of a truck. They have been generalized to simulate other kinds of horizontal impacts, including railcar switching impacts, conveyor jam-ups and pallet marshaling. A newer type of horizontal impact test has more controllable parameters, and includes programmable devices that can vary the shock duration, which will vary depending on the event to be simulated. The tumbling/revolving drum test is an old test, no longer favored by the technical community because it is not repeatable and severely abuses a package, somewhat akin to throwing it down an endless flight of stairs.

Since most dynamic testing attempts to reproduce what is in reality a highly variable environment, an important question in shock and vibration testing concerns test levels and intensities. How much vibration time equals a transportation cycle, and what percentage of impacts (and from what height) does a package need to survive? There is no simple answer.

It is important to identify the objectives of the test before an appropriate method, intensity and time can be chosen. Specific information about damage is particularly useful when developing package tests. Tests to solve specific problems are often designed to reproduce the package's specific damage characteristics, e.g. the ends of bags burst in side drops and provide a measure of the difference between the performance of alternative packages, i.e. how much input energy is required to burst, rather than simply pass/fail. Dynamic testing addresses the forces that packages encounter during transportation and handling. Packages also need to withstand stacking forces during storage, reproduced by laboratory compression testing.

4.4.2 *Compression testing*

Most compression testing research has been conducted on corrugated fiberboard boxes, since the walls of boxes are often expected to carry the load of a stack. Factors that are known to influence corrugated fiberboard box compression strength include the material properties, the board combining method, the box dimensions, manufacturing defects, interior partitions, temperature, humidity, stacking/loading method, and time.

It should be noted, however, that in most cases the product inside the box helps to support the stacked load. For example, packages for products like canned goods, glass jars and plastic bottles may not need the shipping container to provide any compression strength at all, and instead use the strength of the product itself.

There are two types of tests. ASTM D642 and ISO 12048 apply an increasing load until failure in a test machine. ASTM D4577 and ISO 2234 use a constant load over a specified time period, and may use a test machine or a simple stationary apparatus.

Some food manufacturers have learned that a common source of damage is clamp-truck side-to-side compression damage. (A clamp truck squeezes a unit load in order to handle it.) In response, they have developed sideways compression tests.

Since corrugated fiberboard loses a great deal of its strength when it gets wet, compression tests are usually conducted at standard conditions of 23°C (73°F) and 50% relative humidity. But there may also be a need for testing boxes in high humidity conditions or preconditioning boxes before testing.

4.5 Packaging materials and systems

The common materials and systems used for logistical packaging are relatively simple. They include corrugated fiberboard boxes, shrink-film bundles, reusable totes and unitization materials like pallets and stretch film. This section describes the properties and forms of each and illustrates when and where they are commonly used.

4.5.1 Corrugated fiberboard boxes

Corrugated fiberboard boxes are well known for their good stacking strength (when dry), easy availability and inexpensive cost. Corrugated fiberboard is the most common material used for shipping containers, and the regular slotted container (RSC) shown in Figure 4.2 is the most common design.

In a well-designed box, the load bearing panels have their flutes parallel to the direction of the anticipated load: for stacking strength the flutes should run vertically. When side-to-side strength is more important (in clamp handling, for example), it may be better for the flutes to run horizontally.

Corrugated fiberboard is easy to recycle, both from a technical and a logistical point of view. Used boxes are generally discarded in large, homogenous piles by factories, warehouses and retail stores – businesses which have an

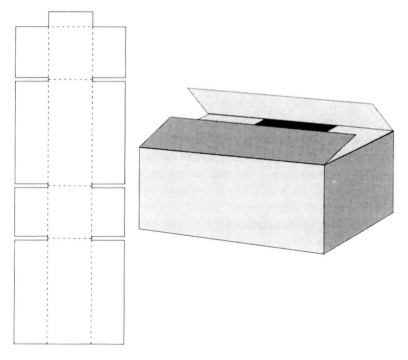

Figure 4.2 Regular slotted container (RSC).
Source: *Packaging Materials* (Twede & Goddard, 1998). Reprinted with permission from Pira International.

incentive to reduce their disposal cost by recycling. As a result, corrugated board has a very high recycling rate. Corrugated fiberboard has been used to make shipping containers for almost 100 years. A series of standard grades have been adopted by most countries. It is categorized in three ways: by the thickness and spacing of the fluted *medium*, by the weight of the *facings*, and by the quality of paper used.

The most widely used flute configurations are known simply as A, B, C, and E. The first corrugated materials were either coarsely fluted A-flute or fine B-flute. The intermediate grade, C-flute has now become the most commonly used type, being a compromise of the best qualitites of the other two. E-flute has very small flutes, and there are even finer grades called microflute, which are used as alternatives to solid fiberboard. Dimensions given by the British Standards Institute and US Fibre Box Association are given in Table 4.1.

Kraft liners to face the board range from 125 to $400\,g\,m^{-2}$ with 150, 200 and 300 grades predominating in Europe. In the US, where the materials are specified by basis weight, liners range from 26 to 90 lb (per $1000\,ft^2$) with 26#,

Table 4.1 Common forms of corrugated fiberboard

Flute	Flutes/metre	Flutes/ft	Flute height
A	105–125	36 ± 3	4.8 mm or 3/16″
B	150–185	50 ± 3	2.4 mm or 3/32″
C	120–145	42 ± 3	3.6 mm or 9/64″
E	290–320	94 ± 4	1.2 mm or 3/64″

Source: BS 1133, Section 7 and Fibre Box Association Handbook.

33# and 42# predominant. The corrugated medium is generally $127 \, \mathrm{g \, m^{-2}}$ (26# in the US).

Single wall board (with 2 facings) is the most common form used for cases and trays. Double and triple wall boards are used for palletload-sized intermediate bulk containers, used for some dry ingredients in the food industry. At one extreme, single face board is soft and used for wrapping items like light bulbs and glass bottles. The other extreme is multi-wall laminated structures made into lightweight pallets.

Corrugated board has an important drawback: it can lose much of its strength (indeed, all of its compression strength) when it is wet. Further, the commonly used starch-based adhesives are also moisture sensitive. It makes good design sense, where possible, to design the box with minimal head space, allowing the inside products to help support the load. This will prevent the uneven collapsing of containers which can topple a palletload. Wax dipping or coating has been used for particularly wet contents, like broccoli which is shipped with ice, but this practice is diminishing because the wax causes problems during recycling.

Corrugated fiberboard boxes are increasingly being used as advertising media in point-of-purchase displays, and so higher quality printing is demanded. There are three options: direct printing, preprinted liners and litho lamination. The uneven surface of the board limits direct printing to relatively simple one or two color flexography. Ink jets can also print directly on a box, and ink jet printing is particularly well suited for variable short-run information like lot codes. Preprinted liners – high quality flexo printed facing materials – can be built in to the corrugated board at the point of manufacture. Litho lamination can produce the highest quality printing, including full color halftones. It is made by laminating lithograph printed paper to the already converted board.

Corrugated fiberboard shipping containers have become a standard element of most logistical systems. In the USA, transport carriers required their use until transportation was deregulated in 1980. They remain popular because they are easy to purchase, perform well, and are recyclable. The technology of mechanical case packing is well developed. However, there is increasing competition from plastic alternatives, like shrink-wrap and reusable totes, which are lower cost in some situations.

4.5.2 Shrink bundles

Shrink bundle shipping containers are increasingly popular for products that do not require the compression strength of corrugated fiberboard. The products, like cans or bottles, are staged in a corrugated fiberboard tray (for stability), and the array is wrapped with a thin layer of film such as linear low density polyethylene film (LLDPE) and then conveyed through a shrink tunnel which tightens the wrap.

The advantages of a shrink bundle, over the comparable corrugated fiberboard shipping container, are that it uses less material and is less expensive. Whether there is more damage is a matter for debate. A shrink bundle is certainly less objectively protective. But many times less damage occurs because the people handling it can see the contents, and therefore handle the package more gently. Like fiberboard, LLDPE shrink-film can be easily recycled, along with the stretch film and plastic bags that are also discarded by warehouses, factories and retailers.

4.5.3 Reusable totes

As the cost of disposal grows, and as some countries have added incentives for waste reduction, the use of multi-trip packaging has also grown. The most common uses are for inter-plant shipments of ingredients, for retail warehouse to store totes, and for fresh produce from the farm to the retail shelf. Most reusable packages are plastic, although some firms reuse corrugated fiberboard boxes, wooden boxes and pallet boxes. Most of the growing reusable-packaging applications have one thing in common: a short, well-managed supply chain with steady predictable demand. The primary participants are either integrated by corporate ownership, contracts, partnership or administration under the control of one firm.

Good supply chain management is important because of the need to control the movement of reusable containers and the need to share the benefits. All partners in a reusable system must cooperate to maximize container use, and an explicit relationship is required for coordination and control. Otherwise, containers are easily lost or misplaced. The shipment cycle should be short, in time and space, in order to minimize the investment in the container pool and to minimize return transportation costs. The demand for products should be steady with little variation because the number of containers needed depends on the number of days in the cycle including during peak demand.

Deciding to invest in a reusable packaging system is a very different task from purchasing *expendable* containers. Many packers are tempted to justify the purchase only in terms of the savings in expendable container costs. The decision should consider all explicit relevant costs – the investment determined by the number of packages in the cycle, as well as the costs of handling, sorting, tracking, cleaning and managing the container pool – versus the purchase and disposal costs for expendables. Intangible benefits like improved factory housekeeping, ergonomics and cube utilization, and decreased damage should be considered.

The experience of UK retailers, who have utilized reusable totes for produce for over 10 years, is that the big savings accrue to the retailer. The standardized modular nature of their reusable totes (the standard footprint is 600 mm × 400 mm) allows containers to be automatically sorted in a distribution center. Modularization facilitates more tidy mixed-loads. They streamline in-store retail operations where the produce is displayed in the totes, and a full one can quickly be swapped for an empty one.

Given the difficulties of container management, there is increasing interest in outsourcing the management, logistics and/or ownership of reusable containers. This is an emerging industry of container pool agencies that organize the participants, assess costs, manage the exchange procedure and container inventory, clean and repair, assess fees and track containers. The containers may be owned by the participants or simply rented to them by the third party. Some pools have been successful in fresh produce and mixed grocery channels in Europe and the US. Such third-party service providers can develop economies of scale to justify a network of depots and vehicles to be able to collect the empties at a reasonable cost.

4.5.4 Unitization

Materials used for unitization vary, but they usually include a pallet or other platform. Wooden pallets predominate. Despite the increasing use of alternatives made from plastic or fiberboard, the wooden pallet has not been matched for versatility, reusability and repairability. Wooden pallets are ubiquitous in food marketing channels, where most handling is done by forklift.

Most pallets in the food industry are reused, to varying degrees of success based on their construction. The choice of wood species has a great impact on cost and durability. The denser and stiffer the wood, the greater its durability and cost. Hardwoods (like oak) are the most durable and costly, and they are used for pallets that are rented as part of *pallet pools*.

Pallets can be designed to be entered on two sides or all four. The vertical members can be blocks in the corners and centers, or *stringers* that extend from front to back, as shown in Figure 4.3. They can have decking only on top or on both faces. Decking on both faces is recommended when palletloads are doublestacked, to spread the weight of the upper load more evenly over the lower one. The decks can have varying numbers of horizontal boards. Wide gaps can cause damage since boxes can deform into them. Other platforms include slipsheets, carts and racks. Carts and racks are often used to deliver orders from a warehouse to a store. They may have shelves and restraint systems to minimize damage due to the mixed nature of the loads.

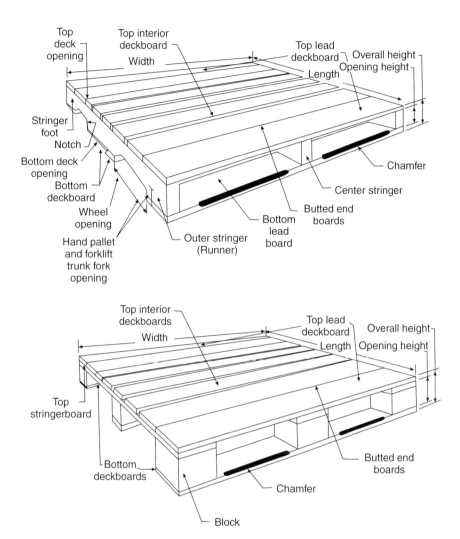

Figure 4.3 Four-way entry pallets with stringers (top) or blocks.
Source: National Wooden Pallet and Container Association, Alexandria, VA, 1998. Reprinted with permission.

A slipsheet is a flexible platform made from a sheet of heavy fiberboard or plastic. They are used because of their low cost and minimum cube usage. However, special handling equipment is required to be used by both the shipper and consignees. A special kind of lift truck attachment pulls the sheet onto a polished steel platform which carries it on the front of the lift truck. To set it down, the attachment pushes the load off the platform. Another special kind of equipment is a clamp truck. It grabs a load between platens and squeezes it with enough pressure to be able to lift it as a unit and carry it in its clamps.

This is used for lightweight sturdy unit loads of products like toilet paper and breakfast cereal cartons.

It is important to note that slipsheet and clamp handling are often used for the benefit of the shipper who wants to eliminate the cost and cube of pallets. But many wholesalers and RDCs can only handle conventional pallets with their forklifts, and do not have specialized equipment to receive the load. They use the forks of an ordinary forklift to scrape unit loads off the floor or to separate stacked units, and then place them on pallets. This causes damage and inefficiency. Such specialized types of unit loads are best used in a more integrated supply chain where all participants can be persuaded to use the same handling methods.

There are a number of materials which can be used to restrain a unit load. Stretchwrap is the most common. Stretchwrap can be applied either manually or mechanically. The turntable type of stretchwrapper is the most common mechanical method: the load spins while a roll of film is played out in a spiral manner up and down the load. The manual applications are usually found in warehouses, where the order pickers wrap mixed loads as they stage them. Whichever method is used, the wrap should be as thin and tight as possible without crushing the load.

Other unit load restraint materials include shrinkwrap, strapping or adhesives. Since shrinkwrapping requires more energy than stretchwrapping, it is less popular. It is used mostly as thick covering for unit loads intended to be stored outside. Strapping or tape, sometimes used around the top two layers of a load adds a minimum level of stability. Pallet stabilizer adhesives applied between layers of boxes provide sheer strength, but allow the boxes to easily pop apart when lifted. Tape and adhesives should be used in cases where they do not damage the outer surface of the shipping container, or where such damage is not considered a problem.

The pallet pattern is generally determined by the dimensions of the shipping containers. To maximize transport and handling efficiency, there should be no wasted space or under-hang; and over-hang is also to be avoided because it causes damage. Cubic efficiency can be improved by designing shipping containers to better fit the footprint: most grocery pallets in the US are $40'' \times 48''$; in Europe, most conform to the ISO standard of $1200\,mm \times 1000\,mm$. There is less than an inch of difference between the two standards, and, for a given box dimensions, some same pallet patterns can be used.

A number of computer programs are available for optimizing a pallet pattern, given the size of shipping containers. They can also be used to optimize the size of the shipping container itself as well as the primary packages inside, given the dimensions or volume of a product. Sometimes a very small dimensional change can result in dramatic transport savings. The pallet cube optimization programs have been found to be especially valuable for long distance shipments, exports and fast moving consumer goods, where transport savings add up quickly.

4.6 Conclusion

This chapter has shown the importance of logistical packaging to a supply chain's operations. Packaging should be considered at an early stage of product design. Ideally, the primary package, the shipping container and the unit load should be considered as an integrated design problem.

The size, shape, weight, and properties of the primary package and the nature of its contents will determine the requirements for the logistical packaging, given the characteristics of a particular distribution channel. Changes to the product or primary package for reasons of cost savings or marketing improvements will very likely have an impact on the performance required as well as dimensions and unit load efficiency.

The objective of logistical systems is to deliver food products and raw materials to the right place, at the right time, at the most effective total cost. Packaging adds value only when it serves these objectives.

References

American Society for Testing and Materials (1999) *Standards and Other Documents Related to Performance Testing of Shipping Containers*, 3rd edn, ASTM, Philadelphia.
Bowersox, D.J., Closs, D.J. and Cooper, M.B. (2002) *Supply Chain Logistics Management*, McGraw Hill Irwin, Burr Ridge, IL.
Brody, A.L. and Marsh, K.S. (eds) (1997) *The Wiley Encyclopedia of Packaging Technology*, 2nd edn, John Wiley & Sons, New York.
Fibre Box Association (1999) *Fibre Box Handbook*, FBA, Rolling Meadows, IL.
Fiedler, B.M. (ed.) (1995) *Distribution Packaging Technology*, Institute of Packaging Professionals, Herndon, VA.
Friedman, W.F. and Kipnees, J.J. (1977) *Distribution Packaging*, 2nd edn, John Wiley & Sons, New York.
Hanlon, J.F., Kelsey, R.J. and Forcino, H.E. (1998) *Handbook of Package Engineering*, 3rd edn, Technomic, Lancaster, PA.
International Safe Transit Association (2001) *ISTA Resource Book*, ISTA, Lansing, MI.
Maltenfort, G.C. (1988) *Corrugated Shipping Containers: An Engineering Approach*, Jelmar, Plainview, NY.
Maltenfort, G.C. (ed.) (1989) *Performance and Evaluation of Shipping Containers*, Jelmar, Plainview, NY.
McKinlay, A.H. (1998) *Transport Packaging*, Institute of Packaging Professionals, Warrenton, VA.
National Motor Freight Traffic Association (annual) *National Motor Freight Classification*, NMFTA, Alexandria, VA.
National Wooden Pallet and Container Association (1998) *Uniform Standard for Wood Pallets*, Alexandria, VA.
Schaffner, D.J., Schroder, W.R. and Earle, M.D. (1998) *Food Marketing: An International Perspective*, McGraw-Hill, Boston.
Twede, D. and Parsons, B. (1997) *Distribution Packaging for Logistical Systems: A Literature Review*, Pira, UK.

5 Metal cans

Bev Page, Mike Edwards and Nick May

5.1 Overview of market for metal cans

The total world market for metal containers is estimated at 410 billion units per annum. Of this, drink cans account for 320 billion and processed food cans account for 75 billion. The remainder are aerosol and general line cans. Drink cans may be divided into those for non-carbonated drinks (liquid coffee, tea, sports drinks etc.) and carbonated beverages (soft drinks and beer), many of which pass through a pasteurisation process.

5.2 Container performance requirements

Metal packages for food products must perform the following basic functions if the contents are to be delivered to the ultimate consumer in a safe and wholesome manner:

- preserve and protect the product
- resist chemical actions of product
- withstand the handling and processing conditions
- withstand the external environment conditions
- have the correct dimensions and the ability to be practically interchangeable with similar products from other supply sources (when necessary)
- have the required shelf display properties at the point of sale
- give easy opening and simple/safe product removal
- be constructed from recyclable raw materials.

In addition, these functions must continue to be performed satisfactorily until well after the end of the stated shelf life period. Most filled food and drink containers for ambient shelf storage are subjected to some form of heat process to prolong the shelf life of the product. For food cans, this will normally provide a shelf life of up to 2–3 years or more. The heat process cycles used to achieve this are particularly severe and the containers must be specifically designed to withstand these conditions of temperature and pressure cycles in a steam/water atmosphere. Following heat processing, when the can temperature has returned to ambient, there will normally be a negative pressure in the can, i.e. a vacuum. Under these conditions, the food product itself does not provide any strength to the can to resist external loads.

In the case of carbonated beverage cans, which form the bulk of drink cans filled, once the container is closed, the carbonation pressure continues to provide significant physical support to the container until the moment of opening. In the case of still liquids, such as juices, juice drinks and wine, nitrogen gas may be used to provide the necessary internal pressure for rigidity and compression strength of thin-walled DWI containers.

5.3 Container designs

Regardless of the particular can-forming process used, the shapes of metal containers are very relevant to their cost, physical performance and compatibility with the filled product.

For most metal food and drink containers the cost of the metal itself is 50–70% of the total container cost. The amount of metal in any particular container is, therefore, the most significant cost item, and this is related to the metal thickness, temper and its surface area. In can design, metal thickness is determined by the need for physical performance in handling, processing and storage of the filled container. Surface area is determined by the volume contents and the chosen shape of the container. For ease of manufacture, handling, filling and closing, most food and drink cans have a circular cross section. However, for different physical performance, cost and product uses, cans may vary from shallow (height less than the diameter) to tall (height greater than the diameter). Figure 5.1 demonstrates typical circular food and drink container shapes.

Non-round cross section containers are typically used for fish and meats that are heat processed, as well as for products such as edible oils, which do not need to be processed. These are described in Figure 5.2.

Open trays of round or non-round section are used for baked food products or with lids as *take away* food containers. Powder products such as dried milk, instant coffee and infant formulae are packed into circular cans with lever lids and diaphragm seals.

Closure systems for food and drink cans are by necessity very different in their mode of operation. Food cans require an aperture with either total or virtually full internal diameter of the container through which to remove the product, whereas the aperture for drink cans is designed to suit the method of consumption. Historically, food cans have required a can-opening tool to remove the plain lid. In more recent years, full aperture easy-open ends (FAEOs) have been developed based on designs originally used for drink products. Whether plain or easy-open ends are used, the end panel for virtually all food and drink cans is mechanically seamed-on to produce a double seam that is capable of withstanding all the heat-processing cycles in use. Heat-sealing of foil lids onto metal containers also withstands full heat process cycles, provided overpressure is applied to the retort to reduce the expansion load on the foil

122 FOOD PACKAGING TECHNOLOGY

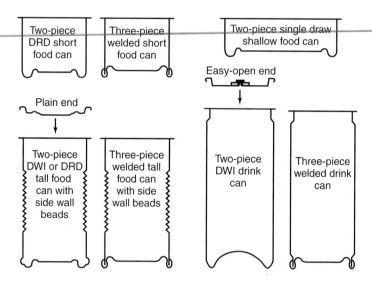

Figure 5.1 Typical round section food and drink cans.

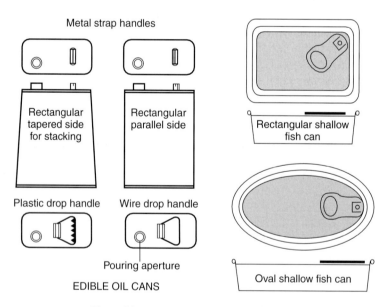

Figure 5.2 Typical non-round section food cans.

membrane. Other pack sealing systems commonly used for less demanding products have screw top closures with wads or sealant material to ensure adequate performance.

5.4 Raw materials for can-making

Steel and aluminium are used for metal container and closure construction for food and drink products. Both are relatively low-cost materials that are non-toxic, having adequate strength and are capable of being work hardened.

5.4.1 Steel

Steel is used in the form of a low-carbon steel which is initially produced as blackplate. This is then converted into tinplate or tin-free steel (TFS) for container and closure manufacture.

Tinplate is created by electrolytically coating blackplate with a thin layer of tin. The tin is coated on both sides of the plate in thickness to suit the internally packed product and the external environment. Different thicknesses of tin may be applied to each side of the plate. Tin, plated in sufficient thickness, provides good corrosion-resisting properties to steel, and is suitable for direct contact with many products including specific foodstuffs such as *white* fruits (e.g. peaches, apricots, pineapple and pears) and certain tomato-based products (e.g. tomatoes in brine and beans in tomato sauce). However, for most foods and drinks it is necessary to apply an organic coating to the inside surfaces of the tinplate container to provide an inert barrier between the metal and the product packed. This barrier acts to prevent chemical action between the product and container and to prevent taint or staining of the product by direct contact with the metal (see later). The tin surface assists in providing good electrical current flow during welding processes. Being a very soft metal, it also acts as a solid lubricant during the wall ironing process of forming two-piece thin wall cans.

TFS, also referred to as electrolytic chromium/chrome oxide coated steel (ECCS), is created by electrolytically coating blackplate with a thin layer of chrome/chrome oxide. This must then be coated with an organic material to provide a complete corrosion-resistant surface. The metallic layer of ECCS provides an excellent key for adhesion of liquid coatings or laminates to the surface. ECCS is usually marginally less expensive than tinplate. However, being a matt surface, after coating with clear lacquer it does not provide a reflective surface like tinplate. ECCS in its standard form is not suitable for welding without prior removal of the chrome/chrome oxide layer. The Japanese steel makers have developed modified tin-free metallic coatings for steel that do permit satisfactory welding of this material.

5.4.2 Aluminium

Aluminium for light metal packaging is used in a relatively pure form, with manganese and magnesium added to improve the strength properties. This material cannot be welded by can-making systems and can only be used for seamless (two-piece) containers. The internal surfaces of aluminium containers are always coated with an organic lacquer because of the products normally packed.

5.4.3 Recycling of packaging metal

Both aluminium- and steel-based packaging materials are readily re-melted by the metal manufacturers. Waste materials arising during the can-making processes may be returned for recycling through third party merchants. Post-consumer metal packaging waste is collected and, after automatic separation from other waste materials, is ultimately returned to the metal manufacturers for re-melting. Aluminium and steel suffer no loss of quality during the re-melting process so may be reused an unlimited number of times for the production of first-quality packaging material. Certain recycling processes permit the tin to be separated from the steel base prior to re-melting.

5.5 Can-making processes

Food and drink cans may be constructed either as three-piece or two-piece containers. Three-piece cans consist of a cylindrical body rolled from a piece of flat metal with a longitudinal seam (usually formed by welding) together with two can ends, which are seamed onto each end of the body. The three-piece can-making process is very flexible, as it is possible to produce almost any practical combination of height and diameter. This process is particularly suitable for making cans of mixed specifications, as it is relatively simple to change the equipment to make cans of different dimension. Container size flexibility facilitates the use of pack promotions offering free extra product.

Two-piece cans are made from a disc of metal that is reformed into a cylinder with an integral end to become a seamless container. To this is seamed a loose end to finally close the can. Drawing is the operation of reforming sheet metal without changing its thickness. Re-drawing is the operation of reforming a two-piece can into one of smaller diameter, and therefore greater height, also without changing its thickness. Drawn and re-drawn containers are often referred to as DRD cans.

Ironing is the operation of thinning the wall of a two-piece can by passing it through hardened circular dies. The draw and wall ironing process (DWI) is very economical for making cans where the height is greater than the diameter and is particularly suited for making large numbers of cans of the same basic specification.

5.5.1 Three-piece welded cans

Three-piece welded food cans are only constructed from steel, as aluminium is not suitable for welding by this particular process. Coils of steel, after delivery from the steel maker, are cut into sheets approximately $1\,m^2$. The cut sheets are then coated, and printed if necessary, to protect and decorate the surfaces. Areas where the weld will be made on the can body are left without coating or print to ensure the weld is always sound. The coatings and inks are normally dried by passing the sheets through a thermally heated oven where the temperature is in the range 150–205°C. Alternatively, for some non-food contact uses, ultraviolet (UV)-sensitive materials may be applied. These are cured instantaneously by passing the wet coating/ink under a UV lamp.

The sheets are next slit into small individual blanks, one for each can body, each blank being rolled into a cylinder with the two longitudinal edges overlapping by approximately 0.4 mm. The two edges are welded by squeezing them together whilst passing an alternating electric current across the two thicknesses of metal (see Fig. 5.3). This heats up and softens the metal sufficiently for a sound joint to be made. If the can is internally coated with lacquer it is generally necessary to apply a repair side stripe lacquer coat to the inside of the weld to ensure coating continuity over the whole can.

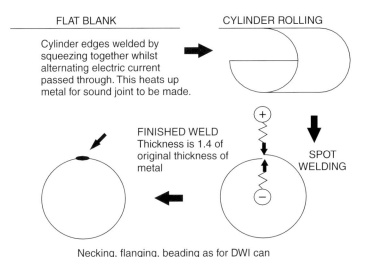

Figure 5.3 Three-piece can welding principles.

For food cans, the can body now passes through a flanging machine where the top and bottom of the can body are flanged outward to accept the can ends. For drink cans, the top and bottom edges of the can body are necked-in to reduce the diameter prior to the creation of the flanges. This permits ends to be fitted which are smaller in diameter than that of the can body, reducing the cost of the end and the space taken up by the seamed can.

For both food and drink cans, one end is then mechanically seamed-on to the bottom of the can body. This end is commonly referred to as the maker's end (ME). Where easy-open ends are fitted to three-piece cans, it is common practice for this end to be fitted at this point, leaving the plain end (non-easy-open) to be fitted after filling. This practice allows the seamed easy-open end to pass through the finished can testing process. The end applied by the packer/filler after can filling is commonly referred to as the canner's end (CE).

At this stage, tall food cans (height-to-diameter ratio more than 1.0) pass through a beading machine where the body wall has circumferential beads formed into it. The beads provide additional hoop strength to prevent implosion of the can during subsequent heat process cycles. All cans finally pass through an air pressure tester, which automatically rejects any cans with pinholes or fractures. This completes the manufacture of empty three-piece food and drink cans.

5.5.2 Two-piece single drawn and multiple drawn (DRD) cans

Pre-coated, laminated and printed tinplate or TFS is fed in sheet or coil form in a reciprocating press that may have single or multiple tools. At each tool station the press cycle cuts a circular disc (blank) from the metal and whilst in the same station draws this in to a shallow can (cup). During the drawing process the metal is reformed from flat metal into a three-dimensional can without changing the metal thickness at any point. After this single draw, the can may be already at its finished dimension. However, by passing this cup through a similar process with different tooling, it may be re-drawn into a can of smaller diameter and greater height to make a draw–redraw can (DRD). This process may be repeated once more to achieve the maximum height can. At each of these steps, the can base and wall thickness remain effectively unchanged from that of the original flat metal. These processes are shown in Figure 5.4. Following this body-forming operation, necking, flanging and beading operations follow according to the end use and height-to-diameter ratio of the can (as for three-piece welded cans).

For all two-piece cans pinhole and crack detection on finished cans is carried out in a light-testing machine. This measures the amount of light passing across the can wall using high levels of external illumination. One advantage of two-piece cans is that there is only one can end instead of two, meaning that one major critical control hazard point is eliminated.

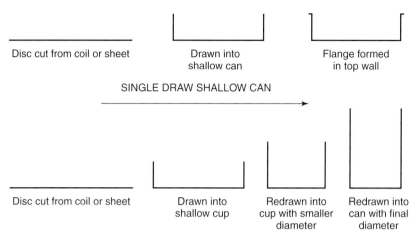

Figure 5.4 Two-piece drawn can forming.

The single drawing process is also used to make aluminium- or steel-tapered shallow trays for eventual heat-sealing with coated metal foil. The container bodies are constructed from metal laminated with organic film. The single drawing process is also used for the manufacture of folded aluminium baking trays and *take away* containers. In this process the aluminium is allowed to fold, as the metal is converted from a flat sheet into a shaped container.

5.5.3 *Two-piece drawn and wall ironed (DWI) cans*

The DWI cans are constructed from uncoated tinplate or aluminium. However, DWI cans for processed food are only made from tinplate as thin wall aluminium cans do not have sufficient strength to withstand the heat process cycles.

For this process, which is described in Figure 5.5, the coiled metal, as it is unwound, is covered with a thin film of water-soluble synthetic lubricant before being fed continuously into a cupping press. This machine blanks and draws multiple shallow cups for each stroke, as described under the section entitled *Drawn cans* above. The cups are then fed to parallel body-making machines which convert the cups into tall cans. This is the drawing and ironing process where the cups are first redrawn to the final can diameter and then rammed through a series of rings with tungsten carbide internal surfaces which thin (iron) the can walls whilst at the same time increasing the can height. During this process the can body is flooded with the same type of lubricant used in the cupping operation. In addition to assisting the ironing process, the lubricant cools the can body and flushes away any metallic debris. No heat is applied to the can during this process – any heat generated being from the cold working of the metal as it is thinned. After the forming of the can body the uneven top edge of the can is

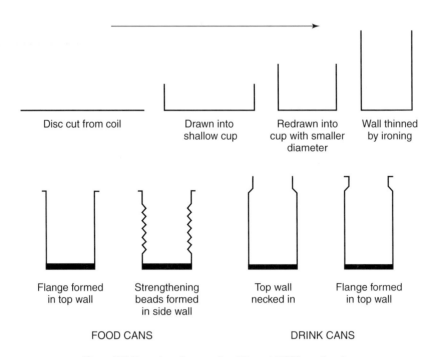

Figure 5.5 Two-piece drawn and wall ironed (DWI) can forming.

trimmed to leave a clean edge and a can of the correct overall height. Trimmed can bodies are passed through chemical washers and then dried. This process removes all traces of lubricant and prepares the metal surfaces for internal and external coating and ultimately external decoration (drink cans only).

For food cans, which will ultimately receive a paper label, an external coating is applied by passing them under a series of waterfalls of clear lacquer which protects the surface against corrosion. The lacquer is dried by passing the cans through a heated oven. Following this the can body now passes through a flanging machine where the top of the can is flanged outwards to accept the can end, which will be fitted after the can is filled with product. The flanged can is next passed through a beading machine which forms circumferential beads in the can wall, to give added strength to the can. After all the mechanical forming operations have been completed, every can is tested by passing through a light tester which automatically rejects any cans with pinholes or fractures. The inside of each can is then coated with lacquer using an airless spray system. The special lacquer is applied to protect the can itself from corrosion and prevent its contents from interacting with the metal. This lacquer is finally dried in a thermal oven at a temperature of about 210°C.

For drink cans, the clean cans are coated externally with a clear or pigmented base coat that forms a good surface for the printing inks. The coating is then dried by passing the cans through a thermally heated oven. The next step is a high-speed printer/decorator which applies the printed design around the outside of the can wall in up to eight colours plus a varnish. A rim-coater coat applies a heavy varnish to the base of each can in order to provide added protection against scuffing during distribution and external corrosion, especially as such products are often kept in the cold humid conditions of chilled refrigerators. The cans now pass through a second oven to dry the ink and varnish. The inside of each can is coated with lacquer using an airless spray system. The special lacquer is applied to protect the can itself from corrosion and prevent its contents from interacting with the metal. This lacquer is finally dried in an oven at a temperature of about 210°C. Following this, the can body now passes through a necker/flanger machine where the diameter of the top wall is first reduced (necked-in) before the top edge is flanged outwards to accept the can end. After all the mechanical forming operations have been completed, every can is tested by passing it through a light tester which automatically rejects any cans with pinholes or fractures.

5.6 End-making processes

Can ends for mechanical double seaming are constructed from aluminium, tinplate or TFS. Aluminium and TFS are always coated on both sides with organic lacquer or film laminate whilst the metal is still in coil or flat sheet form. For tinplate these coatings are optional, depending upon the product being packed in the container and the specified external environmental conditions.

The base of a three-piece can will always be a plain end (non-easy-open). For food cans, the top may be either plain (requiring an opening tool) or full aperture easy-open (FAEO). Rectangular solid meat cans employ a key opening device to detach both the scored body section and ME. For drink cans, the top is usually referred to as a Stay-on Tab (SOT), enabling the opening tab and pierce-open end section to be retained on the can. The SOT end has largely superseded the traditional ring-pull end.

All ends for processed food cans have a number of circular beads in the centre panel area to provide flexibility. These allow the panel to move outwards, as internal pressure is generated in the can during the heating cycle of the process and so reduce the ultimate pressure achieved in the can. During the cooling process, this flexibility permits the centre panel to return to its original position.

Ends for beer and carbonated drink cans do not require the above feature as the can's internal pressure is always positive. The plate thickness and temper have to be appropriate to the level of carbonation of the product and, if

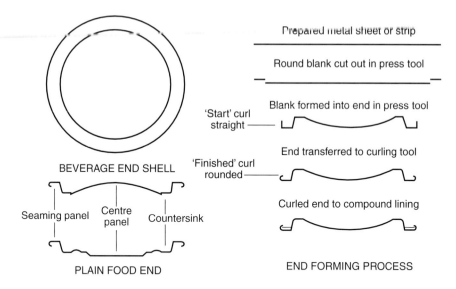

Figure 5.6 Plain end forming.

applicable, pasteurisation treatment; otherwise excessive internal pressure may cause can ends to peak or distort.

5.6.1 *Plain food can ends and shells for food/drink easy-open ends*

The initial processes for making plain food can ends and easy-open ends for food and drink cans are the same. The body of an end that will be ultimately converted into an easy-open end is referred to as a shell.

Plain ends/shells may be stamped directly from wide coils of metal or from sheets cut from coils. Whether from coil or sheet, the metal is fed through a press that produces multiple stampings for every stroke. After removal from the forming tool, the edges of the end shells are then curled over slightly to aid in the final operation of mechanical seaming the end onto the flange of the filled can. After curling, the end shells are passed through a lining machine that applies a bead of liquid-lining compound around the inside of the curl. This process is described in Figure 5.6.

The compound lining is a resilient material that, during mechanical forming, will flow into the crevices of the double seam and thereby provide a hermetic seal.

5.6.2 *Conversion of end shells into easy-open ends*

The principles used in the conversion of end shells are the same for both full aperture food easy-open ends and small aperture drink easy-open ends. The

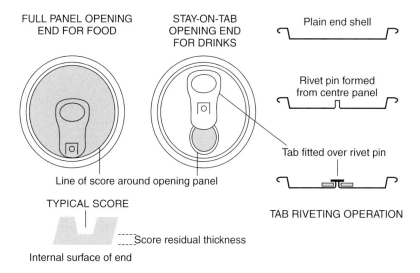

Figure 5.7 Easy-open end conversion.

conversion operations comprise scoring (partially cutting through) the perimeter of the opening panel and attaching a metal tab with which to tear-open the panel. These operations are described in Figure 5.7. Scoring is necessary to reduce the force required to open the end to an acceptable level.

The pull-tab is made from a narrow strip of pre-coated aluminium or steel, which is in coil form. The strip is first pierced and cut, and then the tab is formed in two further stages before it is ready to be joined to the end shell.

The shells pass through a series of dies that score them and form a hollow upstanding rivet in the centre panel of the shell. The tab is then placed over the upstanding rivet on the shell, and the rivet is deformed to make a joint between the two components. The finished ends, ready for capping the filled cans, are packed into paper sleeves and palletised for shipment to the can filler.

5.7 Coatings, film laminates and inks

Organic materials are used to provide barrier or decorative coatings to metal containers and closures. These may be in the form of liquid-applied coatings and inks or film laminates. For three-piece cans, two-piece drawn containers and can ends, the metal is coated and printed while it is flat, in coil or sheet form, prior to the can or end forming operations. For two-piece drawn and wall-ironed containers all coating and decoration is carried out after the can body has been formed.

The coating of metal coil or sheet is always done by roller-coating. For three-piece welded cans with an internal coating, it is usually necessary to apply a coating to the inside of the weld area after the body has been made. This may be done by roller-coating or powder/liquid spray.

The internal surfaces of two-piece DWI cans are coated by airless spray. Although lacquer coatings provide a barrier to metal pick-up, there may be defects present such as micro-channels, micro-cracks or fissures through which metallic ions can transfer to the product. The degree of metal exposure in a lacquered DWI can may be tested by conductivity measurements using an electrolyte solution. For DWI food cans, where paper labels are normally applied, the outside surface is flood coated with clear lacquer with the can in the upturned position. Drink cans have an optional external coating which is applied by roller. This is used to enhance the can decoration and for this reason is usually white.

Metal printing of these products may be onto flat sheet or circular cans, as appropriate. The processes used are, respectively, offset lithography or dry offset. Inks and coatings are formulated for curing by either thermal oven or UV lamp, depending on the particular chemistry of the material.

5.8 Processing of food and drinks in metal packages

It is beyond the scope of this text to discuss the production of the full range of food and drink products that are put into metal containers. This section focuses on the production of foods and drinks in cans, rather than, for example, aluminium trays, though equivalent process stages can be assumed. The text covers stages that will be common to many processes, highlighting generalised good practice where the process challenges the container.

5.8.1 Can reception at the packer

The suitability of a given can and end for an application should be confirmed with the supplier before use. The level of inspection of containers and ends when delivered to the food/drink packer depends upon the working relationship with the supplier. Some packing companies with well-established relationships do no inspection, relying upon certificates of conformance. Other packers carry out inspections according to long established practice, e.g. a small sample per bulk delivery, whilst others use statistical sampling plans, e.g. BS6001-1:1999.

Each unit of cans delivered, e.g. pallet, should be identified to allow traceability to the production lot. Inkjet coding of can bodies aids traceability. It is a good practice to use can lots in production sequence, so that when defects are identified at packing, the whole lot can be put on hold more easily. Likewise, the time of use of a lot of cans should be recorded, so that when a defect is identified after packing, the affected finished product can be recalled.

Table 5.1 Summary of quality assurance checks on in-coming containers at food packers

External inspection of can bodies	Internal inspection of can bodies	Inspection of can ends
Dimensions	Cleanliness	Dimensions
Seam defects on makers end (three-piece cans)	Internal lacquer: presence, continuity and adhesion	Compound: presence and distribution
Welds (non-drawn cans)	Oil	Defective curl
Fractured plate		Damaged
Pin holed (especially for embossing)		Fractured
Damaged		Distorted
Defective flange		Corrosion
External lacquer: presence, continuity and adhesion		Easy-open score: dye penetration
Rust		Easy-open tab
Print quality of lithographed cans		

Packers should maintain a library of defective containers, and at the earliest possible opportunity, classify defects into critical, major or minor in order to keep findings in perspective. Samples of defects are commonly retained for examination by agents of the can suppliers, to aid in decision-making on the fate of a lot.

Inspections of empty cans may include the checks in Table 5.1. Cans awaiting use should be kept in appropriate dry, clean conditions and free from potential mechanical damage, particularly to can flanges. Extra care is required for partially used pallets or containers of stock, to ensure they are adequately re-protected.

5.8.2 Filling and exhausting

Cans should be de-palletised outside of food production areas to minimise the risk of secondary packaging contaminating the food. It is common to see ends being removed from sleeves in food production areas, near the can seamers, but it is not the best practice. Cans should be inverted and cleaned prior to filling. The cleaning may be carried out by an air jet, steam or water (or a combination), depending on the anticipated contamination risk.

Filling should be carried out accurately, as consistency is important for the performance of the container through later process stages (food safety and legality are also considerations, though outside of the scope of this text). The operation should also take place without damage to the can. Product should not be filled in such a manner as to physically prevent the placement of the end on the can. The operation should avoid external contamination of the can and the flange area. Food contamination of can flanges can affect the formation of the seam at closing. In some industry sectors, contamination of the flange is difficult to avoid, e.g. canning of small fish, shredded beetroot and bean-sprouts, where it is difficult to prevent product from overlapping the flange. Manufacturing controls should be in place to minimise the possibility of such cans being seamed, e.g. pre-seamer visual inspection.

It is also important that filling allows a reasonable headspace, as it can influence final vacuum level (residual oxygen affects internal corrosion and product quality) and also helps minimise internal pressure on can ends during heat processing and cooling.

In order to achieve a vacuum, cans may be (a) hot-filled with product and closed with or without direct steam injection into the headspace during double seaming (steam flow closure); (b) filled with product (ambient or hot-fill) and then exhausted (with or without end clinched on) prior to double seaming (with or without steam flow close); or (c) closed in a vacuum chamber. Exhausting involves passing filled cans through a steam-heated chamber (exhaust box) at $c.$ 90°C. Exhausting serves to remove entrapped air from the product (e.g. in a fruit salad), raise initial temperature of the product prior to heat processing (thereby shortening process time/cooking effect) and help ensure that a good vacuum level is achieved. The closure method employed – (a), (b) or (c) – depends on a number of considerations such as cost, product characteristics, production efficiency and vacuum level required.

A finished can vacuum between 10 and 20 in.Hg (70–140 kPa) is common for canned foods. Alternative levels, however, of between 0 and 6 in.Hg (0–40 kPa) may be seen for containers with low headspace, or 26 in.Hg (180 kPa) for high vacuum packs such as sweet-corn in a little brine. The vacuum created is multifunctional. For some packs the vacuum level must be tightly controlled to prevent container damage due to gas expansion during the sterilisation process. The extent to which this is an issue will depend upon the tendency of the product to expand, particularly due to trapped or dissolved air in the product, and the sensitivity of the container, e.g. cans with easy-open ends or weak plate. Where pack damage is likely the can vacuum will be routinely monitored after seaming.

Historically, the presence of a vacuum in food cans has been used as an indicator of freedom from pack leakage and/or spoilage of the contents. Loss of vacuum is, therefore, sometimes detected by dud-detection systems (e.g. Taptone sound detector) as an indicator of pack failure prior to finished product labelling.

Carbonated beverage filling speeds of 2000 DWI cans per minute can be achieved using a rotary filler and seamer. Pressure control is important to achieve the right level of carbonation of the product and avoid excessive foaming, leading to high spillage and wastage, due to under-filled cans, and high levels of air in cans (effecting internal corrosion and deterioration of product quality). The objectives for pressure control in beverage and beer cans are quite different from that in food cans, as the aim is a positive pressure in the finished pack. However, excessively high pressures must be avoided. Control involves de-aeration of ingredients and using low filling temperatures (often 5°C or lower) to give greater stability of the carbonation. The use of low filling temperatures does create a risk of external condensation after closure.

For carbonated beverages at the filling head, there is a balance in the force required to hold the can against the head and that required to restrain pressures developed during filling.

Effective washing of cans after double seaming is often required to minimise external carry-over of food/drink components that may contribute to container corrosion at a later process stage, e.g. acid syrups or brine.

5.8.3 Seaming

The most important closure type used for metal containers is the double seam. The seam is formed in two operations from the curl of the can end and the flange of the can body (Fig. 5.8).

For heat-processed foods, the operation of the can seamer is critical to process safety and would normally be regarded as a critical control point in a hazard analysis for such products. The control is the correct set-up of the can seamer for the can being closed. This critical control point will be monitored by periodic visual inspection. For example, sampling every 30 min and making measurements from seam tear-downs or cut sections (e.g. 1 can per seamer head every 4 h of continuous production) (Department of Health, 1994). For beverage cans, the safety risks are less significant, but seam defects can have huge financial implications, so standards are equally high. Additional inspections should be carried out after jams where cans have become trapped in the seamer. A successful operation requires well-trained staff to maintain and operate the seamers, and to assess seam quality.

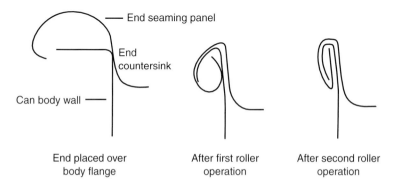

Figure 5.8 Stages in the formation of a double seam.

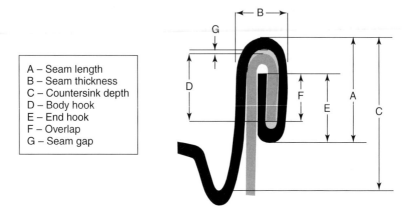

Figure 5.9 The main measured parameters in a double seam.

A – Seam length
B – Seam thickness
C – Countersink depth
D – Body hook
E – End hook
F – Overlap
G – Seam gap

Ideally, seam specifications are provided by the can maker, though under some circumstances in-house standards are used (though it would be unwise to allow greater tolerances than those given by the can maker). To aid the comparison between suppliers, general seam standards are available from organisations such as the Metal Packaging Manufacturers Association (MPMA, 1993, 2001). These define seam parameters, and clarify those that are considered critical for acceptable can seams. In the UK a typical list would be the following parameters:

- seam thickness
- seam height (length)
- overlap
- free Space
- body hook butting
- % wrinkle or tightness rating (this parameter is unusual as it is not measured from the seam section but by visual inspection of the removed cover hook).

These parameters are measures of the extent of folding of the hook created by the can body flange into the hook created by the end curl, ensuring that there is sufficient overlap and that the five layers of metal are sufficiently compressed together (Fig. 5.9).

In other parts of the world different parameters may be emphasised. For example, in the USA, body hook butting is commonly not considered, while greater emphasis is placed upon overlap and the presence of a visible and continuous pressure ridge.

It should be noted that although there is some interchangeability between seamer tooling and can suppliers, it is good practice to recheck seamer performance when changes are planned.

When assessing a seam it is normal to take multiple sections. For cans with a soldered or welded side-seam, the weld is taken as 12 o'clock, and sections would typically be taken from 2 and 10 o'clock which is historically where most problems occur. For drawn cans, two points opposite each other are used. For more complex shapes, which are more difficult to seam, a greater number of sections are taken, e.g. two long sides and all four corners of a rectangular can.

It is good practice to produce control charts of the critical seam parameters so that drift or step changes in values can be quickly identified, and corrected.

5.8.4 Heat processing

The temperatures to which cans are exposed during food and drink sterilisation (typically, 115–135°C) and pasteurisation processes (typically, 90–105°C) are relatively low compared to those used in the can manufacturing process so will not generally be a limiting factor. However, the combined temperature/mechanical conditions in a retort are challenging to containers.

Campden & Chorleywood Food Research Association is an internationally recognised source of guidelines on thermal process evaluation for a wide range of heat preserved food. It is always necessary, however, to conduct in-plant heat penetration tests on canned product in order to establish recommended process parameters of time and temperature.

When cans undergo sterilisation or pasteurisation by heating, internal pressure developed inside the container can be sufficient to cause distortion of the container, or opening of easy-open scores on can ends. This potential is greatest at periods of maximum pressure differential between the process medium and the can interior, for example, during the change from heating to cooling. The design of cans allows for the pressure developed during normal food manufacturing operations; however, this still relies upon correct operation of the heating equipment to ensure that pressure differentials are not excessive. For example, for a classic batch steam retort, when cooling commences it is common for the operator to introduce both water and compressed air into the retort so that as the steam collapses, the air provides the external pressure to prevent the can from distending. Failure to control the air pressure profile correctly will cause significant damage to batches of cans, typically referred to as peaking (distension) or panelling (collapse). Likewise, in continuous sterilisation systems such as reel and spiral retorts that operate with transfer valves between pressurised shells, these valves must maintain the correct pressure.

Loading of batch heat processing systems may be as layers of cans into a basket, commonly onto a false bottom which drops with successive layers (a Busse system), or scramble packed. Generally, scramble packing will present greater challenges to the container than layered loading due to container container impacts. The effect of these impacts can be minimised by loading the baskets in a water tank.

Rotary batch retort systems require layers of cans to be clamped down into baskets using a pneumatic or mechanical press, preventing can movement during the heat process that will result in impacts/scuffing. However, balanced against the desire to restrain movement is the need to avoid crushing of any part of the load.

The heating environment can create extra difficulties in terms of prevention of external container corrosion as electrochemical reactions can be established between the machinery and container, through process water (Mannheim *et al.*, 1983).

It is vitally important that saturated steam retorts are properly vented to remove air which may lead to external corrosion of cans or, more seriously, under-processing due to poor heat transfer. In addition, carry-over of boiler chemicals into the retort may also have some impact upon corrosion. Older retorts tend to be constructed of mild steel which gradually corrodes leaving rust particles present on cans at the end of the heat process that will add to corrosion problems on containers.

Some plant and retort designs are inherently damaging to the lacquers on can ends because the cans are rolled on their seams. Embossed codes used by packers can also challenge can integrity and corrosion protection.

Beverage cans that require pasteurisation are typically heat processed in tunnel pasteurisers, or hot filling may be sufficient.

5.8.5 *Post-process can cooling, drying and labelling*

The drying of sterilised food containers after cooling is critical to minimise microbiological recontamination risks. For both food and beverage cans drying of cans is important to prevent external container corrosion during storage.

It is well established that can double seams are not completely settled immediately after heat processing, to the extent that very small volumes of water may be sucked back through the seams i.e. micro-suction. If water entering the can carries microorganisms, the recontamination can lead to either food poisoning or spoilage (CCFRA, 1980). This possibility can be minimised by ensuring that seams are of good quality (and that the pack is free from other puncture defects), by ensuring good hygienic conditions immediately after sterilisation and by drying containers. One means of minimising microbiological recontamination risks is to use cooling water containing biocides such as free chlorine, though levels should be controlled to prevent accelerated container corrosion. It is vitally important that cans are not handled whilst still wet and hot as, in the worst case scenario, micro-suction of pathogenic microorganisms such as *Clostridium botulinum* spores into the can will cause a potentially fatal food poisoning.

The quality of the can cooling water should be routinely monitored as a high salt content can leave hygroscopic salt residues on can surfaces and cause rusting. A small amount of money spent on a simple water conductivity meter to routinely monitor total dissolved solids can save product and financial loss by preventing spoilage.

To a large extent, food-can drying may be achieved by removing containers from the heat process at a temperature sufficiently high to drive off residual moisture. A (shaken can) temperature of about 40°C is recommended for cans leaving a retort. There is, however, an upper temperature limit, as commercially sterile products should not be held within the growth temperature range of thermophilic microorganisms which will survive the heat process. Some products, such as raw mushrooms, may have a high natural thermophile count. Depending on the likely ambient temperature during distribution, a more severe sterilisation process may be necessary.

Various types of drying aide can be used including

- tipping devices to empty water from countersunk ends
- air knives
- hot beds upon which cans are rolled
- surfactant dips.

Labelling, shrink-wrapping of trays and stretch-wrapping of pallets, when cans are wet is particularly problematic because of prolonged entrapment of water. Use of the correct label paper quality and recommended adhesive is important. For example, highly acidic or alkaline starch-based adhesive can effect external can corrosion.

5.8.6 Container handling

Any visual defect on a container should be regarded as significant, simply because it may influence a customer's decision to purchase that container. Where the defect is severe this may threaten the integrity of the container either by puncture or allowing corrosion. The US National Food Processors Association have published guidelines for sorting of defective food cans (NFPA, 1975).

Failure to re-engineer container handling systems that repeatedly cause damage must be regarded as a bad practice. Impacts during handling after the heat process, especially on seam areas, play a significant part in allowing microbiological recontamination of can contents. Guide-rails on conveyor systems should avoid contact with sensitive seam areas. Necked-in designs for cans reduce seam to seam impacts.

5.8.7 Storage and distribution

Both processed and empty cans should be stored in controlled conditions. Condensation that can lead to corrosion (see Section 5.10) is a major issue if inappropriate combinations of container temperature, air temperature and air humidity are allowed to occur. The risk of condensation at a given set of storage conditions can be predicted using psychrometric charts. Historically, the use of tempering rooms to avoid sudden changes of temperature has been recommended. Where the risk is severe and climate control difficult, e.g. shipping across the equator, the use of desiccants within secondary packaging might be considered. Warehouse storage should ensure that there are no draughts, windows are closed and air movement is minimised e.g. door flaps. Stretch-wrapping of pallets and shrink-wrapping of trays help to reduce the risk of condensation as well as the accumulation of salt-laden dust which is hygroscopic and can effect rusting. Efficient stock control will serve to minimise the risk of corrosion. Coastal canneries are particularly at risk from corrosion due to airborne salt from sea spray. Extra precautions may be necessary, e.g. externally lacquered can components. Can ends are frequently externally lacquered to provide extra protection against external corrosion, particularly at the top edge of the double seam.

Paperboard divider sheets or pallet layer pads are often re-used and salt residues can accumulate leading to rusting and possible perforation of double seams. Paper should be within the recommended pH and salt (chloride and sulphate) content range. Freezing of filled cans should also be avoided to prevent deformation of cans due to ice expansion.

Absence of water is important to minimise external corrosion and also to prevent ingress of bacteria through the can seams. For example, shipping containers and any building used for storage should be water-tight.

Pallet systems used for transporting cans should restrain the load in a manner that minimises the risk of damage during distribution. This may involve using a frame on top of the pallet, straps and/or shrink-wrapping/stretch-wrapping. The pallets themselves should have the correct performance specification and be in good mechanical condition. Protruding nails can present a food-safety risk by can puncture; it is also bad practice to use staple guns in close proximity to finished cans for the same reason. The moisture content and chemistry of pallet wood and other secondary packaging materials should be suitable to prevent corrosion issues.

Can specifications should include the axial loads that they can sustain, which enables makers to advise on maximum stacking heights for storage of finished goods, to prevent compression failure of lower can layers in the pallet stack. Warehouse practices should ensure even distributions of weight on supporting pallets. Containers should be held off the floor, and away from walls, to prevent moisture build up. Division of stocks into smaller units will aid in ventilation.

Stocks of canned goods should be regularly inspected to remove damaged containers, as one defective container can trigger a chain reaction through the spread of moisture and corrosive product spills through a stack.

The decoration of litho-printed cans is sensitive to scratching on conveyor systems and measures, such as using plastic-coated guide rails, may need to be taken.

Many instances of food poisoning associated with post process leakage of cans have resulted from case cutter damage at the point of opening for retail display (Stersky *et al.*, 1980).

5.9 Shelf life of canned foods

Canning of heat preserved foods is a method of food preservation that relies upon the hermetic sealing of foods inside a metallic container and the sterilisation or pasteurisation of the food by heat treatment. No preservatives are therefore necessary to prevent the food spoiling due to the growth of microorganisms. Some chemical reactions can, however, continue to take place inside the can, albeit slowly; these include breakdown of colour, flavour and other natural food components. In addition, the food interacts with the container.

The shelf life of canned foods is determined by a variety of factors but all relate to deteriorative reactions of some form or another, either those introduced during production or processing activities or those occurring during storage. In order to understand these processes better it is important to be able to define shelf life more exactly. Shelf life can be defined in two ways: minimum durability and technical shelf life.

- *Minimum durability* is defined as the period of time under normal storage conditions during which a product will remain fully marketable and will retain any specific qualities for which express claims have been made. However, beyond this point the food may still be satisfactory for consumption.
- *Technical shelf life* is defined as the period of time under normal storage conditions after which the product will not be fit to eat.

For example :

- A canned fruit product is being sold with the claim 'contains 10 mg/100 g of vitamin C'. On production, the product contained more than 10 mg/100 g but after 18 months storage, the vitamin C content has been reduced to only 7.5 mg/100 g. The minimum durability has therefore been exceeded but this loss of vitamin does not make the food unfit to eat. The technical shelf life has therefore not been reached.

- A second product is found after two years to contain 250 mg kg^{-1} of tin. This level is above the maximum UK legal level of 200 mg kg^{-1} tin (Tin in Food Regulations, 1992) and so the technical shelf life has been exceeded.

Three main factors affect the shelf life of canned foods and are implicated in deteriorative reactions:

- sensory quality of the foodstuff, including colour, flavour (plus taints) and texture
- nutritional stability
- interactions with the container.

The first two of these are outside the scope of this chapter, the remainder of which will concentrate on container interactions, both with the can contents and with the external environment.

5.9.1 Interactions between the can and its contents

All foods interact with the internal surface of the can in which they are packed. The most common form of this interaction is corrosion. In plain tinplate containers, this takes the form of etching or pitting corrosion, and staining of the surface may also occur. However, as described earlier in this chapter, internal lacquers are available which reduce this effect by providing a barrier between the food and the metal can wall. This also allows the use of other forms of metal container (e.g. tin-free steel or aluminium) which would otherwise be corroded very quickly.

In the unlacquered form, only tinplate has any corrosion resistance to the acids found in foods; all the other metals must be lacquered. Even tinplate must be lacquered where particularly aggressive products are packed, such as tomato purée, or where there is a danger of pitting corrosion or surface staining (for example, in meat products).

5.9.2 The role of tin

Conventional food cans are composed primarily of steel with a thin layer of tin applied to the internal and external surfaces. The tin coating is an essential component of the can construction and plays an active role in determining shelf life. The most significant aspect of the role of the tin coating is that it protects the steel base-plate which is the structural component of the can. Without a coating of tin, the exposed iron would be attacked by the product and this would cause serious discoloration and off-flavours in the product and swelling of the cans; in extreme cases the iron could be perforated and the cans would lose their integrity. The second role of tin is that it provides a chemically reducing environment, any oxygen in the can at the time of sealing being rapidly consumed by the dissolution of tin. This minimises product oxidation and prevents colour loss and flavour loss in certain products. It is this positive aspect of tin that makes it appropriate for particular product types to be packed in tinplate containers with internally plain (i.e. unlacquered) can component(s) – body and/or ends. Several attempts to replicate this effect of quality preservation with certain products – for example, by introducing tin into lacquers and adding permitted tin salts – have been made, but none are as effective as the normal tinplate can. The increasing use of fully lacquered cans for some of these products in recent years, in order to reduce the tin content of the product, is generally believed by the food manufacturers to have resulted in some loss of quality in the food product.

In order to confer these positive attributes the tin must dissolve into the product. The rate of dissolution is normally relatively slow and shelf life is specified such that the level of tin remains below the UK legal limit of 200 mg kg^{-1} within the anticipated shelf life. Container and product specifications are defined to ensure that this is achieved.

Tin corrodes preferentially off tinplate surfaces due to the ability of tin to act as a sacrificial anode in the corrosion process. The corrosion of tin is, however, relatively slow due to the large hydrogen over-potential which exists on its surface. This protects the steel from corrosion and explains why a relatively thin layer of tin is able to provide such good corrosion protection, see Figure 5.10.

Most food materials contain very low levels of tin (<10 mg kg^{-1}), although foods packed in containers which have internally exposed tin may, under certain conditions, contain much higher levels.

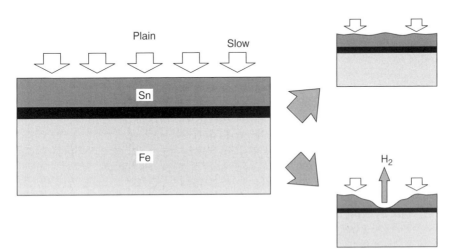

Figure 5.10 Corrosion processes in a plain, unlacquered tinplate can. Normally the tin dissolves evenly. If localised detinning occurs, however, the underlying iron is attacked and hydrogen is evolved.

5.9.3 The dissolution of tin from the can surface

Tin in canned food is derived from the tin coating which dissolved into the product during storage. This time dependence, together with the many other factors that control tin content, make the concept of mean tin levels difficult to deal with, even for a single product. One of the few generalisations that can be made is that tin levels in products packed in fully lacquered cans are very low. In cans with an unlacquered component, however, corrosion is essential in that it confers electrochemical protection to the iron which makes up the structural component of the can. Tin pick-up is normally relatively slow – typically, 3–4 mg kg^{-1} month^{-1} for a 73×111 mm plain-bodied can of peach slices in syrup – and should not give rise to excessive tin levels within the expected storage life of a product. Under certain and unusual circumstances, however, dissolution of the tin is more rapid than it should be and high tin levels can be reached. Many factors interact in a complex way to affect the rate at which tin concentration increases in the food. It is because of these complex interactions that the only way of reliably predicting the rate of tin pick-up, and therefore shelf life, is through packing trials and previous experience of the product. There are numerous factors that influence the rate of tin pick-up and these factors are well established:

- *Time and temperature.* Tin is dissolved over time at a rate influenced by storage temperature, initially at a higher rate than later in storage.
- *Exposure of the tinplate.* The area of exposure of tin is less important than the presence or absence of exposed metal. Containers with no exposed tin will give low tin levels, whereas products where there is even partial exposure, e.g. asparagus cans with one plain end or a tin fillet, will dissolve tin at significant levels.
- *Tin-coating weight.* Although the thickness of the tin coating will ultimately limit the maximum possible level of tin, the rate of tin pick-up is increased when thinner tin coatings are used. Many other aspects of container specification are also important, e.g. tin crystal size, passivation treatment etc.
- *Type and composition of the product.* Several factors such as acidity/pH have a direct influence on the rate of tin dissolution. Certain compounds such as specific organic acids and natural pigments may complex metals to alter the corrosivity of the product in respect of tin and iron.
- *Presence of certain ions.* Certain ions such as nitrates can greatly increase the rate of corrosion. These can arise from the product itself, from ingredients such as water and sugar, or from contaminants such as certain fertiliser residues.
- *Vacuum level.* Two chemical factors that increase the rate of tin pick-up are residual oxygen and the presence of chemical compounds such as nitrates (sometimes referred to as cathodic depolarisers) (Fig. 5.11).

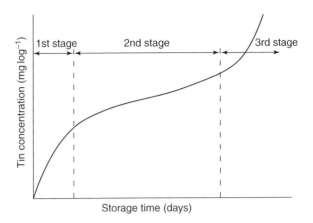

Figure 5.11 Tin dissolution in acidic foods, schematic rate curve (adapted from Mannheim & Passey, 1982).

These accelerators are used up as tin is dissolved and therefore primarily influence the earlier stages of corrosion (phase 1). This means that as storage continues and the tin concentration increases, the rate at which tin is dissolved normally falls and the level of tin tends to plateau off (phase 2). This reduced level of de-tinning continues until most of the tin is dissolved and significant iron exposure occurs, when the rate at which tin is dissolved accelerates again (phase 3).

This third phase of the plot is normally outside the normal shelf life and is therefore seldom of any significance until high tin levels are reached as most of the tin coating has been removed and significant iron exposure occurs.

5.9.4 Tin toxicity

High concentrations of tin in food irritate the gastrointestinal tract and may cause stomach upsets in some individuals, with symptoms which include nausea, vomiting, diarrhoea, abdominal cramps, abdominal bloating, fever and headache. These are short-term symptoms with recovery expected soon after exposure. These effects may occur in some individuals at tin concentrations above 200 mg kg^{-1} (the legal limit) with an increased risk of effects at concentrations above 250 mg kg^{-1}. A wide range of foods in internally plain tinplate cans have been consumed for many years without any long-term health effects having been identified.

Tin corrosion occurs throughout the shelf life of the product. It is therefore imperative to take steps to reduce the rate of corrosion. Accelerating factors include heat, oxygen, nitrate, some chemical preservatives and dyes, and certain particularly aggressive food types (e.g. celery, rhubarb). A high vacuum level is one effective method of reducing the rate of tin pick-up in cans with un-lacquered components. There is a legal maximum level of 200 mg kg^{-1} of tin in food products in the UK (Tin in Food Regulations 1992, S.I. 1992 No. 496), and this level becomes the limiting shelf life time point in most cases.

5.9.5 Iron

There is no recommended maximum level or legal limit for the iron content of foods. Iron is an essential element in the diet, and so this aspect plays no part in limiting the legally permitted shelf life of food products. However, high levels of iron in the food will make it unpalatable. Dissolution of iron does occur from tinplate and from TFS containers although its rate is limited by physical factors such as the amount of steel baseplate which is exposed through the tin layer or through the lacquer. All tinplate containers have microscopic pores in the tin layer exposing the steel beneath. Normally these corrode at a slow rate but under certain situations pitting corrosion may occur, leading to preferential attack on the steel with deep *craters* or *pits* being produced which could lead to perforation and product spoilage (Fig. 5.12).

High iron corrosion usually only occurs towards the end of tin corrosion when significant areas of steel become exposed. Once the base steel is exposed,

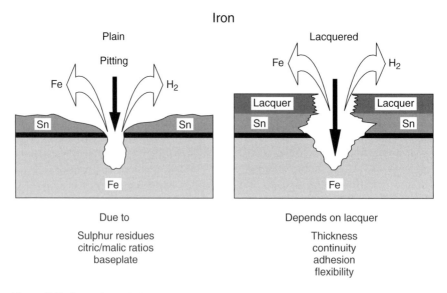

Figure 5.12 Comparison of pitting process in unlacquered (left) and lacquered (right) tinplate food cans.

components of the product (e.g. fruit acids) may corrode the iron and yield hydrogen gas which causes cans to swell.

Iron levels bring about an end to shelf life when they affect the product through flavour changes or colour changes. Even at relatively low levels, iron pick-up in lacquered cans may cause metallic taints for canned products, e.g. certain lager beers and colas. Certain wines cannot be successfully be packed in DWI tinplate cans due to their very high iron sensitivity. Instead, they need to be packed into aluminium DWI cans. Dissolved iron can also cause colour changes in certain products and chelating agents may be used to counter this effect.

5.9.6 Lead

Lead was a problem with older, soldered cans but levels are now very low. However, some tinplate is contaminated with minimal amounts of lead, and certain environmental pressure groups in the USA are pressing for reduction in these levels. The manufacture of lead soldered cans may still be found in the developing world.

5.9.7 Aluminium

All aluminium cans have very good lacquer systems to prevent contact of the food with the metal. Therefore, aluminium levels are generally very low, but occasionally even this low level may affect sensitive products such as beer, causing cloudiness or haze.

5.9.8 Lacquers

The presence of lacquer or enamel very effectively limits dissolution of tin into the product, and so the use of lacquers is becoming increasingly common, even with those products which were previously packed in plain tinplate cans.

There are several different types of lacquer in common use today. By far the most common type is the Epoxy Phenolic group, which are suitable for packing meat, fish, vegetable and fruit products. These have largely replaced the Oleoresinous group, which had a similar wide range of application. Some canners use cans lacquered with vinyl resins, which have the important quality of being free from any taste and odour, and are therefore particularly suitable for dry packs such as biscuits and powders, but also some drinks. White vinyl lacquers have been used where staining of the underlying metal caused by reaction with the product is a problem. Also, white vinyl lacquers have been used for marketing reasons in order to present a hygienic/clinical appearance and not the aesthetically undesirable corrosion patterns on tinplate. The Organosol group are also free from any taste or smell, and have also found applications for beverage cans.

In a three-piece can, it is often desirable to protect the exposed metal at the side-seam, even if the rest of the inside can body is not to be lacquered. This strip of lacquer is sometimes known as the side-stripe. A number of the lacquers mentioned above are used for this purpose, but some powdered coatings are also used.

Coatings are well screened before release for canned food applications. As part of due diligence, the coating application and cure conditions have to be strictly adhered to and full cure regularly confirmed.

5.10 Internal corrosion

In addition to the gradual dissolution of tin or iron from the internal surface of cans during their shelf life, described above, failure of cans may also be caused by internal corrosion as a result of mechanical damage to the cans or a manufacturing fault, or an unusually aggressive reaction between the can and its contents.

Mechanical damage to cans, such as denting caused by poor handling, can result in cracking of the internal lacquer. This will allow the product to gain access to the underlying metal, and may result in quite rapid localised corrosion, depending on the can and the product.

The formation of beads in the can body or rings in the can ends (see above) can sometimes result in either cracking of the internal lacquer at these points or loss of adhesion between the lacquer and the metal. Either may eventually result in local corrosion of the metal by the product. The cause of the problem often lies in insufficiently flexible lacquers, caused either by an excessive lacquer thickness or incorrect stoving (curing) of the lacquer. Similarly, the formation of embossed codes on can ends may also result in cracking of the lacquer, leading to local corrosion.

Occasionally, internal corrosion may result from an unusually aggressive reaction between the can and its contents, causing the lacquer to peel away from the can surface. The causes of these reactions are often very complex, and sometimes the only solution is to use a different lacquer.

5.11 Stress corrosion cracking

Stress corrosion is the acceleration of corrosion in certain environments when metals are externally stressed or contain internal tensile stresses due to cold working. Stress corrosion is one of the most important types of corrosion because it can occur in so many metals. Because the conditions that cause cracking in one metal may not cause cracking in another, it is very difficult to predict where attack will occur. Stress corrosion cracking is sometimes seen in steel cans in the beaded area of the body, where cracks occur in the metal and are preferentially corroded.

5.12 Environmental stress cracking corrosion of aluminium alloy beverage can ends

The aluminium alloy used for the manufacture of easy-open ends for drink cans is specially developed to give the required mechanical properties. This alloy is however subject to environmental stress cracking corrosion due to reaction with moisture. This process is also greatly accelerated by the presence of contaminants such as residual salts, notably chlorides and other halides. The score areas on both pull-tab and stay-on-tab easy-open ends are particularly susceptible to this form of cracking corrosion because of the tensile stress to which this part of the end is subjected. This problem cannot occur without the presence of moisture, so great care must be taken after can filling that easy-open can ends are thoroughly washed with clean water and dried before being put into store. Even during filled can storage, humidity conditions need to be controlled by provision of adequate ventilation etc.

5.13 Sulphur staining

Sulphur staining or *Sulphide* staining is characterised by blue-black or brown marks on the inside of tinplate or tin-free steel cans. In lacquered cans, this headspace phenomenon occurs during processing, and is caused by sulphur compounds from the proteins in the product reacting, in the presence of residual oxygen, with iron in solution which usually originates from base steel exposure at a cut edge or other point (pores, worked radii on expansion beads of ends) where iron exposure is increased. The black deposit formed is a complex of iron sulphides, oxides and hydroxides. Sulphur staining can occur with protein-containing products, e.g. peas, sweet corn, fish or meat. It is most obvious in the headspace. It is regarded as no more than a cosmetic problem, as it is not harmful in any way, and does not normally lead to further corrosion. However, it does look unsightly, and often results in consumer complaints. For this reason, when products susceptible to sulphur staining are packed, a can lacquer is usually selected which will either resist the penetration of the sulphur compounds, or mask the problem. These latter lacquers are generally grey in colour and contain zinc or aluminium compounds, which react with the sulphur compounds to produce white metal sulphides which are harmless and not readily visible. However, this approach is not suitable for acid products, where the acids may attack the coating to produce zinc or aluminium salts which could be harmful to health.

5.14 External corrosion

Any problem which causes external damage to the container may terminate its shelf life earlier than intended. Particularly important is the avoidance of external corrosion. Since rusting requires the presence of metal, oxygen and moisture, it can be prevented by removing any one of the factors. Outside the can, moisture is the only factor that can be readily controlled. External corrosion may be exacerbated by any of the following factors:

Condensation due to	temperature fluctuations, humidity changes, draughts, poor stacking
Labels and paperboard divider sheets	high chloride or sulphate
Incomplete drying	free water
Hygroscopic deposits	moisture from humid environment
Low external tin coating and/or lacquer coverage	high metal exposure
Physical damage	damage to lacquer and tincoating of metal
Rusty retorts	rust particles
Poor venting of retorts	oxygen and water
Boiler water carry-over into retort	alkali detinning
Label adhesive	too acidic or alkaline

External corrosion often occurs at specific points on the can, such as the end seams, or the score lines on easy-open ends. This is sometimes interpreted as a fault in the can, whereas in many cases, the problem is due to poor drying or storage of the cans. In these circumstances, corrosion has simply begun at the weakest point on the can.

External corrosion can also be caused by leaking product from neighbouring cans. This can be a particular problem with beverage cans, where a single leaking can, possibly caused by mechanical damage, results in product leaking all over the other cans in the store.

5.15 Conclusion

In general, the shelf life of a can will depend upon the product, the can specification and the storage conditions in which it is held. Every can is a unit of sale and as such each and every can must comply with the legislation relevant to it. For this reason, and because of the many variables associated with the product, from the grower through to the retailer, it is very difficult to assign a specific shelf life to any product. The experience of the packer is most critical in arriving at sensible conclusions. Overall, however, due to the many different factors affecting shelf life, it is often impossible to predict, and tests with the actual product are the best course of action.

Recent innovations in the design and manufacture of metal packaging for food products include: large opening stay-on-tab ends for drink cans, widgets to provide a foam head to beer and chilled coffee, self-heating and self-chilling drink cans, full aperture food can ends which are easier to open, square section processed food cans for more efficient shelf storage, peelable membrane ends for processed food cans, two-piece draw and wall iron as well as two-piece draw redraw cans made from steel with plastic extrusion coatings.

These innovations and many others not specifically referred to here will ensure that metals will continue to have an extremely important part to play in the cost efficient packaging of foods for short or long term ambient storage conditions. The inherent strength of metal containers and the fact that they are impervious to light contribute to a high level of protection for the contained product over long shelf life periods.

References and further reading

Page, Bev (2001) *Metal Packaging, An introduction*, Pira International Ltd, Leatherhead, Surrey KT22 7RU, UK.

CCFRA (1980) *Post Process Sanitation in Canneries*, Technical Manual No. 1, Campden & Chorleywood Food Research Association, Chipping Campden, Gloucestershire, UK.

CCFRA (1984) *Visual Can Defects*, Technical Manual No. 10, Campden & Chorleywood Food Research Association, Chipping Campden, Gloucestershire, UK.

Department of Health (1994) *Guidelines for the Safe Production of Heat Preserved Foods*, HMSO, London.

Downing, D.L. (1996) Metal containers for canned foods, in *A Complete Course in* Canning and Related Processes. Book 2: Microbiology, Packaging, HACCP and Ingredients, CTI Publications.

Mannheim, C.H., Adan, H. and Passy, N. (1983) External corrosion of cans during thermal treatment. *Journal of Food Technology*, **18**, 285–293.

Mannheim, C.H. and Passy, N. (1982) Internal corrosion and shelf-life of food cans and methods of evaluation, *CRC Critical Reviews in Food Science and Nutrition*, **17**(4), 371–407.

Metal Packing Manufacturers' Association (1993) *Recommended Industry Specifications for Open Top Processed Food Cans*, Metal Packaging Manufacturers' Association, Maidenhead, Berkshire SL6 1NJ, UK.

Metal Packing Manufacturers' Association (2001) *Recommended Industry Specifications for Beer and Carbonated Soft Drink Cans*, Metal Packaging Manufacturers' Association, Maidenhead, Berkshire SL6 1NJ, UK.

Morgan, E. (1985) Tinplate and modern canmaking technology, Oxford: Pergamon Press.

NFPA (1975) *Safety of Damaged Canned Food Containers – Guidelines for Consumers, Regulatory Officials, Canners, Distributors and Retailers*. Bulletin 38-L, National Food Processors Association 1133 20th Street, N.W. Washington DC 20036.

Pilley, Kevin, P. (1994) Lacquers, varnishes and coatings for food & drink cans and for the metal decorating industry. Birmingham: ICI Packaging Coatings.

Stersky, A., Todd, E. and Pivnick, H. (1980) Food poisoning associated with post-process leakage (ppl) in canned foods. *Journal of Food Protection*, **43**(6), 465–476.

Tin in Food Regulations (1992) S.I. 1992 No. 496, London: The Stationary Office.

6 Packaging of food in glass containers

P.J. Girling

6.1 Introduction

6.1.1 Definition of glass

The American Society for Testing Materials defined glass as 'an inorganic product of fusion which has cooled to a rigid state without crystallizing' (ASTM, 1965).

Chemically, we know that glass is made by cooling a heated, fused mixture of silicates, lime and soda to the point of fusion. Morey says that, after cooling, it attains a condition which is continuous with, and analogous to, the liquid state of that substance, but which, as a result of a reversible change in viscosity, has attained so high a degree of viscosity as to be for all practical purposes solid (Morey, 1954).

We know that the atoms and molecules in glass have an amorphous random distribution. Scientifically this means that it has failed to crystallize from the molten state, and maintains a liquid-type structure at all temperatures. In appearance it is usually transparent but, by varying the components, this can be changed-as also can important properties such as thermal expansion, colour and the pH of aqueous extracts. Glass is hard and brittle, with a chonchoidal (shell-like) fracture.

6.1.2 Brief history

Glass beads and arrow heads have been found that date back to the bronze age, which started in the eastern end of the Mediterranean area around 3000 BC. Ornamental glass has been found in excavations in Egypt and Mesopotamia. The invention of the blow stick in Roman times led to the manufacture of hollow glass containers. Glass became one of the earliest forms of packaging. Container manufacture was mechanized in the United States in the late 19th century.

6.1.3 Glass packaging

The two main types of glass container used in food packaging are bottles, which have narrow necks, and jars and pots, which have wide openings. Glass closures are not common today, but were once popular as screw action stoppers with rubber washers and sprung metal fittings for pressurized bottles, e.g. for carbonated beverages, and vacuumized jars, e.g. for heat preserved fruits and vegetables. Ground glass friction fitting stoppers were used for storage jars, e.g. for confectionery.

6.1.4 Glass containers market sectors for foods and drinks

A wide range of foods is packed in glass containers. Examples are as follows: instant coffee, dry mixes, spices, processed baby foods, dairy products, sugar preserves (jams and marmalades), spreads, syrups, processed fruit, vegetables, fish and meat products, mustards and condiments etc. Glass bottles are widely used for beers, wines, spirits, liqueurs, soft drinks and mineral waters. Within these categories of food and drinks, the products range from dry powders and granules to liquids, some of which are carbonated and packed under pressure, and products which are heat sterilized. Table 6.1 gives an overview of the proportions of containers made for the various usage sectors in the UK.

In the categories listed, there has been an overall increase of 28% approximately, in the number of glass containers made in the UK in the period 1993–2002. Within the range listed in Table 6.1, there has been a significant decrease (of around 50%) in the number of milk bottles mainly due to the reduction in doorstep milk delivery and its replacement with plastic and paperboard containers, sold through supermarkets, garage forecourts etc. The sector that has expanded significantly is for flavoured alcoholic drinks (Crayton, 2002).

6.1.5 Glass composition

6.1.5.1 White flint (clear glass)

Colourless glass, known as white flint, is derived from soda, lime and silica. This composition also forms the basis for all other glass colours. A typical composition would be: silica (SiO_2) 72%, from high purity sand; lime (CaO) 12%, from limestone (calcium carbonate); soda (Na_2O) 12%, from soda ash;

Table 6.1 Proportions of glass containers made in UK for various market sectors

Product	%	Trend in consumption
Beer	30	Rising steadily from early 1990s
Food	24	Steady in 1990s, now falling slowly
Spirits and liqueurs	18.5	Steady
Flavoured alcoholic beverages	16	Rapid from start in 1997 and rising
Soft drinks	6.5	Steady but below 1990s average
Milk	2	Steady
Wine	2	Steady
Cider	1	Steady

Source: Derived from data supplied by Rockware Glass.

alumina (Al_2O_3), present in some of the other raw materials or in feldspar-type aluminous material; magnesia (MgO) and potash (K_2O), ingredients not normally added but present in the other materials. Cullet, recycled broken glass, when added to the batch reduces the use of these materials.

6.1.5.2 Pale green (half white)
Where slightly less pure materials are used, the iron content (Fe_2O_3) rises and a pale green glass is produced. Chromium oxide (Cr_2O_3) can be added to produce a slightly denser blue green colour.

6.1.5.3 Dark green
This colour is also obtained by the addition of chromium oxide and iron oxide.

6.1.5.4 Amber (brown in various colour densities)
Amber is usually obtained by melting a composition containing iron oxide under strongly reduced conditions. Carbon is also added. Amber glass has UV protection properties and could well be suited for use with light-sensitive products.

6.1.5.5 Blue
Blue glass is usually obtained by the addition of cobalt to a low-iron glass. Almost any coloured glass can be produced either by furnace operation or by glass colouring in the conditioning forehearth. The latter operation is an expensive way of producing glass and commands a premium product price. Forehearth colours would generally be outside the target price of most carbonated soft drinks.

6.2 Attributes of food packaged in glass containers

The glass package has a modern profile with distinct advantages, including:

- *Quality image* – consumer research by brand owners has consistently indicated that consumers attach a high quality perception to glass packaged products and they are prepared to pay a premium for them, for specific products such as spirits and liqueurs.
- *Transparency* – it is a distinct advantage for the purchaser to be able to see the product in many cases, e.g. processed fruit and vegetables.
- *Surface texture* – whilst most glass is produced with a smooth surface, other possibilities also exist, for example, for an overall roughened ice-like effect or specific surface designs on the surface, such as text or coats of arms. These effects emanate from the moulding but subsequent acid etch treatment is another option.

- *Colour* – as indicated, a range of colours are possible based on choice of raw materials. Facilities exist for producing smaller quantities of non-mainstream colours, e.g. Stolzle's feeder colour system (Ayshford, 2002).
- *Decorative possibilities*, including ceramic printing, powder coating, coloured and plain printed plastic sleeving and a range of labelling options.
- *Impermeability* – for all practical purposes in connection with the packaging of food, glass is impermeable.
- *Chemical integrity* – glass is chemically resistant to all food products, both liquid and solid. It is odourless.
- *Design potential* – distinctive shapes are often used to enhance product and brand recognition.
- *Heat processable* – glass is thermally stable, which makes it suitable for the hot-filling and the in-container heat sterilization and pasteurization of food products.
- *Microwaveable* – glass is open to microwave penetration and food can be reheated in the container. Removal of the closures is recommended, as a safety measure, before heating commences, although the closure can be left loosely applied to prevent splashing in the microwave oven. Developments are in hand to ensure that the closure releases even when not initially slackened.
- *Tamper evident* – glass is resistant to penetration by syringes. Container closures can be readily tamper-evidenced by the application of shrinkable plastic sleeves or in-built tamper evident bands. Glass can quite readily accept preformed metal and roll-on metal closures, which also provide enhanced tamper evidence.
- *Ease of opening* – the rigidity of the container offers improved ease of opening and reduces the risk of closure misalignment compared with plastic containers, although it is recognized that vacuum packed food products can be difficult to open. Technology in the development of lubricants in closure seals, improved application of glass surface treatments together with improved control of filling and retorting all combine to reduce the difficulty of closure removal. However, it is essential in order to maintain shelf life that sufficient closure torque is retained, to ensure vacuum retention with no closure back-off during processing and distribution.
- *UV protection* – amber glass offers UV protection to the product and, in some cases, green glass can offer partial UV protection.
- *Strength* – although glass is a brittle material glass containers have high top load strength making them easy to handle during filling and distribution. Whilst the weight factor of glass is unfavourable compared with plastics, considerable savings are to be made in warehousing and distribution costs. Glass containers can withstand high top loading with minimal secondary packaging. Glass is an elastic material and will absorb energy,

up to a point, on impact. Impact resistance is improved by an even distribution of glass during container manufacture and subsequent treatment.
- *Hygiene* – glass surfaces are easily wetted and dried during washing and cleaning prior to filling.
- *Environmental benefits* – glass containers are returnable, reusable and recyclable. Significant savings in container weight have been achieved by technical advances in design, manufacture and handling.

6.2.1 Glass pack integrity and product compatibility

6.2.1.1 Safety
Migration studies on glass have shown it to be an inert material as regards its application to packaged foods and, from a health and hygiene viewpoint, it is regarded as an optimal material for containing food and drinks.

6.2.1.2 Product compatibility
Glass containers are noted for the fact that they enable liquid and solid foods to be stored for long periods of time without adverse effects on the quality or flavour of the product.

6.2.2 Consumer acceptability

Market research has indicated that consumers attach a high quality perception to glass packaged products. Recent findings of a report on consumer perceptions carried out by The Design Engine, on behalf of Rockware Glass, concluded that there are five key and largely exclusive benefits for food packaging in glass (The Design Engine, 2001), namely:

1. aesthetic appeal
2. quality perception
3. preferred taste
4. product visibility and associated appetite appeal
5. resealability.

6.3 Glass and glass container manufacture

6.3.1 Melting

Glass is melted in a furnace at temperatures of around 1350°C (2462°F) and is homogenized in the melting process, producing a bubble-free liquid. The molten glass is then allowed to flow through a temperature controlled channel (forehearth) to the forming machine, where it arrives via the feeder at the correct temperature to suit the container to be produced. For general containers, suitable for foods and carbonated beverages, this would be in the region of 1100°C (2012°F).

6.3.2 Container forming

In the feeder (Fig. 6.1) the molten glass is extruded through an orifice of known diameter at a predetermined rate and is cropped into a solid cylindrical shape. The cylinder of glass is known in the trade as a *gob* and is equivalent in weight to the container to be produced. The gob is allowed to free-fall through a series of deflectors into the forming machine, also known as the IS or individual section machine, where it enters the parison. The parison comprises a neck finish mould and a parison mould, mounted in an inverted position. The parison is formed by either pressing or blowing the gob to the shape of the parison mould. The parison is then reinverted, placed into the final mould and blown out to the shape of the final mould, from where it emerges at a temperature of approximately 650°C (1200°F). A container is said to have been produced by either the *press and blow* or *blow and blow* process (Fig. 6.2).

In general terms, the press and blow process is used for jars and the blow and blow process for bottles. An alternative, for lightweight bottles, is the *narrow neck press and blow* process. The press and blow process is generally

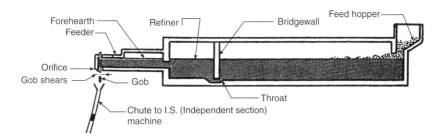

Figure 6.1 The feeder – molten glass is extruded through the orifice at a predetermined rate and is cropped into a solid cylinder known as a *gob* (courtesy of The Institute of Packaging).

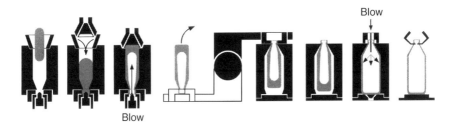

Figure 6.2 The blow and blow forming process (courtesy of Rockware Glass).

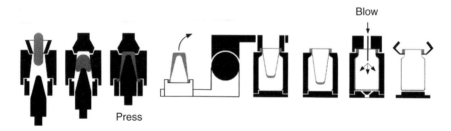

Figure 6.3 The wide mouth press and blow forming process (courtesy of Rockware Glass).

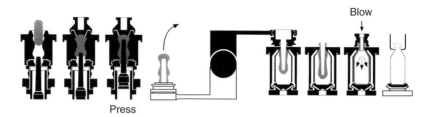

Figure 6.4 The narrow neck press and blow forming process (courtesy of Rockware Glass).

best suited to produce jars with a neck finish size of ≥35 mm (≥1.25″); the other two processes are more suited to produce bottles with a neck finish size of ≤35 mm (≤1.25″) (Fig. 6.3).

The narrow neck press and blow process offers better control of the glass distribution than the blow and blow process, allowing weight savings in the region of 30% to be made (Fig. 6.4).

6.3.3 Design parameters

One of the design parameters to be borne in mind when looking at the functionality of a glass container is that the tilt angle for a wide-mouthed jar should be ≥22° and that for a bottle ≥16°. These parameters are indicative of the least degree of stability that the container can withstand. (For other design parameters, see Figs 6.5 and 6.6.)

6.3.4 Surface treatments

Once formed, surface treatment is applied to the container in two stages: hot end and cold end treatment, respectively.

6.3.4.1 Hot end treatment

The purpose of hot end surface treatment is to prevent surface damage whilst the bottle is still hot and to help maintain the strength of the container. The

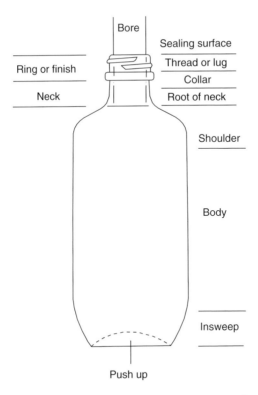

Figure 6.5 The parts of a glass container. (Reproduced, with permission, from Giles, G.A. (1999), *Handbook of Beverage Packaging*, Blackwell Publishing (Sheffield Academic Press), Oxford.)

most common coating material deposited is tin oxide, although derivatives of titanium are also used. This treatment tends to generate high friction surfaces; to overcome this problem, a lubricant is added.

6.3.4.2 Cold end treatment

The second surface treatment is applied once the container has been annealed. Annealing is a process which reduces the residual strain in the container that has been introduced in the forming process. The purpose of the cold end treatment is to create a lubricated surface that does not break down under the influence of pressure or water, and aids the flow of containers through a high speed filling line. Application is by aqueous spray or vapour, care being taken to prevent entry of the spray into the container, the most commonly used lubricants being derivatives of polyester waxes or polyethylene. The surface tension resulting from this treatment can be measured by using Dynes indicating pens.

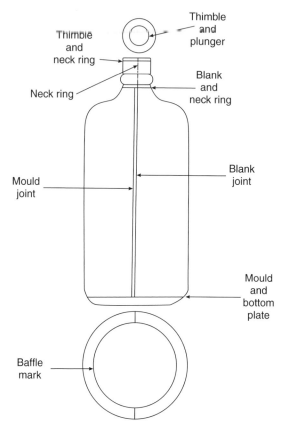

Figure 6.6 The positions of the moulding joints of the glass container. (Reproduced, with permission, from Giles, G.A. (1999), *Handbook of Beverage Packaging*, Blackwell Publishing (Sheffield Academic Press), Oxford.)

Labelling compatibility should be discussed with either the adhesive supplier or the adhesive label supplier depending on the type of label to be used.

6.3.4.3 *Low-cost production tooling*

The tooling cost for a glass container is approximately one-fifth that of a plastic container. Whilst the numbers produced per cavity are lower than for plastic, this can be advantageous, because the design can be modified or completely revamped in a much shorter time-span than plastic; thus, the product image can be updated and the product marketability kept alive. The numbers produced per mould cavity vary depending on the number of production runs required, the complexity of the shape and the embossing detail. In general, 750 000 pieces can be produced from a complex mould and 1 000 000 pieces from a mould of a simple round shape. There can be upwards of 20 moulds per production set.

6.3.4.4 Container inspection and quality

As with packaging in general, quality assurance is needed to ensure that consumer safety, brand owner's needs and efficiency in handling, packing, distribution and merchandising are achieved.

Quality assurance needs are defined and incorporated into the specification of the glass container at the design stage and by, consistency in manufacture, thereby meeting the needs of packing, distribution and use. Quality control, on the other hand, comprises the procedures, including on-line inspection, sampling and test methods used to control the process and assess conformity with the specification.

The techniques used can broadly be defined as chemical, physical and visual.

Chemical testing by spectrophotometry, flame photometry and X-ray fluorescence is used to check raw materials and the finished glass. Small changes in the proportions and purity of raw materials can have a significant effect on processing and physical properties.

Physical tests include checking dimensional tolerances, tests for colour, impact strength, thermal shock resistance and internal pressure strength. Visual tests check for defects that can be seen (Sohani, 2002).

The list of possible visually observable defects is quite long and though most of them are comparatively rare, it is essential that production be checked by planned procedures. The categories of these defects comprise various types of cracks, glass strands (bird swings and spikes), foreign bodies and process material contamination from the process environment, mis-shapes and surface marks of various kinds. For a comprehensive list, see Hanlon *et al.*, Handbook of Package Engineering pp. 9-24–9-26.

Defects are classified as being,

- critical, e.g. defects which endanger the consumer or prevent use in packing
- major, e.g. defects which seriously affect efficiency in packing
- minor, e.g. defects that relate to appearance even though the container is functionally satisfactory.

Visual inspection on manufacturing and packing lines is assisted today by automatic monitoring systems, as shown in Figure 6.7, where this is appropriate. Systems are available for container sidewall inspection using multiple cameras that detect opaque and transparent surface defects (Anon, 2002). Infrared cameras can be used in a system to examine containers directly after formation (Dalstra & Kats, 2001). On the packing line, claims have been made that foreign bodies can be detected in glass containers running at speeds up to 60 000 capped beer bottles per hour (Anon, 2001).

An X-ray system such as that from Heimann Systems Corp. (Eagle Tall™) is designed for the automatic inspection of products packed in jars. The system

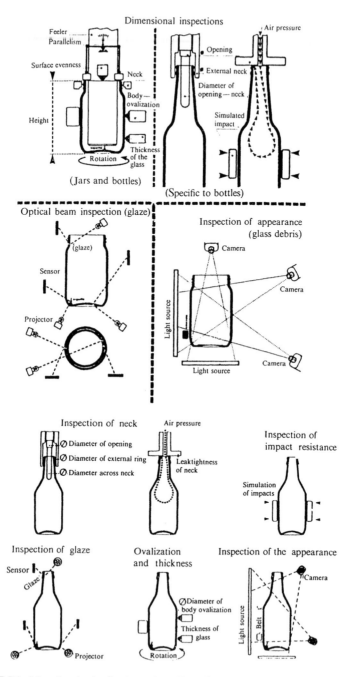

Figure 6.7 Principles of production line inspections. (Reproduced, with permission, from John Wiley & Son Inc.)

can detect foreign materials such as ferrous and stainless metals, glass particles, stone, bone and plastic materials. This equipment runs at 100 m min^{-1}.

6.4 Closure selection

Closures for glass packaging containers are usually metal or plastic, though cork is still widely used for wines and spirits. Effecting a seal is achieved either by a tight fitting plug, a screw threaded cap applied with torque in one of several ways or a metal cap applied with pressure and edge crimping. Hermetic or airtight sealing can be achieved by heat sealing a flexible barrier material to the glass usually with an overcap for protection and subsequent reclosing during use. The aluminium foil cap applied to a milk bottle is one of the simplest forms of closure.

All these closures are applied to what is known as the *finish* of the container. This may seem an odd name for the part of the container which is formed first but in fact this name goes back to the time of blowing and forming glass containers by hand when the rim was the last part to be formed and therefore called the *finish*.

Four key dimensions determine the finish as shown in Figure. 6.8. Industry-wide standards for these dimensions have been agreed upon. The contour of glass threads are round, and closures, both metal and plastic, with symmetrical threads will fit the appropriate containers.

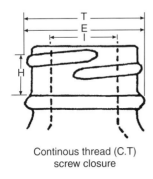

Continous thread (C.T) screw closure

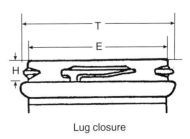

Lug closure

I	Diameter at smallest opening inside finish
T	Thread diameter measured across the threads
E	Thread root diameter
H	Top of finish to top of bead or to intersection with bottle shoulder on beadless designs

Figure 6.8 Standard finish nomenclature (courtesy of The Institute of Packaging).

Careful choice of closure is essential. Too large a closure can create leakage due to the force generated upon it either from internal gas pressure or from heating during processing of the product. Too small a closure may well introduce an interference fit between the minimum through bore on the glass container and the filler tube. The types of closure available fall into three main categories:

1. normal seal
2. vacuum seal
3. pressure seal.

6.4.1 Normal seals

Normal seals, that is those used for non-vacuum/non-pressure filled products, comprise composite closures of plastic/foil, for products such as coffee, milk powders, powder and granular products in general and for mustards, milk and yoghurts.

Glass lends itself both to induction and conduction sealing without prior treatment of the glass finish, but is considered suitable only for dry powders and products, such as peanut butter and chocolate spreads, which do not require a further heating process.

Crimped seals using foils, used for instance on milk containers, have been used for a long time, and are cost effective.

6.4.2 Vacuum seals

Vacuum seals are metal closures with a composite liner to seal onto the glass rim. They can be pressed or twisted into place, at which time a vacuum is created by flushing the headspace with steam. They lend themselves quite readily to in-bottle pasteurization and retort sterilization and sizes range from 28 to 82 mm. For beverages, sizes are usually in the 28–40 mm range.

6.4.3 Pressure seals

Pressure seals can be metal or plastic with a composite liner to make the seal, and can either be pressed or twisted into place. They include:

- preformed metal, e.g. crown or twist crown
- metal closures rolled-on to the thread of the glass
- roll-on pilfer proof (ROPP)
- preformed plastic screwed into position with or without a tamper evidence band.

Selecting the correct glass finish to suit the closure to be used is essential. Advice on suitability should be sought from both the closure and the glass manufacturers before the final choice is made.

6.5 Thermal processing of glass packaged foods

Glass containers lend themselves to in-bottle sterilization and pasteurization for both hot and cold filled products. Subject to the headspace volume conditions being maintained and thermal shock ground rules being observed, no problems will be experienced.

In general terms, hot-fill products filled at 85°C and then cooled will require a minimum headspace of 5%, whilst a cold filled product requiring sterilization at 121°C will require a 6% minimum head space. In all cases the recommendations of the closure supplier should be obtained before preparing the design brief. It should be noted that the thermal shock of glass containers is twice as high when cooling down as when warming up. To avoid thermal shock, cool down differentials should not exceed 40°C and warm up differentials should not exceed 65°C.

Internal pressure resistance. A well-designed glass container can withstand an internal pressure of up to 10 bar (150 pounds per square inch), although the norm required rarely exceeds 5 bar. It is also capable of withstanding internal vacuum conditions and filling of thick concentrates, with steam-flushing of the headspace to produce the initial vacuum requirements for the closure seal.

Resealability. Preformed metal, rolled-on metal and preformed plastic closures can all be readily applied to the neck finish of a glass container. Prise-off crown closures offer no reseal, whilst the twist-off crown satisfies reseal performance within reason.

6.6 Plastic sleeving and decorating possibilities

Glass containers can accept a wide range of decorative formats, i.e. labelling, silk screen printing with ceramic inks, plastic sleeving, acid etching, organic and inorganic colour coating and embossing (with good definition, especially for carbonated products). The rigidity of the container offers a good presentation surface for decorating, which is not subject to distortion from internal pressure or internal vacuum (Ayshford, 2002).

When plastic sleeving the container, it is essential to test the sleeving film under in-bottle pasteurization temperatures to ensure that no secondary movement of the sleeve occurs. Care should also be taken not to exceed the stretch limits of the film by ensuring the maximum and minimum diameters of the area to be sleeved do not fall outside the stretch ratios of the film specification.

Sleeving also offers fragment retention properties, should the container become damaged in use.

6.7 Strength in theory and practice

The theoretical strength of glass can be calculated and it is extremely high. In practice, the strength is much lower due to surface blemishes, such as microcracks, which are vulnerable stress points for impacts such as occur during handling and on packing lines. Work has therefore been concentrated on:

- improving the surface to reduce defects
- improving surface coatings in manufacture
- avoiding stress during manufacture and use.

Major investigations of packing line performance, noting all breakages, and using techniques such as high speed video, can lead to improvements in performance by eliminating stress points.

Broken bottles can be reconstructed and thereby demonstrate the type of impact which caused the failure, such as whether it occurred at a slow or fast rate or whether it was caused by an external, or internal, pressure related fault. The strength of a glass container is also dependent on shape and thickness. The inter-relationship can be subjected to computer modeling as a design-aid to:

- identify vulnerable features in a proposed design
- calculate the effect of modifying the design
- simulate the effect of lightweighting by reducing thickness.

Specific tests can be carried out on containers to check:

- vertical crushing (relevant to stacking)
- internal pressure (relevant where proposed contents are packed under pressure as with carbonated drinks)
- thermal shock (relevant for hot-filling, pasteurization and sterilization).

Thermal shock relates to heat transfer and as glass is a very good insulator, heat is conducted slowly across the walls when a hot liquid is filled. Another important heat related property is the dimensional change per degree change in temperature, which is low for glass. This property is also subject to the glass formulation, e.g. Pyrex is a well known type of glass with an even lower heat expansion compared with standard white flint soda glass. This is achieved by replacing some of the soda with boric oxide and increasing the proportion of silica.

It has been recognized that achieving an even distribution of glass in the walls of a container is the major factor in successfully reducing weight whilst maintaining adequate strength.

6.8 Glass pack design and specification

6.8.1 Concept and bottle design

Leading glass manufacturers have state-of-the-art design expertise and systems that can be readily integrated with design house concepts to design a container which meets the requirements of branding, manufacture, filling and distribution under recommended good manufacturing practices and procedures.

The brand manager/packaging technologist can now quite readily bring together all the expertise necessary to produce a food container of the ultimate design, cost and quality to meet all their needs.

An understanding of the product specification and the filling line requirements is essential at the concept design stage. The information required includes:

- type and density of the product
- carbonation level, if required, in the product
- closure type/neck specification required
- quantity to be filled
- type of filling process (hot-fill/cooled, hot-fill/pasteurized, ambient-fill/sterilized or any combination of these processes)
- is the container to be a specified volume-measuring container?
- what type of filler is to be used (volumetric or vacuum-assisted)?
- what is the filler tube size/diameter?
- is the container to be refillable or single-trip?
- speed of the filling operation, i.e. bottles per minute
- impact forces on the process line (for ultra lightweight designs, line impact speed should not exceed 25 inches per second)
- what pallet size is to be used in the distribution of filled stock?
- is the depalletizer operation sweep-off or lift-off?

From this information, the glass manufacturer can select the correct finish and closure design, surface treatment requirements, the type of pack to be used for distribution to the filling line and the handling systems. Wherever possible, the body size of the container should ensure a snug fit to the pallet, since any overhang of the glass beyond the edge of the pallet could result in breakage in transit, whilst underhang on the pallet could lead to instability. Compression, tension strapped packs can be accommodated together with live bed deliveries. This creates a highly efficient delivery system with minimal stock-holding on site, by means of just-in-time (JIT) deliveries.

Ever more challenging briefs are demanding more from packaging materials. It is well known that consumers have an innate, high quality perception of glass packaging. This emotional connection between consumer and brand is highly valued by food and drink consumers. Add to this the ability of glass to be formed into unique shapes with a wide range of decoration techniques and it is clear why glass is also the preferred choice for designers. With increased emphasis on production speed and efficiency, design freedom decreases.

Low volume. For low volume, limited or special edition products, the design freedom is high, as hand operated or semi-automated processing lines are used. Bottles may be produced using single gob machines and have a high (+0.8) capacity to weight ratio.

Main stream. For main stream production volumes, design freedom decreases, with automatic filling lines and bulk distribution being important. Bottles will be produced using larger double gob machines and have capacity to weight ratios of around 0.6–0.7.

High volume. For high volume brands, which probably have multinational distribution, the design freedom is strictly controlled to ensure compatibility with very high speed (+1000 bpm) filling lines. These brands will be produced using the largest double and triple gob NNPB machines and have capacity to weight ratios down to 0.5. A *full circle* design process creates a range of radical design options and a common sense view on the likely costs and implications of each concept. This ensures all design options are fully explored and the best design solutions are rapidly brought to the market.

Concept design. A concept design team focuses on the packaging as a brand communication tool. Using brand analysis they ensure the pack is as active as possible, at the point of sale and in use, in communicating the brand's value and positioning. Concept designers are able to work very closely with a customer's design agency, supporting the design process so that a wide range of creative options are explored, yet at the same time highlighting the practical consequences of the design options. This allows realistic, balanced decisions to be made at the earliest possible stage of the project.

Product design. Taking computer information from the concept designers, or any design agent, product designers apply a series of objective tests to the design to ensure it is fit for the purpose. These include stress analysis to check retention of carbonated products, packing line stability, and impact analysis to assess the containers' filling line performance. Strength for stacking and distribution is also checked. On completion of these tests a detailed specification of the design is issued and a 3D computer model displayed. The 3D computer model is used to create exact models for market research and to seek approval for new designs.

Mould design. The mould design team translates the product specification into mould equipment that will reproduce the container millions of times. Depending on the manufacturing plant and the process to be used, the mould equipment will vary. The level of precision required for modern glass container production is extremely exacting and has a direct effect on product quality.

The product design computer model is used to control all aspects of the design, ensuring exact replication of the design into glass. The design is now ready to be transferred to the mould makers.

Production. Quality information from each production run is fed back to the product and mould design teams to ensure best practice is used on all designs and that design teams are up to date with improvements in manufacturing capability. This closes the *full circle*.

6.9 Packing – due diligence in the use of glass containers

Receipt of deliveries. Glass containers are usually delivered on bulk palletized shrink-wrapped pallets. A check should be made for holes in the pallet shroud and broken glass on the pallet, and any damaged pallets rejected. The advice note should be signed accordingly, informing the supplier and returning the damaged goods.

Storage/on-site warehousing. Pallets of glass must not be stored more than six high, they must be handled with care and not shunted. Fork-lift trucks should be guarded to prevent the lift masts contacting the glass. Where air rinser cleaning is used on the filling line, the empty glass containers should not be stored outside. Pallets damaged in on-site warehousing must not be forwarded to the filling area until they have been cleared of broken glass.

Depalletization. A record should be made of the sequence and time of use of each pallet and the product batch code. Plastic shrouds must be removed with care to prevent damage to the glass; if knives are used, the blade should be shrouded at all times, so as not to damage the glass. It is necessary to ensure that the layer pads between the glass containers are removed in such a way as to prevent any debris present from dropping onto the next layer of glass. Breakages must be recorded and clean-up equipment provided to prevent any further contamination.

Cleaning operation

- *Air rinse.* The glass must be temperature-conditioned to prevent condensate forming on the inside, which would inhibit the removal of cardboard debris. The air pressure should be monitored to ensure that debris is not suspended and allowed to settle back into the container.
- *On-line water rinse.* Where hot-filling of the product takes place, it is essential to ensure that the temperature of the water is adequate to prevent thermal shock at the filler, i.e. not more than 60°C (140°F) differential.
- *Returnable wash systems.* The washer feed area must be checked to ensure that the bottles enter the washer cups cleanly. A washer-full of bottles must not be left soaking overnight. In the longer term this would

considerably weaken the container and could well create a reaction on the bottle surface between the hot end coating and the caustic in the washer. Where hot-filling is taking place, it is necessary to ensure that the correct temperature is reached to prevent thermal shock at the filler.

Filling operation. Clean-up instructions should be issued and displayed, so that the filling line crew know the procedure to follow should a glass container breakage occur and the need to record all breakages. It is essential to ensure that flood rinsing of the filler head in question is adequate to prevent contamination of further bottles. It is necessary to ensure that filling levels in the container comply with trading standards' requirements for measuring containers.

Capping. Clean-up instructions on the procedure to follow should breakage occur in the capper should be issued and displayed, and all breakages recorded. The application torque of the caps and vacuum levels must be checked at prescribed intervals, as must the cap security of carbonated products.

Pasteurization/sterilization. It is necessary to ensure that cooling water in the pasteurizer or sterilization retort does not exceed a differential of more than 40°C (104°F), to prevent thermal shock situations. The ideal temperature of the container after cooling is 40°C, which allows further drying of the closure and helps prevent rusting of metal closures. Air knives should be used to remove water from closures to further minimize the risk of rusting.

Labelling. Where self-adhesive labels are to be used, all traces of condensate must be eliminated to obtain the optimum conditions for label application. Adhesives must not be changed without informing the glass supplier, since this could affect the specification of adhesives/surface treatments.

Distribution. It is essential to ensure that the arrangement of the glass containers in the tray, usually plastic or corrugated fibreboard, is adequate to prevent undue movement during distribution, that the shrink-wrap is tight and that the batch coding is correct and visible.

Warehousing. The pallets of filled product must be carefully stacked to prevent isolated pockets of high loading that might create cut through in the lining compound of the container closures, as this would result in pack failures.

Quality management. The procedures of good management practice in the development, manufacture, filling, closing, processing (where appropriate), storage and distribution of food products in glass containers discussed in this chapter have been developed to ensure that product quality and hygiene standards are achieved along with consumer and product safety needs. Their application indicates *due diligence* in meeting these needs. It is essential that all procedures are clearly laid down, training is provided in their use and that regular checks are made on their implementation.

Companies can demonstrate *due diligence* by achieving certification under an accepted Quality Management Standard, such as ISO 9000. In the UK, the British Retail Consortium (BRC) and The Institute of Packaging (IOP) have cooperated in the publication of a Technical Standard and Protocol (see 'Further reading' section), which can be integrated with their ISO 9000 procedures. The BRC is a trade association representing around 90% of the retail trade and the IOP is the professional membership body, established in 1947, for the packaging industry. The IOP has amongst its objectives the education and training of people engaged in the packaging industry. This Technical Standard and Protocol requires companies to:

- adopt a formal Hazard Analysis System
- implement a documented Technical Management System
- define and control factory standards, product and process specifications and personnel needs.

6.10 Environmental profile

6.10.1 Reuse

Glass containers can be reused for food use. However, there is only one well established household example in the UK – that of the daily doorstep delivery of fresh milk in bottles and the collection of the empty bottles. There are wide disparities in the number of trips that can be expected depending on the location, with around 12 trips per bottle being the national average. The decline of doorstep delivery has been rapid over the last decade but the system of reuse is well established. In the licensed trade, and in most places where drinks are served to customers, the drinks manufacturers operate returnable systems.

6.10.2 Recycling

Glass is one of the easiest materials to be recycled because it can be crushed, melted and reformed an infinite number of times with no deterioration of structure. It is the only packaging material that retains all its quality characteristics when it is recycled. Using recycled glass (cullet), in place of virgin raw materials, to manufacture new glass containers reduces

- the need to quarry and transport raw materials
- the energy required to melt the glass
- furnace chimney emissions
- the amount of solid waste going into landfill.

In order to recycle glass, it must first be recovered. In the UK, glass is brought by consumers to bottle banks. Currently, approximately 600 000 t/a are recovered – a figure which must increase sharply if the UK is to meet increased European Union targets for glass recovery at currently generated levels of glass container waste. Currently the recycled content of the average glass container is around 33%. The recycled proportion is higher for green than for clear containers and reflects the proportions of clear and green glass taken to bottle banks by the public. Green glass may now contain as much as 85–90% recycled glass.

6.10.3 Reduction – lightweighting

In the period 1992–2002, it is claimed that the average weight of glass containers has been reduced by 40–50% (*source*: Rockware Glass). This is an average reduction. Some brand owners still retain heavy containers, e.g. spirits and liqueurs, and this causes the progress made by the glass industry to reduce the weight of packaging to be understated.

6.11 Glass as a marketing tool

Glass packaging supports brand differentiation and product identification by the use of:

- creative and unique shapes and surface textures
- ceramic printing, acid etching and coating
- labelling, both conventionally and by plastic shrink sleeving.

Glass can be readily formed into a multiplicity of shapes to provide shelf appeal. Jars may be designed to be table presentable and have convenience in handling features, and bottles have been redesigned to reflect changing drinking habits. Printed pressure sensitive plastic labels using adhesives which are as clear, or transparent, as glass can be used to give a *no label* effect. Precision in manufacturing and subsequent rigidity of glass containers enable them to meet EC measuring container regulations in terms of capacity (volume) and product give-away through overcapacity or container expansion.

Current developments include the use of metallic, thermochromic, photochromic finishes, UV activated fluorescent and translucent inks and the ability to incorporate embossed, foiled, velvet textured and holographic materials. These finishes are compatible with laser etching and offer the possibility of permanent traceability coding.

References

American Society for Testing Materials, ASTM (1965) C 162-56, Standard Definition of Terms Relating to Glass Products, ASTM Stand., Part 13, 145–159.

Anon (2001) Packaging Today, Heineken and GEI launch bottle inspection system, **23**(9), p. 18.

Anon (2002) Packaging Innovation, Inspecting for bottle defects, **6**(11), p. 6.
Ayshford, H. (2002) Packaging magazine, www.dotpackaging.com/features/glass/
Crayton, C. (2002) Packaging magazine, www.dotpackaging.com/features/glass/
Dalstra, J. and Kats, H. (2001) Glass International, Relocation inspection tasks to the hot end, **24**(6), pp. 35–37.
The Design Engine (2001) Qualitative Research Report.
Sohani, S. (2002) Packaging India. Glass containers: tests procedures and their significance for quality evaluation, **35**(1), pp. 19–20.
Morey, G.W. (1954) The properties of glass, 2nd edn, Reinhold Publishing Corp., New York, p. 28.

Further reading

Tech 7 Strength and Performance Standards for the Manufacture and Use of Carbonated Beverage Containers. Sheffield, UK: British Glass Publications.
Tech 9 General Guide Lines for the Use of Glass Containers, Sheffield, UK: British Glass Publications.
Closure Manual, London, UK: British Soft Drinks Association (BSDA).
Safe Packing of food and drink in Glass Containers, Chipping Campden, UK: Campden & Chorleywood Food Research Association.
Moody, B. Packaging Glass, London, UK: Hutchinson Bentham Publishers.
Handbook of Beverage Packaging (ed. G. Giles), Sheffield Academic Press.
Food Packaging Technology (1996), (eds G. Bureau and J.-L. Multon), **1**, VCH Publishers (John Wiley & Son).
Fundamentals of Packaging Technology by Walter Soroka revised UK edition by Anne Emblem and Henry Emblem, published by The Institute of Packaging.
Technical Standard and Protocol for companies manufacturing and supplying Food Packaging Materials for Retailer Branded Products, ISBN 0 11 702842 8, Published by The Institute of Packaging, www.iop.co.uk.

7 Plastics in food packaging

Mark J. Kirwan and John W. Strawbridge

7.1 Introduction

7.1.1 Definition and background

The most recent EU Directive relating to 'plastic materials and articles intended to come into contact with foodstuffs' (reference 2001/62/EC) defines *plastics* as being: 'organic macromolecular compounds obtained by polymerisation, polycondensation, polyaddition or any similar process from molecules with a lower molecular weight or by chemical alteration of natural macromolecular compounds'.

Plastics are widely used for packaging materials and in the construction of food processing plant and equipment, because:

- they are flowable and mouldable under certain conditions, to make sheets, shapes and structures
- they are generally chemically inert, though not necessarily impermeable
- they are cost effective in meeting market needs
- they are lightweight
- they provide choices in respect of transparency, colour, heat sealing, heat resistance and barrier.

Referring again to the Directive, *molecules with a lower molecular weight* are known as monomers and the *macromolecular compounds* are known as polymers – a word derived from Greek, meaning *many parts*.

The first plastics were derived from natural raw materials and, subsequently, in the first half of the 20th century, from coal, oil and natural gas. The most widely used plastic today, polyethylene, was invented in 1933 – it was used in packaging from the late 1940s onwards in the form of squeeze bottles, crates for fish replacing wooden boxes and film and extrusion coatings on paperboard for milk cartons.

In Europe, nearly 40% of all plastics is used in the packaging sector, and packaging is the largest sector of plastics usage (Association of Plastics Manufacturers in Europe, APME). About 50% of Europe's food is packed in plastic packaging (British Plastics Federation, BPF).

Plastics have properties of strength and toughness. For example, polyethylene terephthalate (PET) film has a mechanical strength similar to that of iron, but under load the PET film will stretch considerably more than iron before breaking.

Specific plastics can meet the needs of a wide temperature range, from deep frozen food processing (−40°C) and storage (−20°C) to the high temperatures of retort sterilization (121°C), and reheating of packaged food products by microwave (100°C) and radiant heat (200°C). Most packaging plastics are thermoplastic, which means that they can be repeatedly softened and melted when heated. This feature has several important implications for the use and performance of plastics, as in the forming of containers, film manufacture and heat sealability.

Thermosetting plastics are materials which can be moulded once by heat and pressure. They cannot be resoftened, as reheating causes the material to degrade. Thermosetting plastics such as phenol formaldehyde and urea formaldehyde are used for threaded closures in cosmetics, toiletries and pharmaceutical packaging but are not used to any great extent for food packaging.

Plastics are used in the packaging of food because they offer a wide range of appearance and performance properties which are derived from the inherent features of the individual plastic material and how it is processed and used.

Plastics are resistant to many types of compound – they are not very reactive with inorganic chemicals, including acids, alkalis and organic solvents, thus making them suitable, i.e. inert, for food packaging. Plastics do not support the growth of microorganisms.

Some plastics may absorb some food constituents, such as oils and fats, and hence it is important that a thorough testing is conducted to check all food applications for absorption and migration.

Gases such as oxygen, carbon dioxide and nitrogen together with water vapour and organic solvents permeate through plastics. The rate of permeation depends on:

- type of plastic
- thickness and surface area
- method of processing
- concentration or partial pressure of the permeant molecule
- storage temperature.

Plastics are chosen for specific technical applications taking the specific needs, in packing, distribution and storage, and use of the product into consideration, as well as for marketing reasons, which can include considerations of environmental perception.

7.1.2 Use of plastics in food packaging

Plastics are used as containers, container components and flexible packaging. In usage, by weight, they are the second most widely used type of packaging and first in terms of value. Examples are as follows:

- rigid plastic containers such as bottles, jars, pots, tubs and trays
- flexible plastic films in the form of bags, sachets, pouches and heat-sealable flexible lidding materials
- plastics combined with paperboard in liquid packaging cartons
- expanded or foamed plastic for uses where some form of insulation, rigidity and the ability to withstand compression is required
- plastic lids and caps and the wadding used in such closures
- diaphragms on plastic and glass jars to provide product protection and tamper evidence
- plastic bands to provide external tamper evidence
- pouring and dispensing devices
- to collate and group individual packs in multipacks, e.g. Hi-cone rings for cans of beer, trays for jars of sugar preserves etc.
- plastic films used in cling, stretch and shrink wrapping
- films used as labels for bottles and jars, as flat glued labels or heat-shrinkable sleeves
- components of coatings, adhesives and inks.

Plastic films may be combined with other plastics by coextrusion, blending, lamination and coating to achieve properties which the components could not provide alone. Coextrusion is a process which combines layers of two or more plastics together at the point of extrusion. Lamination is a process which combines two or more layers of plastics together with the use of adhesives. Different plastic granules can be blended together prior to extrusion. Several types of coating process are available to apply plastic coatings by extrusion, deposition from either solvent or aqueous mixtures or by vacuum deposition.

Plastics are also used as coatings and in laminations with other materials such as regenerated cellulose film (RCF), aluminium foil, paper and paperboard to extend the range of properties which can be achieved. Plastics may be incorporated in adhesives to increase seal strength, initial tack and low temperature flexibility.

Plastics can be coloured, printed, decorated or labelled in several ways, depending on the type of packaging concerned. Alternatively, some plastics are glass clear, others have various levels of transparency, and their surfaces can be glossy or matte.

Plastics are also used to store and distribute food in bulk, in the form of drums, intermediate bulk containers (IBCs), crates, tote bins, fresh produce trays and plastic sacks, and are used for returnable pallets, as an alternative to wood.

The main reasons why plastics are used in food packaging are that they protect food from spoilage, can be integrated with food processing technology, do not interact with food, are relatively light in weight, are not prone to breakage, do not result in splintering and are available in a wide range of packaging structures, shapes and designs which present food products cost effectively, conveniently and attractively.

7.1.3 Types of plastics used in food packaging

The following are the types of plastics used in food-packaging

- polyethylene (PE)
- polypropylene (PP)
- polyesters (PET, PEN, PC) (note: PET is referred to as PETE in some markets)
- ionomers
- ethylene vinyl acetate (EVA)
- polyamides (PA)
- polyvinyl chloride (PVC)
- polyvinylidene chloride (PVdC)
- polystyrene (PS)
- styrene butadiene (SB)
- acrylonitrile butadiene styrene (ABS)
- ethylene vinyl alcohol (EVOH)
- polymethyl pentene (TPX)
- high nitrile polymers (HNP)
- fluoropolymers (PCTFE/PTFE)
- cellulose-based materials
- polyvinyl acetate (PVA).

Many plastics are better known by their trade names and abbreviations. In the European packaging market, PE constitutes the highest proportion of consumption, with about 56% of the market by weight, and four others, PP, PET, PS (including expanded polystyrene or EPS) and PVC, comprise most of the remaining 46% (source BPF). The percentages may vary in other markets, but the ranking is similar. The other plastics listed meet particular niche needs, such as improved barrier, heat sealability, adhesion, strength or heat resistance.

These materials are all thermoplastic polymers. Each is based on one, or more, simple compound or monomer. An example of a simple monomer would be ethylene, which is derived from oil and natural gas. It is based on a specific arrangement of carbon and hydrogen atoms. The smallest independent unit of ethylene is known as a molecule, and it is represented by the chemical formula C_2H_4.

Polymerisation results in joining thousands of molecules together to make polyethylene. When the molecules join end to end, they form a long chain. It is possible for molecules to proliferate as a straight chain or as a linear chain with side branches. The length of the chain, the way the chains pack together and the degree of branching affect properties, such as density, crystallinity, gas and water vapour barrier, heat sealing, strength, flexibility and processability.

The factors which control polymerisation are temperature, pressure, reaction time, concentration, chemical nature of the monomer(s) and, of major significance, the catalyst(s). A catalyst controls the rate and type of reaction but is not, itself, changed permanently. The recent introduction of metallocene (cyclopentadiene) catalysts has resulted in the production of high-performance plastics and has had a major impact on the properties of PE, PP and other plastics, such as PS. In some cases, the resulting polymers are virtually new polymers with new applications, e.g. breathable PE film for fresh produce packing, and sealant layers in laminates and coextrusions.

It is appropriate to consider PE as a family of related PEs which vary in structure, density, crystallinity and other properties of packaging importance. It is possible to include other simple molecules in the structure, and all these variables can be controlled by the conditions of polymerisation – heat, pressure, reaction time and the type of catalyst.

All PEs have certain characteristics in common, which polymerisation can modify, some to a greater and some to a lesser extent, but all PEs will be different from, for example, all polypropylenes (PP) or the family of polyesters (PET).

Similar considerations apply to all the plastics listed; they are all families of related materials, with each family originating from one type or more types of monomer molecule.

It is also important to appreciate the fact that plastics are continually being developed, i.e. modified in the polymerisation process, to enhance specific properties to meet the needs of the:

- manufacture of the film, sheet, moulded rigid plastic container etc.
- end use of the plastic film, container etc.

In the case of food packaging, end use properties relate to performance properties, such as strength, permeability to gases and water vapour, heat sealability and heat resistance, and optical properties, such as clarity.

Additionally, the way the plastic is subsequently processed and converted in the manufacture of the packaging film, sheet, container etc., will also have an effect on the properties of that packaging item.

7.2 Manufacture of plastics packaging

7.2.1 Introduction to the manufacture of plastics packaging

The plastic raw material, also known as resin, is usually supplied by the polymer manufacturer in the form of pellets. Plastics in powder form are used in some processes. Whilst some plastics are used to make coatings, adhesives or additives in other packaging related processes, the first major step in the

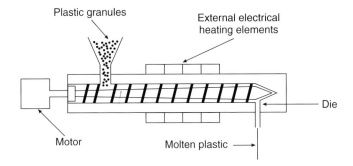

Figure 7.1 Extruder.

conversion of plastic resin into films, sheets, containers etc., is to change the pellets from solid to liquid or molten phase in an extruder.

The plastic is melted by a combination of high pressure, friction and externally applied heat. This is done by forcing the pellets along the barrel of an extruder using specially designed, polymer-specific, screw under controlled conditions that ensure the production of a homogeneous melt prior to extrusion (Fig. 7.1).

In the manufacture of film and sheet, the molten plastic is then forced through a narrow slot or die. In the manufacture of rigid packaging, such as bottles and closures, the molten plastic is forced into shape using a precisely machined mould.

7.2.2 Plastic film and sheet for packaging

Generally, films are by definition less than 100 μm thick (1 micron is 0.000001 metres or 1×10^{-6} m). Film is used to wrap product, to overwrap packaging (single packs, groups of packs, palletised loads), to make sachets, bags and pouches, and is combined with other plastics and other materials in laminates, which in turn are converted into packaging. Plastic sheets in thicknesses up to 200 μm are used to produce semi-rigid packaging such as pots, tubs and trays.

The properties of plastic films and sheets are dependent on the plastic(s) used and the method of film manufacture together with any coating or lamination. In film and sheet manufacture, there are two distinct methods of processing the molten plastic which is extruded from the extruder die. In the *cast* film process, the molten plastic is extruded through a straight slot die onto a cooled cylinder, known as the chill roll (Fig. 7.2).

In the *blown*, or tubular, film process, the molten plastic is continuously extruded through a die in the form of a circular annulus, so that it emerges as a tube. The tube is prevented from collapsing by maintaining air pressure inside the tube or bubble (Fig. 7.3).

In both the processes, the molten polymer is quickly chilled and solidified to produce a film which is reeled and slit to size.

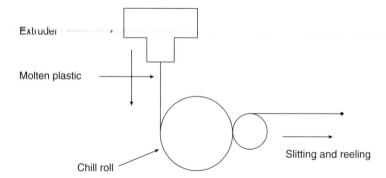

Figure 7.2 Production of cast film.

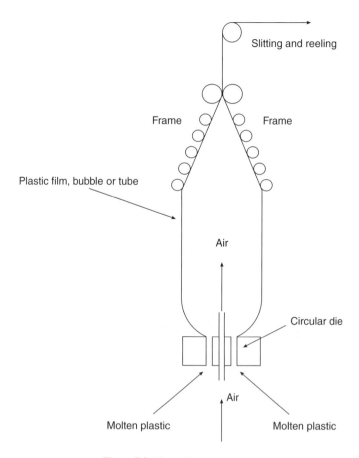

Figure 7.3 Blown film manufacture.

For increased strength and improved barrier properties, film can be stretched to realign, or orient, the molecules in both the machine direction (MD), and across the web in the transverse (TD) or cross direction.

In the Stenter-orienting process, transverse stretching of the cast flat sheet is carried out using clips which grip and pull the film edges, so that the width increases. Stretching in the MD can be achieved with several sets of nip rolls running at faster speeds.

With the blown, or tubular, film process, orienting is achieved by increasing the pressure inside the tube to create a tube with a much larger diameter (Fig. 7.4).

Film stretched in one direction only is described as being *mono-oriented*. When a film is stretched in both the directions, it is said to be *biaxially orientated*. Packing the molecules closer together improves the gas and water vapour barrier properties. Orientation of the molecules increases the mechanical strength of the film.

Cast films and sheets which are not oriented are used in a range of thicknesses and can be thermoformed by heat and either pressure or vacuum to make the bottom webs of pouches and for single portion pots, tubs, trays or blister packs.

Cast films are also used in flexible packaging because they are considered to be tough; if one tries to tear them, they will stretch and absorb the energy, even though the ultimate tensile strength may be lower than that with an oriented equivalent.

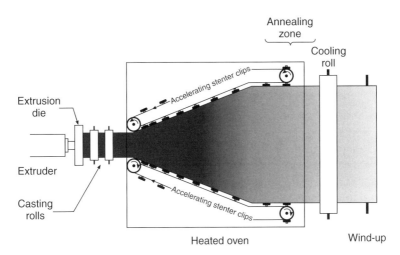

Figure 7.4 TD orientation by Stenter and MD orientation by acceleration in machine direction (courtesy of The Institute of Packaging).

Oriented films are brought close to their melting point to anneal or release stresses in them and to minimise the amount of shrinkage which may take place when being heated in a post-production process such as printing or heat sealing. Failure to anneal heat set films will ensure that they have very unstable thermal characteristics and allow the films to shrink tightly onto cartons or bottles when heated.

It is difficult to puncture or initiate a tear in an oriented film, but once punctured, the alignment of the molecules allows easy proliferation of the rupture and tear. This feature is made use of to assist the opening of film sachets by incorporating a tear-initiating notch mechanically into the sealing area.

Oriented films may have as little as 60% elongation before breaking, whereas cast polypropylene, for example, may extend by 600% before finally breaking. This property is exploited to great effect with linear low-density polyethylene (LLDPE), in the application of stretch wrapping, because the non-branching polymer chains allow easy movement of the polymer molecules past each other. By adding special long-chain molecules in the manufacturing process, it is possible to ensure that the film clings to itself.

The majority of plastic films are transparent and not easily coloured by dyeing or adding pigments. In order to develop opacity, films can be cavitated during film manufacture. Cavitation causes internal light scattering, which gives a white or pearlescent appearance. A simple analogy for the light scattering effect is to consider the example of beating and blending egg white with sugar to produce a meringue, which has a white appearance due to the bubbles trapped inside the beaten egg white. With some plastics, such as cast PE, a chemical compound can be added to the plastic resin, which gives off a gas such as nitrogen or carbon dioxide, when heated in the film manufacturing process. The small gas bubbles in the plastic cause light scattering, which gives the film a pearlescent appearance.

However, because oriented films are thin, there is the possibility of the bubbles being so large that the film may be ruptured. So instead of using gas bubbles, a shearing compound or powder is added to the polymer, causing internal rupturing of the plastic sheet as it is being stressed. This causes voids in the film and light is scattered across the whole spectrum. Incident white light is reflected inside the film as a result of the differing refractive index between the plastic and free air. The process lowers the density of the film and may give more cost-effective packaging as a result of the increased area yield.

The technique of pigmenting plastics has been developed using white compounds such as calcium carbonate or, more usually, titanium dioxide, to give a white appearance. The addition of such an inorganic filler, however, increases the density by up to 50%, lowering the yield and increasing the risk of mechanically weakening the film. Early attempts to pigment film produced an abrasive surface, and the practice today is to ensure that there is a skin of pure resin on the outer layers which acts as an encapsulating skin to give the film a smooth and glossy surface. White pigmented cast sheet material is used in thermoforming pots and dishes for dairy-based products.

Metallising with a very thin layer of aluminium is an alternative way to achieve opacity by causing a high proportion of incident light to be reflected off the surface away from the film. This technique has the added benefit of improving barrier properties.

Transparency, the opposite of opacity, depends on the polymer concerned and on the way the film has been produced. If the film is allowed to cool down slowly, then large crystals may be formed and this gives the film a hazy appearance due to the diffraction and scattering of incident light by the crystals. Transparency improves as polymer crystallinity decreases and is also affected by additives in the film. If the size of the additive particle is too large or if, as with slip agents, they migrate to the surface, the film becomes hazy.

The surface of a film needs to be as smooth as possible to enhance the surface for printing. A rough surface will give a matte appearance to the final printed effect, which is usually considered to be less attractive than a shiny, mirror smooth appearance. Furthermore, a rough surface may give packaging machine runnability problems, as it may be difficult to make the film slide over machine parts without creating static electricity in the film. This is overcome by incorporating food grade additives in the film. Films will also tend to block and become adhered layer to layer in the reel. Waxes, such as carnauba wax, are added to minimise the blocking. The action of a slip additive, such as silica, depends on the particles of silica migrating to the surface of the film where they act like ball bearings holding the surfaces apart.

For marketing purposes, it may be desirable to create a unique impact on the shelf at the point of sale, and hence films have been developed which are matte on one side and have a gloss surface on the other. This is done by casting the film against the matte surface of a sand-blasted chill roll.

It is possible to combine streams of molten plastic from separate extruders in the die to make coextrusions. Higher productivity is achieved for a given thickness of film if the same plastic is extruded in two or more layers and combined in the die to form a single film. Coextrusion is an area of rapid development, with extruders capable of combining up to seven layers of differing plastics to achieve specific properties and characteristics.

7.2.3 Pack types based on use of plastic films, laminates etc.

Single films, coextruded films and coated and laminated films in reel form are used to make plastic bags, sachets, pouches and overwraps.

Plastic bags are made by folding, cutting and sealing with welded seams which are also cut in the same operation. Pouches are usually made from laminates. They may be formed on the packing machine either from one reel by folding, or from two reels and sealing, inside face to inside face on three sides prior to filling and closing. The pouches travel horizontally on these machines with the product filled vertically (Fig. 7.5).

Pouches can have a base gusset or a similar feature, which enables them to stand when filled and sealed. Pouches can be made separately, and they can be filled manually or fed from magazines on automatic filling machines. (Small four-side sealed packs are also referred to as sachets, though the industry is not consistent in naming – the small four side heat sealed packs for tea are referred to as *tea bags*.)

Free-flowing products such as granules and powders can also be filled vertically on form, fill, seal machines where the film is fed vertically from the reel (Fig. 7.6). These packs are formed around a tube, through which the previously apportioned product passes. A longitudinal heat seal is made either as a fin seal, with inside surface sealing to inside surface, or as an overlap seal, depending on the sealing compatibility of the surfaces. The cross seal is combined with cutting to separate the individual packs.

Solid products such as chocolate bars are packed horizontally on form, fill, seal machines (Fig. 7.7). Biscuits can be packed in this way, provided they are collated in a base (plastic) tray, though they are also packed at high speed on roll-wrapping machines with the ends of the film gathered together and heat sealed.

Products packed in cartons are often overwrapped with plastic film, e.g. chocolate assortments and tea bags. The cartons are pushed into the web of

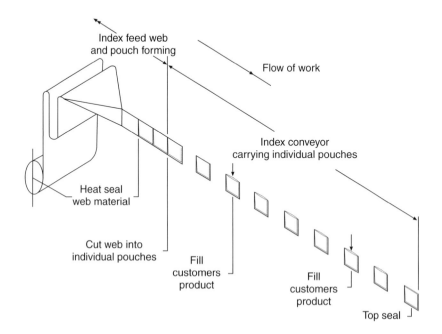

Figure 7.5 Horizontal form/fill/seal sachet/pouch machine.

PLASTICS IN FOOD PACKAGING 185

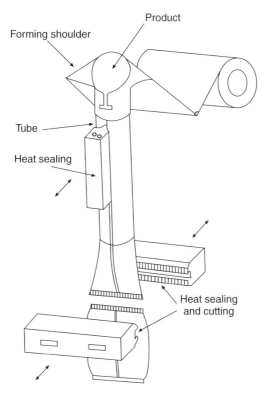

Figure 7.6 Vertical form, fill, seal (f/f/s) machine.

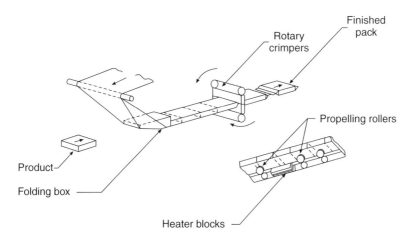

Figure 7.7 Horizontal form, fill, seal (f/f/s), flowpack type machine.

film, a longitudinal seal is made and the end seals are neatly folded, envelope style, prior to sealing with a hot platen which presses against the folded ends.

Shrink wrapping is similar to the overwrapping described above, except that the packs pass through the heated tunnel once the cross seal is made – there are no end seals. The film shrinks over the ends of the pack, the extent depending on the width of the film used.

Another packaging format results in either flexible or semi-rigid packaging, depending on the films used, where the film is fed horizontally and cavities are formed by thermoforming. The plastic sheet, such as PET/PE or PA/PE, is softened by heat and made to conform with the dimensions of a mould by pressure and/or vacuum. Where more precise dimensions for wall thickness or shape are required, a plug matching the mould may also be used to help the plastic conform with the mould. The plastic sheet may be cast, cast coextruded or laminated film, depending on the heat sealing and barrier needs of the application. Products packed in this way are typically cheese or slices of bacon. This form of packaging may be sealed with a lidding film laminate under vacuum or in a modified atmosphere (MAP). The various methods of heat sealing are discussed in Section 7.9.2.

7.2.4 Rigid plastic packaging

Bottles are made by extrusion blow moulding. A thick tube of plastic is extruded into a bottle mould which closes around the tube, resulting in the characteristic jointed seal at the base of the container (Fig. 7.8). Air pressure is then used to

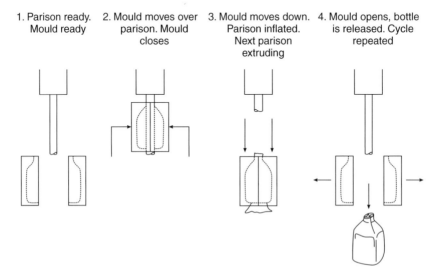

Figure 7.8 Extrusion blow moulding (courtesy of The Institute of Packaging).

force the plastic into the shape of the mould. After cooling, the mould is opened and the item removed. (The bottle will show a thin line in the position where the two parts of the mould are joined.) Blow moulding is used for milk bottles (HDPE) and wide mouth jars.

It is possible to apply coextrusion to extrusion blow moulding so that multi-layered plastic containers can be made with a sandwich of various plastics. An example would be where high oxygen barrier, but moisture sensitive, EVOH is sandwiched between layers of PP to protect the oxygen barrier from moisture. This construction will provide for a 12–18 month shelf life for oxygen-sensitive products such as tomato ketchup, mayonnaise and sauces.

If more precision is required in the neck finish of the container, injection blow moulding, a two-stage process, is used. Firstly, a preform or parison, which is a narrow diameter plastic tube, is made by injection moulding (Fig. 7.9). An injection mould is a two-piece mould where the cavity, and the resulting moulded item, is restricted to the actual, precise, dimensions of the preform. This is then blow moulded in a second operation, whilst retaining the accurate dimensions of the neck finish. This process also provides a good control of wall thickness.

A variation of injection and extrusion blow moulding is to stretch the preform after softening it at the second stage and then stretching it in the direction of the long axis using a rod (Fig. 7.10). The stretched preform is then blow moulded which results in biaxial orientation of the polymer molecules, thereby increasing strength, clarity, gloss and gas barrier. Injection stretch blow moulding is used to make PET bottles for carbonated beverages.

Screw cap and pressure fit closures with accurate profiles are made by injection moulding (Fig. 7.11). Wide mouth tubs and boxes are also made by injection moulding.

Not only are injection moulded items very accurate dimensionally but they can also be made with a very precise thickness, whether it be thick or thin. It should be noted that coextrusion is not possible with injection moulding.

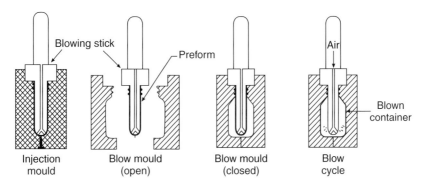

Figure 7.9 Injection blow moulding (courtesy of The Institute of Packaging).

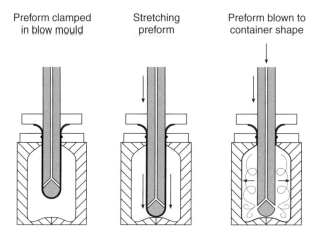

Figure 7.10 Stretch blow moulding – applicable to both extrusion and injection blow moulding (courtesy of The Institute of Packaging).

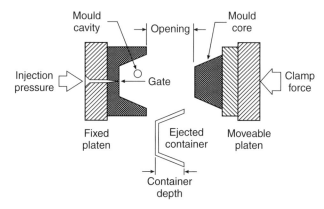

Figure 7.11 Injection moulding (courtesy of The Institute of Packaging).

Injection moulded items are recognised by a pinhead-sized protrusion, known as the gate, on the surface, indicating the point of entry of molten plastic into the mould. With injection blow moulding, the gate mark on the preform is expanded in the blowing action to a larger diameter circular shape.

There are many food applications for rigid and semi-rigid thermoformed containers. Examples include a wide range of dairy products, yoghurts etc. in single portion pots, fresh sandwich packs, compartmented trays to segregate assortments of chocolate confectionery and trays for biscuits. Thermoforming can be combined with packing on in-line thermoform, fill and seal machines. These machines can incorporate aseptic filling and sealing (Fig. 7.12).

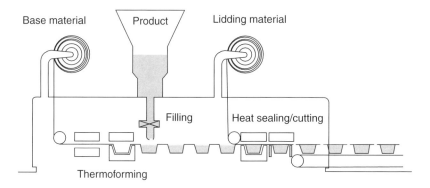

Figure 7.12 Thermoforming, filling and sealing.

Profile extrusion is used to make plastic tubing of constant diameter by inserting a suitably pointed rod in the outlet from the die of the extruder. The tubing can be cut to length and an injection moulded end with closure applied. The tube can be filled through the open end, which is then closed by heat sealing. This type of pack is used for food products such as salad dressings and powders/granules (herbs, spices and seasonings). Where higher barrier properties are required multilayer plastics tubing can be made by coextrusion. An alternative would be to use an end plug and closure, for example, for loosely packed confectionery products. (*Note*: laminated tubes are made with a characteristic heat seal parallel with the long axis.)

Foamed plastics are formed by dispersing a gas in the molten polymer, e.g. EPS. Food trays are made from extruded foam sheet by thermoforming. Insulated boxes for the distribution of fresh fish are made by injection moulding.

Plastic bulk containers are used in the food industry for the distribution of ingredients. They can be made by rotational moulding. This process uses plastics, such as low and high density PE, in powder form. A mould is charged with the right amount of polymer and it is heated and rotated in three axes. This action deposits the plastic on the inside walls of the mould, where it fuses and forms the side walls of the container.

7.3 Types of plastic used in packaging

7.3.1 Polyethylene

PE is structurally the simplest plastic and is made by addition polymerisation of ethylene gas in a high temperature and pressure reactor. A range of low, medium and high density resins are produced, depending on the conditions (temperature, pressure and catalyst) of polymerisation. The processing conditions control the degree of branching in the polymer chain and therefore the density and other properties of films and other types of packaging. Polyethylenes are readily heat sealable. They can be made into strong, tough films, with a good barrier to moisture and water vapour. They are not a particularly high barrier to oils and fats or gases such as carbon dioxide and oxygen compared with other plastics, although barrier properties increase with density. The heat resistance is lower than that of other plastics used in packaging, with a melting point of around 120°C, which increases as the density increases.

Polyethylene is not a conductor of electricity and was first used as an insulator in the 1940s. PE films are therefore highly susceptible of generating a static charge and need to have antistatic, slip agents and anti-blocking compounds added to the resin to assist film manufacturing, conversion and use.

Polyethylene is the most widely used in tonnage terms and is cost effective for many applications. It is the workhorse of the flexible films industry. Polymer plants can be found in all countries around the world, supplying specialist film-making polymers.

LDPE or low density PE is easily extruded as a tube and blown to stretch it by a factor of three times the original area. It is commonly manufactured around $30 \mu m$, with newer polymers allowing down gauging to 20 or $25 \mu m$ within a density range 0.910–$0.925 \, g \, cm^{-3}$.

It is possible to colour the films by blending pigment with the polymer prior to extrusion. Where extruders have more than one die, it is possible to form films with two or more layers of the same material or to produce coextruded films comprised of layers of different plastic materials. With three extruders, it is possible to produce a film where, for example, a moisture-sensitive polymer, EVOH, is sandwiched between protective layers of PE. EVOH provides a gas and odour barrier, and the PE offers good heat-sealing properties and a substrate for printing.

PE film melts at relatively low temperatures and welds to itself when cut with a hot wire, or blade, to form effective seals. For packaging, it is possible to use either premade bags or form/fill/seal machines using flat film in reel form. A major use of white pigmented LDPE film is for making bags for holding frozen vegetables.

By laminating to other substrates with adhesives, or extruding the PE polymer onto another material, or web, it is possible to make strong sachets, pouches and bags with good seal integrity, as the PE flows to fill holes in the sealing area or around contaminants in the seal.

PE and other plastics are used in combination with paperboard to make the base material for liquid packaging cartons which are discussed in Chapter 8. Major uses of PE film are in shrink and stretch wrapping for collating groups of packs and for securing pallet size loads.

LLDPE or linear low-density PE film has a density range similar to that of LDPE. It has short side chain branching and is superior to LDPE in most properties such as tensile and impact strength and also in puncture resistance. A major use has been the pillow pack for liquid milk and other liquid foods.

LDPE and LLDPE can be used in blends with EVA to improve strength and heat sealing. There is a degree of overlap in application between LDPE and LLDPE, due to the fact that there are differences in both, as a result of the conditions of polymer manufacture and on-going product development. The thickness used for specific applications can vary, and this can also have commercial implications.

MDPE or medium-density PE film is mechanically stronger than LDPE and therefore used in more demanding situations. LDPE is coextruded with MDPE to combine the good sealability of LDPE with the toughness and puncture resistance of MDPE, e.g. for the inner extrusion coating of sachets for dehydrated soup mixes.

HDPE or high-density PE is the toughest grade and is extruded in the thinnest gauges. This film is used for boil-in-the-bag applications. To improve heat sealability, HDPE can be coextruded with LDPE to achieve peelable seals where the polymer layers can be made to separate easily at the interface of the coextrusion.

A grade of HDPE film is available with either TD monoaxial orientation or biaxial orientation. This film is used for twist wrapping sugar confectionery and for lamination to oriented PP (OPP). The TD-oriented grade easily tears across the web but is more difficult to tear along the web. Being coextruded, a heat-sealable layer is applied to enable the film to run on conventional form/fill/seal machines. The biaxially oriented film has properties similar to that of OPP but has a higher moisture vapour barrier. It may be coated in the same way as OPP, including metallising, to give a high-barrier performance film with the good sealing integrity associated with PE.

HDPE is injection moulded for closures, crates, pallets and drums, and rotationally moulded for intermediate bulk containers (IBCs). A major application of HDPE is for blow moulded milk containers with a capacity 0.5–3 l.

7.3.2 Polypropylene (PP)

PP is an addition polymer of propylene formed under heat and pressure using Zieger-Natta type catalysts to produce a linear polymer with protruding methyl (CH_2) groups. The resultant polymer is a harder and denser resin than PE and more transparent in its natural form. The usage of PP developed from the 1950s onwards. PP has the lowest density and highest melting point of all the high volume usage thermoplastics and has a relatively low cost. This versatile plastic can be processed in many ways and has many food packaging applications in both flexible film and rigid form.

The high melting point of PP (160°C) makes it suitable for applications where thermal resistance is needed, for example in hot filling and microwave packaging. PP may be extrusion laminated to PET or other high-temperature-resistant films to produce heat-sealable webs which can withstand temperatures of up to 115–130°C, for sterilising and use in retort pouches.

The surfaces of PP films are smooth and have good melting characteristics. PP films are relatively stiff. When cast, the film is glass clear and heat sealable. It is used for presentation applications to enhance the appearance of the packed product. Unlike PE, the cast film becomes brittle just below 0°C and exhibits stress cracking below −5°C and hence has to be used in a laminate if the application requires deep freeze storage. OPP film, on the other hand, is suitable for use in frozen storage.

PP is chemically inert and resistant to most commonly found chemicals, both organic and inorganic. It is a barrier to water vapour and has oil and fat resistance. Aromatic and aliphatic hydrocarbons are, however, able to be dissolved in films and cause swelling and distortion. PP is not subject to environmental stress cracking (ESC). (ESC is a surface phenomenon whereby cracks can appear in moulded plastic as a result of contact with materials which affect the surface structure in critical parts of the design. This can lead to cracking without actually degrading the surface. There are specific tests to check for ESC, and shelf life tests with the actual product to be packed should also be carried out.)

Orientation increases the versatility of PP film. Oriented PP film (OPP or BOPP) was the first plastic film to successfully replace regenerated cellulose film (RCF) in major packaging applications such as biscuit packing. OPP films do not weld or heat seal together easily, as the melting temperature is close to the shrinkage temperature of the film and the surfaces spring apart when being sealed. However, acrylic-coated OPP has good runnability, including heat sealing, on packing machines, designed for RCF, though improved temperature control of the heat-sealing equipment is required. Acrylic coatings also offer good odour-barrier properties. Low temperature sealing water-based coatings are also available to provide improved runnability on packing machines.

An improved barrier film for both gases and water vapour is produced by coating with PVdC. This film is used as a carton overwrap for assortments of chocolate confectionery and for tea bags, providing excellent flavour and moisture barrier protection in both cases. One-side coated EVOH coatings are also available for use in laminations. PP films can be metallised and heat seal coated to produce a high gas and water vapour barrier film.

Many of the PP films are used in the form of laminations with other PP and PE films. This allows for the reverse-side printing of one surface, which is then *buried* inside the subsequent laminate.

Orienting, or extending, the film, typically, by a factor of 5, in both the machine and transverse direction, increases the ultimate tensile strength and some barrier properties such as the barrier to water vapour. Orienting in one direction binds the polymer molecules tightly to each other. However, they behave in a similar way to cotton fibres in a ball of cotton wool, in that they demonstrate low mechanical strength. When cotton fibres are spun, the resulting thread has a high strength. In the same way, orienting the polymer fibres biaxially in two directions results in a film with a high strength. This amount of biaxial orientation increases the area compared with the area extruded by a factor of 25, and the film is, proportionately, reduced in thickness. It is possible to produce oriented film with a consistent thickness of 14 μm for packaging.

PP and PE have the lowest surface tension values of the main packaging plastics and require additional treatment to make them suitable for printing, coating and laminating. This is achieved with a high-voltage electrical (corona) discharge, ozone treatment or by gas jets (Grieg, 2000). These treatments lightly oxidise the surface by providing aldehydes and ketones which increase the surface energy and therefore improve the adhesion, or keying, of coatings, printing inks and adhesives.

OPP film is produced in widths of up to 10 m or more to achieve cost-effective production. The limiting factors in production are either extrusion capacity for the thicker films or winding speed for the very thin films.

Most extrusion units today have more than one extruder, thereby enabling production to run at higher speeds and the use of different polymers feeding into one common die slot. Typically, a film will be made up of three or five layers of resin. The centre layer may be a thick core, either opaque or transparent, secondary polymer layers may have special barrier properties or pigmentation, and the outer layers may be pure PP resin to give gloss to the surface and/or protect the inner resins should they be moisture sensitive, as is the case with EVOH. In addition, thin layers of special adhesion-promoting resins, known as tie layers, may also be extruded.

The range of food products packed in PP films include biscuits, crisps (chips) and snack foods, chocolate and sugar confectionery, ice cream and frozen food, tea and coffee. Metallised PP film can be used for snacks and crisps (chips) where either a higher barrier or longer shelf life is required. PP white opaque PP films and films with twist wrapping properties are available. There are several types of heat seal coating, and in addition, it is possible for converters to apply cold seal coatings on the non-printing side, in register with the print, for wrapping heat-sensitive food products, such as those involving chocolate.

Paperboard can be extrusion coated with PP for use as frozen/chilled food trays which can be heated by the consumer in microwave and steam-heated ovens. Major food applications of PP are for injection-moulded pots and tubs for yoghurt, ice cream, butter and margarine. It is also blow-moulded for bottles and wide mouth jars. PP is widely used for the injection moulding of closures for bottles and jars.

PP can provide a durable *living* hinge which is used for flip top injection-moulded lids which remain attached to the container when opened, e.g. sauce dispensing closure and lid.

It is used in thermoforming from PP sheet, as a monolayer, for many food products such as snacks, biscuits, cheese and sauces. In coextrusions with PS,

EVOH and PE it is used for the packaging of several types of food product including those packed aseptically, by hot filling, and in microwaveable and retortable packs.

7.3.3 Polyethylene terephthalate (PET or PETE)

Polyesters are condensation polymers formed from ester monomers, resulting from the reaction of a carboxylic acid with an alcohol. There are many different types of polyester, depending on the monomers used. When terephthalic acid reacts with ethylene glycol and polymerises, the result is PET.

PET can be made into film by blowing or casting. It can be blow moulded, injection moulded, foamed, extrusion coated on paperboard and extruded as sheet for thermoforming. PET can be made into a biaxially oriented range of clear polyester films produced on essentially the same type of extrusion and Stenter-orienting equipment as OPP. Film thicknesses range from thinner than 12 µm for most polyester films to around 200 µm for laminated composites. No processing additives are used in the manufacture of PET film.

Polyesters have much higher heat resistance than many plastics and, when oriented, have very high mechanical strength. The ester has more radicals which may link with other chemicals, consequently making the surface more reactive to inks and not as resistant to chemicals as compared with polyolefins such as PE and PP.

PET melts at a much higher temperature than PP, typically 260°C, and due to the manufacturing conditions does not shrink below 180°C. This means that PET is ideal for high-temperature applications using steam sterilisation, boil-in-the-bag and for cooking or reheating in microwave or conventional radiant heat ovens. The film is also flexible in extremes of cold, down to −100°C. Heat-sealable versions are available, and it can also be laminated to PE to give good heat-sealing properties. Coated versions using PVdC give a good gas barrier and heat-sealing capability.

PET is a medium oxygen barrier on its own but becomes a high barrier to oxygen and water vapour when metallised with aluminium. This is used for vacuumised coffee and bag-in-box liquids, where it is laminated with EVA on both sides to produce highly effective seals. It is also used in snack food flexible packaging for products with a high fat content requiring barriers to oxygen and ultra violet (UV) light. Metallised PET, either as a strip or as a flexible laminate, is used as a susceptor in microwaveable packaging.

Reverse printed PET film is used as the external ply on f/f/s pouches where it provides a heat-resistant surface for contact with the heat-sealing bars.

The amorphous cast grades can be used as the bottom web in formed applications which are lidded with a heat-sealable grade of PET. These packs can be reheated in microwave and conventional ovens. (*Note*: all polymers are amorphous in the molten state, and rapid cooling fixes the polymer chains in this random state whereas slow cooling allows the molecules to realign themselves in a more formal crystalline state. In the case of PET, the amorphous state is a tough transparent material and the crystalline state is white and brittle.)

PET film is also used as the outer reverse-printed ply in retort pouches, providing strength and puncture resistance, where it is laminated with aluminium foil and either PP or HDPE. PET can be oxide coated with SiO_2 to improve the barrier, whilst remaining transparent, retortable and microwaveable.

Paperboard is extrusion coated with PET for use as ready meal trays which can be reheated in microwave or conventional radiant heat ovens, i.e. dual ovenable. The PET coated side of the paperboard is on the inside of the tray which is erected by corner heat sealing.

PET is the fastest growing plastic for food packaging applications as a result of its use in all sizes of carbonated soft drinks and mineral water bottles which are produced by injection stretch blow moulding. PET bottles are also used for edible oils, as an alternative to PVC.

A foamed coloured PET sheet has been developed under the trade name *Escofoam* which can be laminated and used as the bottom web in thermoformable f/f/s packaging with a printed, peelable seal, top web, e.g. with MAP for fresh meat and fish. A high-barrier laminate, requiring the use of an extruded tie polymer, acting as an adhesion promoter, would comprise PE/EVOH/PE/PET foam.

7.3.4 *Polyethylene naphthalene dicarboxylate (PEN)*

PEN is a condensation polymer of dimethyl naphthalene dicarboxylate (DNC) and ethylene glycol. This polyester polymer has created interest in the last few years due to its improved gas and water vapour barrier and strength properties compared to PET.

It is UV resistant and has higher temperature resistance compared with PET. It can be made into film and blow moulded for bottles.

PEN is a modified polyester resin from BP Chemicals. It is available as either a monopolymer PEN, a copolymer with PET, or a PET/PEN blend. The selection of a specific naphthalate containing resin is dependent on the performance and cost requirements of the particular application.

Suggested applications include one-trip beer/soft drinks bottles, returnable/refillable beer/mineral water bottles, sterilisable baby feeding bottles, hot fill applications, sports drinks, juices and dehydrated food products in flexible packaging. PEN is more expensive than PET, and this has limited its food packaging applications. Because of its relatively high cost, it is likely that PEN containers will only be suitable for use in closed loop returnable packaging systems.

PEN films can be vacuum metallised or coated with Al/Si oxides. It is claimed that the lamination of a metallised PEN film, Hostaphan RHP coated with SiO_x from Mitsubishi Plastic Films, to PE has been used for powdered baby milk. SiO$_x$ coated PEN film may be developed for retorting so that it could be used, for example, for soup in stand-up pouches.

7.3.5 Polycarbonate (PC)

PC is a polyester containing carbonate groups in its structure. It is formed by the polymerisation of the sodium salt of bisphenolic acid with phosgene. It is glass clear, heat resistant and very tough and durable. PC is mainly used as a glass replacement in processing equipment and for glazing applications. Its use in packaging is mainly for large, returnable/refillable 3–6 litre water bottles. It is used for sterilisable baby feeding bottles and as a replacement in food service. (This polycarbonate is not to be confused with the thermosetting polycarbonate used in contact lens manufacture.) It has been used for returnable milk bottles, ovenable trays for frozen food and if coextruded with nylon could be used for carbonated drinks.

7.3.6 Ionomers

Ionomers are polymers formed from metallic salts of acid copolymers and possess interchange ionic crosslinks which provide the characteristic properties of this family of plastics. The best known in food packaging applications is Surlyn™ from Dupont, where the metallic ions are zinc or sodium and the copolymer is based on ethylene and methacrylic acid. Surlyn™ is related to PE. It is clear, tougher than PE, having high puncture strength, and has excellent oil and fat resistance. It is therefore used where products contain essential oils, as in the aseptic liquid packaging of fruit juices in cartons, and fat-containing products such as snack foods in sachets. It has excellent hot tack and heat-sealing properties, leading to increased packing line speeds and output – even sealing when the seal area is contaminated with product. It is used in the packaging of meat, poultry and cheese. It is particularly useful in packing product with sharp protrusions.

Surlyn™ grades are available for use in conventional extrusion and coextrusion blown and cast film, and extrusion coating on equipment designed to process PE. It is also used as a tie or graft layer to promote adhesion between other materials such as PE onto aluminium foil or PET to nylon. An ionomer/ionomer heat seal can be peelable if PE is used, adjacent to one of the ionomer layers and *buried* in the laminate, e.g. PET/PE/Ionomer.

In food packaging, ionomer films, including coextruded films, are used in laminations and extrusion coatings in all the main types of flexible packaging. These include:

- vertical and horizontal f/f/s
- vacuum and MAP packing

- four-side sealed pouches and twin-web pouches with one web thermoformed
- inner ply of paperboard composite cans, e.g. aluminium foil/ionomer
- diaphragm or membrane seals.

Ionomers are used in laminated and coated form with PET, PA, PP, PE, aluminium foil, paper and paperboard.

7.3.7 Ethylene vinyl acetate (EVA)

EVA is a copolymer of ethylene with vinyl acetate. It is similar to PE in many respects, and it is used, blended with PE, in several ways. The properties of the blend depend on the proportion of the vinyl acetate component. Generally, as the VA component increases, sealing temperature decreases and impact strength, low temperature flexibility, stress resistance and clarity increase. At a 4% level, it improves heat sealability, at 8% it increases toughness and elasticity, along with improved heat sealability, and at higher levels, the resultant film has good stretch wrapping properties. EVA with PVdC is a tough high-barrier film which is used in vacuum packing large meat cuts and with metallised PET for bag-in-box liners for wine.

Modified EVAs are available for use as peelable coatings on lidding materials such as aluminium foil, OPP, OPET and paper. They enable heat sealing, resulting in controllable heat seal strength for easy, clean peeling. These coatings will seal to both flexible and rigid PE, PP, PET, PS and PVC containers. (An alternative approach to achieving a peelable heat seal is to blend non-compatible material with a resin which is known to give strong heat seal bonds so that the bond is weakened. Modified EVAs are also available for use in this way.)

Modified EVAs are also used to create strong interlayer tie bonding between dissimilar materials, e.g. between PET and paper, LDPE and EVOH.

EVA is also a major component of hot melt adhesives, frequently used in packaging machinery to erect and close packs, e.g. folding cartons and corrugated packaging.

7.3.8 Polyamide (PA)

Polyamides (PA) are commonly known as nylon. However, nylon is not a generic name; it is the brand name for a range of nylon products made by Dupont. They were initially used in textiles, but subsequently other important applications were developed including uses in packaging and engineering. Polyamide plastics are formed by a condensation reaction between a diamine and a diacid or a compound containing each functional group (amine). The different types of polyamide plastics are characterised by a number which relates to the number of carbon atoms in the originating monomer. Nylon 6 and a related polymer nylon 6.6 have packaging applications. It has mechanical and thermal properties similar to that of PET and therefore similar applications.

PA resins can be used to make blown film, and they can be coextruded. PA can be blended with PE, PET, EVA and EVOH. It can be blow moulded to make bottles and jars which are glass clear, low in weight and have a good resistance to impact.

Biaxially oriented PA film has high heat resistance and excellent resistance to stress cracking and puncture. It has good clarity and is easily thermoformed, giving a relatively deep draw. It provides a good flavour and odour barrier and is resistant to oil and fat. It has a high permeability to moisture vapour and is difficult to heat seal. These features can be overcome by PVdC coating. They can also be overcome by lamination or coextrusion with polyethylene, and this structure is used as the bottom thermoformable web, i.e. deep drawn, for packing bacon and cheese in vacuum packs or in gas-flushed packs (MAP or modified atmosphere packaging). The film can be metallised.

PA film is used in retortable packaging in structures such as PA/aluminium foil/PP. The film is non-whitening in retort processing. PA is relatively expensive compared with, for example, PE, but as it has superior properties, it is effective in low thicknesses.

7.3.9 Polyvinyl chloride (PVC)

If one of the hydrogen atoms in ethylene is replaced with a chlorine atom, the resultant molecule is called vinyl chloride monomer (VCM). Addition polymerisation of vinyl chloride produces PVC.

Unplasticised PVC (UPVC) has useful properties but is a hard, brittle material, and modification is necessary for it to be used successfully. Flexibility can be achieved by the inclusion of plasticisers, reduced surface friction with slip agents, various colours by the addition of pigments and improved thermal processing by the addition of stabilising agents. Care must be exercised in the choice of additives used in film which will be in direct contact with food, particularly with respect to the migration of packaging components into foodstuffs.

Rigid UPVC is used for transparent or coloured compartmented trays for chocolate assortments and biscuits. It is used with MAP for thermoformed trays to pack salads, sandwiches and cooked meats.

Most PVC films are produced by extrusion, using the bubble process. It can be oriented to produce film with a high degree of shrinkability. Up to 50% shrinkage is possible at quite low temperatures. The film releases the lowest energy of the commonly used plastic films when it is heat shrunk around products. It is plasticised, and the high stretch and cling make it suitable for overwrapping fresh produce, e.g. apples and meat in rigid trays using semi-automatic and manual methods.

Printed PVC film is used for heat-shrinkable sleeve labels for plastic and glass containers. It is also used for tamper-evident shrink bands. Thicker grades are thermoformed to make trays which, after filling, are lidded with a heat seal-compatible top web.

PVC has excellent resistance to fat and oil. It is used in the form of blow-moulded bottles for vegetable oil and fruit drinks. It has good clarity. As a film, it is tough, with high elongation, though with relatively low tensile and tear strength. The moisture vapour transmission rate is relatively high, though adequate for the packaging of mineral water, fruit juice and fruit drinks in bottles. PVC softens, depending on its composition, at relatively low temperatures (80–95°C). PVC easily seals to itself with heat, but heat sealing with a hot wire has the disadvantage of producing HCl gas.

The permeability to water vapour and gases depends on the amount of plasticiser used in manufacture. UPVC is a good gas and water vapour barrier, but these properties decrease with increasing plasticiser content. There are grades which are used to wrap fresh meat and fresh produce, where a good barrier to moisture vapour retards weight loss, but the permeability to oxygen allows the product to *breathe*. This allows fresh meat to retain its red colour and products such as fruits, vegetables and salads to stay fresh longer by reducing the rate of respiration, especially when packed in a modified atmosphere (MAP).

7.3.10 *Polyvinylidene chloride (PVdC)*

PVdC is a copolymer of vinyl chloride and vinylidene chloride – the latter forms when two hydrogen atoms in ethylene are replaced by chlorine atoms. PVdC was developed originally by Dow Chemical, who gave it the trade name Saran®.

PVdC is heat sealable and is an excellent barrier to water vapour and gases and to fatty and oily products. As a result of the high gas and odour barrier, it is used to protect flavour and aroma sensitive foods from both loss of flavour and ingress of volatile contaminants. It is used in flexible packaging in several ways:

- *Monolayer film.* A well-known application is the Cryovac range introduced by W.R. Grace and now operated by the Sealed Air Corporation. This includes poultry packing where hot water shrinkable bags are used to achieve a tight wrap around the product. The film can be used in the form of sachets but is less likely to be cost effective compared with other plastic films – some of which may incorporate PVdC as a coating. An interesting use is as sausage and chubb casing.
- *Coextrusions.* PVdC is often used in coextrusion, where, today, extruders incorporate three, five and even seven extrusion layers to meet product protection and packaging machinery needs cost effectively.

- *Coatings*. These may be applied using solutions in either organic solvents or aqueous dispersions to plastic films such as BOPP and PET, to RCF and to paper and paperboard.

Hence, PVdC is a widely used component in the packaging of cured meats, cheese, snack foods, tea, coffee and confectionery. It is used in hot filling, retorting, low-temperature storage and MAP as well as ambient filling and distribution in a wide range of pack shapes.

7.3.11 Polystyrene (PS)

PS is an addition polymer of styrene, a vinyl compound where a hydrogen atom is replaced with a benzene ring. PS has many packaging uses and can be extruded as a monolayer plastic film, coextruded as a thermoformable plastic sheet, injection moulded and foamed to give a range of pack types. It is also copolymerised to extend its properties.

It is less well known as an oriented plastic film, though the film has interesting properties. It has high transparency (clarity). It is stiff, with a characteristic crinkle, suggesting freshness, and has a deadfold property. The clear film is used for carton windows, and white pigmented film is used for labels. The film is printable. It has a low barrier to moisture vapour and common gases, making it suitable for packaging products, such as fresh produce, which need to breathe.

PS is easily processed by foaming to produce a rigid lightweight material which has good impact protection and thermal insulation properties. It is used in two ways. The blown foam can be extruded as a sheet which can be thermoformed to make trays for meat and fish, egg cartons, a variety of fast food packs such as the clam shell-shaped container, as well as cups and tubs. Thin sheets can be used as a label stock. The foam can also be produced in pellet or bead form which can then be moulded with heat and pressure. This is known as expanded polystyrene or EPS. It can be used as a transit case for fresh fish, with thick walls for insulation.

PS so far described is general purpose polystyrene. The main disadvantage as a rigid or semi-rigid container is the fact that it is brittle. This can be overcome by blending with styrene butadiene copolymer, SB or SBC, an elastomeric polymer. The blend is known as high-impact polystyrene or HIPS. Blending produces a tougher material. It is translucent and is often used in a white pigmented form. The sheet can be thermoformed for short shelf life dairy products.

HIPS is also used in multilayer sheet extrusion with a variety of other polymers, each of which contributes to the protection and application needs of the product concerned. Other polymers which may be used in this way with HIPS include PE, PP, PET, PVdC and EVOH. The food products packed with these materials include dairy products such as cream and yoghurt-based desserts,

UHT milk, cheese, butter, margarine, jam, fruit compote, fresh meat, pasta, salads etc. Many of these products are packed aseptically on thermoform, fill and seal machines.

7.3.12 Styrene butadiene (SB)

SB copolymer is also a packaging polymer in its own right – it is tough and transparent, with a high-gloss surface finish. Blown film has high permeability to water vapour and gases. It is used to pack fresh produce. It is heat sealable to a variety of surfaces. The film has good crease retention, making it suitable for twist wrapping sugar confectionery. Injection-moulded containers with an integral locking closure have a flexible hinge, similar in this respect to PP. It is known as K resin in the USA. It can also be used to make thermoformable sheet, injection and blow moulded bottles and other containers with high impact resistance and glass-like clarity. The relatively low density gives SBC a 20–30% yield advantage over other, non-styrenic, clear resins.

7.3.13 Acrylonitrile butadiene styrene (ABS)

ABS is a copolymer of acrylonitrile, butadiene and styrene, with a wide range of useful properties which can be varied by altering the proportions of the three monomer components. It is a tough material with good impact and tensile strength and good flexing properties. ABS is either translucent or opaque. It is thermoformable and can be moulded. A major use is in large shipping and storage containers (tote boxes), and it has been used for thin-walled margarine tubs and lids.

7.3.14 Ethylene vinyl alcohol (EVOH)

EVOH is a copolymer of ethylene and vinyl alcohol. It is related to polyvinyl alcohol (PVOH), which is a water-soluble synthetic polymer with excellent film-forming, emulsifying and adhesive properties. It is a high-barrier material with respect to oil, grease, organic solvents and oxygen. It is moisture sensitive and, in film form, is water soluble. PVOH itself has packaging applications in film form but not in food products, and it is used as a coating for BOPP (see Section 7.4.4).

EVOH was developed to retain the high-barrier properties of PVOH. It is also an excellent barrier to oxygen and is resistant to the absorption and permeation of many products, especially those containing oil, fat and sensitive aromas and flavours. Though it is moisture sensitive to a much lesser degree than PVOH, it is still necessary to *bury* it in multilayered coextruded structures, such as film for flexible packaging, sheets for thermoforming and in blow-moulded bottles, so that it is not in contact with liquid.

The other polymers used depend on the application, i.e. the food product and type of pack. PS/EVOH/PS and PS/EVOH/PE sheets are used for processed cheese, pâtè, UHT milk and milk-based desserts and drinks. It is also used for MAP of fresh meat and for pasta, salads, coffee and hot filled processed cheese, including portion packed cheese and fruit compote.

A higher-barrier sheet can be constructed with PP/EVOH/PP for pasteurisable and retortable products such as fruit, pâté, baby food, sauces like ketchup and ready meals, some of which are reheated by microwave. Coextruded film applications can involve EVOH with nylon, LLDPE and Surlyn™ (ionomer) with food products such as bag-in-box wine, processed and fresh meat.

Extrusion lamination can involve EVOH with PET, LDPE and LLDPE for coffee, condiments and snacks. It is used with PET and PP for tray lidding material. Extrusion lamination of paperboard with EVOH and PE is used for aseptically packed UHT milk and fruit juices where the EVOH layer provides an oxygen barrier as a replacement for aluminium foil. In blow moulding, EVOH is used with PP for sauces, ketchup, mayonnaise and cooking oil and with HDPE for salad dressings and juices. Ketchup and mayonnaise bottles based on EVOH are squeezable.

Small tubes made by profile coextrusion are used for condiments by incorporating EVOH into structures with LDPE and LLDPE. EVOH is an important polymer in many processing applications providing protection for many types of food product.

7.3.15 *Polymethyl pentene (TPX)*

TPX is the trade name for methyl pentene copolymer. It is based on 4-methyl-1-ene and possesses the lowest density of commercially available packaging plastics ($0.83\,cm^3$). It is a clear, heat-resistant plastic which can be used in applications up to 200°C. The crystalline melting point is 240°C (464°F). TPX offers good chemical resistance, excellent transparency and gloss. It can be extruded and injection moulded.

It was introduced by ICI in the 1970s and is now available from Mitsui Chemical of America, Inc. The main food packaging use is as an extrusion coating onto paperboard for use in baking applications in the form of cartons and trays for bread, cakes and other cook-in-pack foods. This packaging is dual ovenable, i.e. food packed in this way may be heated in microwave and radiation ovens. The surface of this plastic gives superior product release compared with aluminium and PET surfaces. TPX coated trays must be formed by the use of interlocking corners as they cannot be erected by heat sealing.

7.3.16 *High nitrile polymers (HNP)*

HNPs are copolymers of acrylonitrile. They are used in the manufacture of other plastics such as ABS and SAN. The nitrile component contributes very good gas and odour barrier properties to the common gases, together with good chemical resistance. HNPs therefore offer very good flavour and aroma protection.

A commercially available range of HNPs approved for food packaging is made under the trade name Barex®, introduced by Sohio Chemical and now BP Chemicals. This is a rubber-modified acrylonitrile–methyl acrylate copolymer. Grades are available for blown and cast film, extrusion blow moulding, injection stretch blow moulding and injection moulding. It is a clear, tough, rigid material with a very good gas barrier and chemical resistance. It is used as the inner layer in blow-moulded bottles coextruded with HDPE. Barex® films can be coextruded as film and sheet or laminated with PE, PP and aluminium foil for flexible packaging applications with food products. Sheet materials can be thermoformed.

7.3.17 Fluoropolymers

Fluoropolymers or fluoroplastics are high-performance polymers related to ethylene where some or all of the hydrogen atoms are replaced by fluorine, and in the packaging polymer a hydrogen is also replaced by a chlorine atom to produce polychlorotrifluoroethylene (PCTFE). The trade names are Aclar® (Honeywell) and Neoflon™ (introduced as Kel-F® by 3Ms and now manufactured by Daikin).

This material has the highest water vapour barrier of all the commercially available packaging polymers, is a very good gas barrier and offers high resistance to most chemicals at low temperatures. In many applications, it is a suitable replacement for aluminium foil. It is available as a film or sheet. It is transparent, heat sealable and can be laminated, thermoformed, metallised and sterilised.

It is relatively expensive and is best known as a thermoformable blister pack material laminated with PVC for pharmaceutical tablets. Food packaging applications are possible but are not highlighted at the present time.

Polytetrafluoroethylene (PTFE), better known as DuPont Teflon®, is a high-melting point, inert and waxy polymer. It is used in the form of tape and coatings on packaging machines to reduce adhesion, where that could be a problem, e.g. heat seal bars, and to reduce friction where packaging materials move over metal surfaces.

7.3.18 Cellulose-based materials

The original packaging film was regenerated cellulose film (RCF). Pure cellulose fibre derived from wood is dissolved and then regenerated by extrusion through a slot, casting onto a drum and following acid treatment, is wound up as film. It is commonly known as *Cellophane*, though this is in fact a trade name. RCF is not a thermoplastic material in that it is not processed in a molten phase or softened by heat. Cellulose is, however, a high molecular weight, naturally derived, polymer.

To make it flexible, it is plasticised with humectants (glycol type). The degree of flexibility can be adjusted to suit the application. The degree of flexibility can range from a fairly rigid level to the most flexible, which is known as *twist-wrap* used to wrap individual units of sugar confectionery. RCF is *dead folding* so that it keeps a folded, or twist wrapped, condition. It is a poor barrier to water vapour, and this property is made use of with products which need to lose moisture, such as pastries and other flour confections, to achieve the correct texture when packed. Plastic films such as PP or PE would keep the relative humidity (RH) too high inside such a package and therefore encourage mould growth. When dry, RCF is a good barrier to oxygen.

Heat sealability and an improved barrier to water vapour and common gases are achieved by coating. The coatings are either nitrocellulose (MS type) or PVdC (MX type). A range of barrier performances are possible by choice and method of coating. It can be coloured and metallised. (Red coloured *breathable* RCF secured with a printed label on the base is used to wrap Christmas puddings.) RCF is printable.

RCF can be laminated with paper, aluminium foil, PET, metallised PET and PE to achieve specific levels of performance and appearance. It is resistant to heat and is used in laminates where reheating involves temperatures in the range 220–250°C. Usage today is much reduced with the availability of the lower cost BOPP, which matches RCF in many properties. The pack design is usually a form/fill/seal pouch type. RCF is used in food packaging for gift packs and for packaging which is specified as biodegradable (compostable).

Uncoated thick RCF is used to demonstrate tamper evidence on a bottle. This is done by moistening a small diameter RCF sleeve, slipping it over the bottle closure and top part of the neck and allowing it to dry when it shrinks tightly to the bottle and closure.

Cellulose acetate is also derived from cellulose. It has high transparency and gloss. It can be printed. It has been used as a laminate with paperboard for confectionery cartons and as a window in carton design. In both applications, it is more expensive than BOPP. It has also been replaced by PVC, PET and PP as a sheet material for the manufacture of transparent cartons.

7.3.19 Polyvinyl acetate (PVA)

PVA is a polymer of vinyl acetate which forms a highly amorphous material with good adhesive properties in terms of open time, tack and dry bond strength. The main use of PVA in food packaging is as an adhesive dispersion in water. PVA adhesives are used to seal the side seams of folding cartons and corrugated fibreboard cases and to laminate paper to aluminium foil.

7.4 Coating of plastic films – types and properties

7.4.1 Introduction to coating

Coatings are applied to the surfaces of plastic films to improve heat-sealing and barrier properties. They are also applied to rigid plastics to improve barrier. Traditionally, the most common method of application to film has been by using an etched roll, as this gives consistent and accurate coating with weights up to around $6\,g\,m^{-2}$. This level of coating is commercially available from film manufacturers. If higher coating weights are required, this is normally carried out by converters.

With environmental concerns being an important factor, water-based coating systems have been developed. It is unusual to find solvent-based systems being used today. Where they are used they are mostly alcohol based, with butanol being the highest boiling point solvent used. Priming coats are applied. Where the coating is to be applied to both sides of the film, the primer is applied to both sides simultaneously using reverse driven gravure rollers. Adhesion to the base film is essential, and hence an adhesive-type coating with antistatic properties is normally used, with a coating weight of less than $1\,g\,m^{-2}$.

Metallised coating using aluminium has been available for some time. This is carried out by converting the aluminium to vapour in a vacuumised chamber and depositing it on the surface of plastic film, paper and paperboard for packaging purposes. More recently, SiO_x, a mixture of silicon oxides, has been deposited in thin layers on several plastic films. It is likely that other materials, e.g. DLC, diamond-like coating based on carbon which has been internally applied to PET bottles, will be developed further for both bottles and film.

Extrusion coating is also a method of applying a plastic coating, though this usually refers to the application of plastics to other materials such as aluminium foil, paper and paperboard. Coating as a technique for improving the properties of plastic film and containers is an active area for innovation.

7.4.2 Acrylic coatings

Acrylic coatings are applied to plastic film, particularly OPP. The coating is glass clear, hard, heat sealable and very glossy. It has an initial sealing temperature of around 100°C. The melting point is sharply defined. This means that the coating can easily slide over hot surfaces without sticking. A typically acceptable lower sealing strength would be $250\,g\,m^{-2}$/25 mm seal width. With a film shrinkage temperature of 150°C, this would give a sealing range of 50°C. It is necessary to have some slip and anti-blocking compounds incorporated in the coating to achieve the best packaging machine runnability. The coating thickness is generally about $1.0\,g\,m^{-2}$, and with a specific gravity close to 1, this gives a thickness of 1 μm.

7.4.3 PVdC coatings

PVdC coatings may be modified to produce either a good heat-sealing polymer or a high-barrier polymer. There is a compromise to be made between the quality of sealing and the barrier properties required. Modification of the polymer to give a wider sealing range lowers the threshold for sealing to around 110°C at the expense of the gas barrier. PVdC coatings are applied to films and paper.

The majority of general-purpose coatings supplied will have sealing properties starting to seal at 120°C and oxygen barrier of around $25\,\text{ml}\,\text{m}^{-2}/24\,\text{h}$. For PET film, PVdC is normally chosen for high oxygen barrier ($10\,\text{ml}\,\text{m}^{-2}/24\,\text{h}$) and as a result may have poor sealing properties.

The formulation needs to incorporate both silica and waxes as slip and anti-blocking agents to prevent the coatings from sticking to the hot sealing surfaces. Typically, film producers apply coating weights of $3\,\text{g}\,\text{m}^{-2}$ or $2\,\mu\text{m}$ thickness. The specific gravity of PVdC is 1.3. Surface coatings can be applied to rigid containers such as the surface of PET beer bottles.

7.4.4 PVOH coatings

With the environmental concern that dioxins may be produced if chlorine-based compounds are incinerated, an alternative high gas barrier has been sought to replace PVdC without needing to modify coating parameters. PVOH emulsions meet this requirement, but they are sensitive to moisture, losing barrier properties if the RH increases to more than 65% RH. Films with PVOH are therefore likely to be used as part of a laminate, with the PVOH on the inside of the web. BOPP with PVOH on the outside can be used, provided it is overlacquered with a protective varnish. PVOH also has no sealing properties. It is, however, an excellent surface to receive printing inks with low absorption or retention of solvents. Coating weights are similar to PVdC, but the specific gravity is nearer to 1.0 and film yield is slightly higher.

7.4.5 Low-temperature sealing coatings (LTSCs)

LTSCs for OPP which seal at lower temperatures and have a wider sealing range are required to meet the demand for faster packaging machine speeds. These coatings, based on ionomer resins, applied in the form of emulsions are an alternative to both acrylic and PVdC coatings. As silica and waxes are likely to raise the sealing temperature threshold of any coatings, they are kept to a minimum with the consequence that friction is higher on LTSC than with the conventional coatings. The LTSC does not stick or block to PVdC or acrylic coatings and hence it is possible to have differentially coated films. The ionomer surface has good ink receptivity and does not retain printing ink solvents.

7.4.6 Metallising with aluminium

Direct vacuum metallising with alumium on plastic films results in a significant increase in barrier properties. This is because these films are smooth and a continuous layer of even thickness can be applied. Films treated in this way are PET, PA and OPP. A major cost factor is the time taken to apply the vacuum after a reel change. This favours 12 μm PET because a large area can be contained in a reel. When applied to PET, the film can be used to metallise paper and paperboard by transfer from the film using a heated nip roll, after which the PET can be reused (Fig. 7.13).

7.4.7 SiO_x coatings

SiO_x has recently been introduced as a coating. This material has excellent barrier properties. It is applied by vacuum deposition. SiO_x coated PET film is commercially available and is used in the retort pouch laminates in Japan. It is transparent, retortable, recyclable and has excellent barrier properties. An alternative method of coating is being developed by Lawson Mardon in Switzerland which uses plasma pretreatment followed by evaporation of the silicon using an electron beam (EB) (Lohwasser, 2001).

SiO_x has also been applied to plastic bottles, giving an oxygen barrier which is 20 times greater than the barrier of an uncoated bottle (Matsuoka *et al.*, 2002). The Glaskin process introduced by Tetrapak also vacuum coats the inside of PET beer bottles. Bottles coated in this way have been used by several leading breweries in Europe (Anon, 2000b,c). Less flavour scalping has been claimed to result in a minimum shelf life of 6 months. The

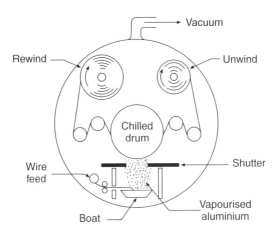

Figure 7.13 Metallising process (courtesy of The Institute of Packaging).

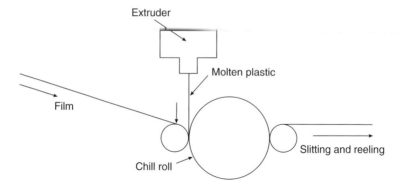

Figure 7.14 Extrusion coating of plastic film.

European use of thermal or EB chemical vapour deposition of silicon oxide and the reactive evaporation of aluminium is discussed in Naegeli (2001).

7.4.8 DLC (Diamond-like coating)

A relatively new coating is known as DLC (diamond-like coating). It comprises a very thin layer of carbon. PET bottles do not give as long a shelf life as glass for the bottled beer market. A DLC coating on the inside of PET bottles has been trialled extensively in Japan. Significant improvements in barrier have been reported (Ayshford, 1998; Anon, 2000a).

7.4.9 Extrusion coating with PE

A heat seal coating can be applied to a heat-resistant film such as PET and PA by extrusion coating the film with PE (Fig. 7.14).

7.5 Secondary conversion techniques

7.5.1 Film lamination by adhesive

The lamination of plastic films combines two or more films using an adhesive which can be water, solvent or 100% solids based. Lamination is an alternative to coextrusion, where two or more layers of molten plastic are combined during film manufacture. It is also possible to laminate using molten polymer as both an adhesive and a barrier layer in an extrusion lamination. It is also possible to laminate without an adhesive, e.g. laser lamination of thermoplastics.

The choice as to whether to use a laminate or a coextrusion can be quite complicated, depending on several factors. These include:

- product needs, in terms of shelf life and barrier properties
- type of pack, how it will be handled at every stage and the run length
- waste during coextruded film manufacture cannot be recycled whereas waste produced during single film manufacture can be recycled
- coextruded film can only be surface printed, whereas one of the films in a laminate may be reverse printed so that the print can be sandwiched during lamination. This ensures rub resistance, gloss and clarity, dependent on the film concerned
- converters may print and laminate in one cost-effective operation; where this is not possible, the cost increase due to the extra conversion process may be difficult to justify. However, the cost of any alternative approach to providing the protection required by the product, either by a single plastic material or by a form of coextrusion, must also be considered
- film thickness to achieve the required barrier must be considered in case it would be thicker than ideal, with increased stiffness giving sealing problems and handling difficulties. Sealing range may be narrowed due to poor thermal transfer, and heat retention after the seal has been made may allow seals to re-open
- on the other hand, a thicker material will be stiffer so that it handles and displays better than packaging made from a thinner material
- lamination will almost always highlight the different tension in each film web, and it is very common for laminates to suffer from curling. For many applications where the film is cut and then pushed or pulled, the cut edge may need to be flat to give trouble-free feeding through the packaging machine.

There is a vast amount of literature published about laminating and coextrusion. Consideration of what approach to take must be assessed on a case by case basis as there are both technical and commercial factors to be considered (Haas, 1996).

A wide range of adhesives is available. PVA and other water-based adhesives which remain flexible and have a long shelf life may not adhere satisfactorily to polyolefines (PE and PP) with their inert surfaces and excellent moisture barrier. Should such adhesives be used, then a long drying period is required before the laminate can be used and, in practice, paper or paperboard should be one of the substrates to allow the water to disperse.

Polyurethanes and other cross-linking adhesives are preferred for barrier plastics. They are normally applied from a gravure roller at the end of the printing press and the films combined under pressure using a coating weight of $1-3\,\mathrm{g\,m^{-2}}$. Careful selection is essential, as carbon dioxide can form as small bubbles and impair the visual properties of the final laminate. In some film structures, there is the possibility of the adhesive reacting with the film coatings to produce discolouration. When this happens, advice should be sought from the adhesive and ink suppliers. Adhesion strength of several hundreds of grams/25 mm is normal and often the bonding is permanent.

There is pressure to move away from adhesive systems based on the use of organic solvents to water-based and 100% solids systems to reduce solvent emissions. The 100% solids adhesives are materials which cross-link as a result of applied heat, UV or EB radiation. Tacky hot melt adhesives with a wax and EVA content and the use of PE in extrusion laminating would also be considered 100% solids adhesives.

With the increasing speed of printing, it has become common to operate a two-stage process of printing followed by lamination. This allows better control of the processing as the systems are independent.

7.5.2 Extrusion lamination

One web of a laminate can be passed through a curtain of molten PE and then combined with a second film layer whilst the PE is still molten. It is possible to use a small weight of PE (typically $7–10\,g\,m^{-2}$) as both an adhesive and as a means of making a laminate much stiffer due to the increased thickness of the total structure. There is often a need to prime the surfaces of the films to receive the PE and to achieve bond strengths above $200\,g/25\,mm^2$ (Fig. 7.15).

For many applications where the laminate structure does not experience high stresses, the strength of a laminate may in practice only need to be about $100\,g/25\,mm$ at the lowest. The danger is always that the laminate bond may

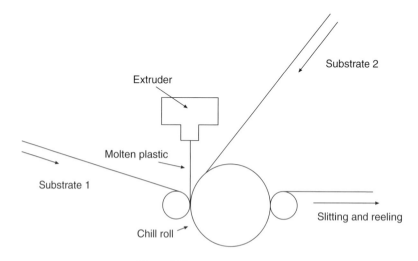

Figure 7.15 Extrusion lamination.

reduce with time, and hence the specification has to be set higher than the known minimum requirement.

7.5.3 Thermal lamination

When two webs each have heat-sealing properties, it is possible to join them together by passing the films through a heated nip roller system. With no adhesive involved, the final weight of the laminate is the same as that of the original components. This process relies on the films each having a low sealing point, as under tension the films may shrink if heated at a high temperature, causing creasing, or stretch under tension. As the films approach the elastic limit, curl may be produced due to the slightly greater shrinkage of one web. Bond strengths should be high depending on the nature of the original coatings. This form of laminating is not common for the production of laminates in the food industry. It is widely used in book cover production. A specific type of thermal lamination is that using a laser to activate the surfaces being bonded (Potente *et al.*, 1995).

7.6 Printing

7.6.1 Introduction to the printing of plastic films

Printing preferences until recently seemed to be geographical, with a tendency for gravure presses being popular in Europe and flexographic presses in North America. This may be historical, as the result of the way the markets developed. As the quality of flexo printing has increased with the introduction of photopolymer plates and with the market requiring shorter and shorter print runs, the flexo process has gained ground in Europe. Combination presses which incorporate flexo and gravure units have also become popular. The number of printing stations has grown in number with up to 10 stations now available on central impression flexographic (CI) machines and on gravure reel fed presses.

7.6.2 Gravure printing

The gravure press consists of a series of printing stations in-line, each applying one colour of liquid ink, applying cold seal latex or PVdC emulsion in-line. A roller is engraved, mechanically, chemically or electrically laser eroded, into a pattern of small cells. These cells hold the ink which is picked up from the ink bath in which the gravure roller rotates. The amount of ink is controlled by the depth and area of the cell, and a doctor blade scrapes off the excess ink. Film is passed over the gravure roller with backing pressure from a lay-on, or impression roll, to pull the ink out of the cells. The inked film is passed into a heated oven to dry off the solvents or water medium. Other ink, or coating, layers are applied in register to achieve the finished design.

The gravure system allows a very large number of prints as the cylinders are hard wearing and accurately reproduce the design. Initial costs are high due to the engraving process, but, for long runs, which can be printed at high speed, the gravure process is cost effective.

7.6.3 Flexographic printing

Flexographic printing may be carried out with a number of printing stations in-line (stack press) or with the printing rollers arranged around a central large diameter drum (central impression). The plates which are now made from a photochemical (plastic) plate material are attached to the printing rolls. Ink is picked up by a cavitated anilox roll and transferred to the printing plate. The ink is then transferred to the film. Because the costs of producing the plates are relatively low, flexographic printing is cost effective, especially for short runs. The quality of reproduction has increased, and has approached that of gravure printing. Productivity on both types of press has increased, and hence the better choice of process for any given print order has become more difficult.

7.6.4 Digital printing

Electronic printing systems have been developed, and with coatings available to receive the new ink systems, it is now possible to create artwork on a computer and transfer the image directly to the packaging film. A design is created on a computer; it may be an individual design or replicated to give several hundreds of impressions. The ink, usually in powder form, is attracted on to the film surface and cured in place. Special coatings are necessary to receive the ink. A standard heat-sealable coating on the reverse side allows the film to be made immediately into packages. The system as yet is only suitable for narrow web widths and is capable of producing test packages for market research or promotional campaigns.

7.7 Printing and labelling of rigid plastic containers

7.7.1 In-mould labelling

Printed labels can be applied to containers and lids during forming. The technique has been adapted for use in thermoforming, e.g. yoghurt pots, and in both blow moulding and injection moulding, e.g. ice cream tubs, lids and large biscuit containers.

Designs in relief can be carried in the walls of the mould. These designs are visible in the moulded item having an embossed or debossed effect. This technique is used to imprint the plastic identification for sorting in waste management schemes, indicate the number of the mould and other manufacturers markings.

7.7.2 Labelling

Several types of printed labelling are common – pressure-sensitive plastic, paper and laminated aluminium foil labels. Sleeve labels are clipped or fixed with adhesive around pots and tubs. Such containers may require features, such as a recessed panel, in their design to facilitate the location of the label. Some packs are designed so that after use a paperboard label and plastic pot may be easily separated to meet waste management needs.

Another popular form of labelling of bottles and jars is the printed plastic shrink sleeve. These are supplied flat and printed in tubular form on reels. After automatic application, the container passes into a heated zone which causes the label to shrink tightly around the container.

7.7.3 Dry offset printing

This method uses a relief plate which after inking transfers the ink to a blanket roll, which in turn applies the design to the plastic surface. This method has especially been developed to print round and tapered containers. The inks are either heat set or UV cured.

7.7.4 Silk screen printing

The design to be printed is carried on a metal or plastic woven mesh. This is placed in contact with the item to be printed, and the thick oil-based ink is forced, or squeezed, through in the design areas with the action of a flexible wiping blade.

7.7.5 Heat transfer printing

The full design is first printed with heat-sensitive inks on a carrier web of PET. This can then be placed in contact with plastic containers at high speed where a heated die transfers the design directly onto the container. Therimage is an example of a well-known form of heat transfer labelling.

Hot foil stamping is a type of heat transfer printing. A heat-resistant ink with an adhesive coating carried on a PET film is placed in contact with the item being printed. A heated metal die with the design in relief is pressed against the PET film, transferring the image. This type of decoration can be used to print a highly reflective metallic image. Hot foil stamping is often used on luxury items such as cartons for chocolate confectionery and labels for bottles of spirits and liqueurs.

7.8 Food contact and barrier properties

7.8.1 The issues

In addition to the maintenance of pack integrity, i.e. efficient closure systems and physical protection during storage and distribution, it is essential that primary food packaging protects the food in such a way that health is not endangered and that the quality is maintained within the expected shelf life. Quality in this context depends on the packaging, the food product and possible interactions between the food and the packaging material. The result may be that detrimental organoleptic and other changes occur, which may be caused by:

- migration of additives, residues and monomer molecules from the packaging material into the food
- permeation of gases, vapours and permeant molecules from the environment into the pack headspace and vice versa
- sorption of components, including volatile flavour compounds and lipids, into the packaging in a process often referred to as *scalping*.

7.8.2 Migration

When food products are packaged, the food is in direct contact with the inside surface of the packaging. It is possible for interaction between the food and the packaging to occur and for components of the packaging to be absorbed by, or react with, the food. In the case of plastic materials, this may involve the basic polymer, and even if this is non-reactive with respect to the ingredients of the food, it is possible that coatings and additives used to facilitate manufacture and use of the plastic may interact with the food. It is therefore essential that the plastic material and associated additives are approved for direct food contact.

To ensure that adequate product safety procedures are carried out, many countries have regulations to maintain safety with respect to plastics in contact with food. In the United States, the regulations emanate from the Federal Food and Drugs Administration.

In the European Union, Directive 89/109/EEC is a Framework Directive concerning the general requirements for all plastic materials and articles in contact with food. The aim is to ensure that they are manufactured in accordance with good practice and that they do not transfer any constituents into food in such a way as to endanger public health or bring about organoleptic or other unacceptable changes in the nature, substance or quality of the food.

Organoleptic in this context refers to the taste, texture, flavour, colour or odour of the food product, and migration is the process whereby chemical components in the packaging material transfer into a food product.

In the European Union, EU Directive 90/128 EEC, and subsequent amendment 2001/62/EC, deal specifically with the use of plastics in contact with food, see website reference. It includes definitions and migration limits, and limits the use of and residues associated with specific substances. A migration limit is either defined in terms of weight released per unit area, e.g. $10\,\text{mg}\,\text{dm}^{-2}$, or where this is either not feasible, e.g. for caps, gaskets, stoppers, or where the volume of the container is from 0.5 to 10 litres, the limit is $60\,\text{mg}\,\text{kg}^{-1}$ of the foodstuff concerned.

Directive 82/711/EEC sets rules for migration testing using specified food simulants, i.e.

- Simulant A: water for aqueous foods
- Simulant B: 3% w/v acetic acid for acidic foods
- Simulant C: 15% v/v ethanol for alcoholic products
- Simulant D: rectified olive oil for fatty/oily foods.

Sophisticated analytical and measurement techniques have been developed to identify and quantify the materials extracted by these techniques. They include the use of GLC (gas liquid chromatography), mass spectrometry and IR analysis. These test procedures are also used, together with sensory testing panels, to evaluate plastic materials and packaging in production and use (Frank, 2001).

7.8.3 Permeation

Permeation through a film is a three-part process:

- solution/absorption of penetrant (vapour or gas) into the polymer surface
- migration/diffusion of penetrant through polymer(s)
- emergence/desorption of penetrant from opposite surface of polymer.

Absorption and desorption depend on the solubility of the permeant, and solubility is greatest when penetrant and material have similar properties.

Other relevant theory comprises:

- Graham's Law (1833) which states that the velocity of diffusion of a gas is inversely proportional to the square root of the density.
- Fick (1855) stated that the quantity of diffusing gas is proportional to concentration and time and inversely proportional to the thickness of the substrate through which it is diffusing.
- Henry's Law (1803) which states that the amount of gas absorbed by a given volume of a liquid at a given temperature is directly proportional to the partial pressure of the gas.

In practice, the film may comprise more than a single polymer, and there may be discontinuities in coatings, pinholes in films, variations in molecular structure and degree of crystallinity. The penetrant molecular size, shape and degree of polarity are relevant and so are ambient conditions. These are all factors which affect diffusion and solubility, which in turn have a direct impact on permeability.

The permeability of plastic films to moisture vapour and common gases such as oxygen, carbon dioxide and nitrogen has been measured by standardised test methods. Oxygen, for example, can cause oxidative rancidity in oil or fat containing food products.

Water vapour permeation into a product may cause a loss of texture, and, on the other hand, the escape of water, in the vapour phase, from a product through the packaging may cause dehydration, textural changes and loss of weight. An example of the latter would be a plastic film-wrapped Christmas pudding which would lose moisture in storage prior to sale, where a compromise has to be made in balancing weight loss in storage with initial weight and the water vapour barrier protection provided by the plastic film. In this example, in addition to flavour retention and texture, the actual weight at the point of sale would also have to meet appropriate regulations.

The results of permeability tests provide guidance with respect to the choice of material(s) for the packaging of specific food products. Some other possible penetrants and the effect of the presence of polymer additives, e.g. plasticisers, can lead to surprising results. It is still necessary to carry out shelf life tests to establish performance in practice with the food under consideration.

7.8.4 *Changes in flavour*

Food manufacturers have to ensure hygiene and freedom from odour and taint in their products. This has implications for the packaging used, in the sense that contaminating material from the external environment must be prevented from changing the flavour, aroma or taste of the food. An example would be the use of PVdC coated BOPP film to overwrap cartons containing tea bags.

Flavour may be lost by *scalping*. This is where organic compounds are absorbed into the packaging or adsorbed on the surface of the packaging material. Flavour may be masked or chemically changed by any ingress of off flavours and aromas from the external environment by permeation (transmission) through, or by migration of contaminating ingredients from, the packaging material.

Food may also be changed by the loss or gain of moisture and by oxidation. Hence, the permeability of the packaging material to the transmission of moisture vapour and oxygen is an important property of the packaging material. The rate of transmission is dependent on the ambient temperature and the principles of permeation.

In addition to the qualitative and quantitative tests referred to in Section 7.8.1 subjective assessment of flavour, odour and taint is carried out by statistically valid sensory testing panels.

7.9 Sealability and closure

7.9.1 Introduction to sealability and closure

The most important function of packaging is to ensure the protection and integrity of the product. This implies that the pack must be securely sealed. With plastic packaging, this can be achieved on the packing line by either heat sealing, application of a closure, such as a screw cap, or by the use of some form of adhesive system.

One of the most overlooked factors in the production line is the efficient performance of the packaging system. Sealing or closing systems are often presumed to perform with little consideration of the material/machine relationship.

The needs of the proposed material or container are seldom discussed with the machinery manufacturer. At the earliest stage, therefore, planning needs to take place between Production, Engineering, Purchasing, Product R&D, Marketing and Packaging Technologists with Machinery and Packaging Suppliers. It may be found that compromises will have to be made to find the optimum solution.

7.9.2 Heat sealing

Product protection and hence effective shelf life are a function of the quality of sealing of the package. Sealing strength is influenced by the thickness of the film web. With the same coating, a doubling of the base film thickness almost doubles the seal strength. Conflictingly, the thicker the material, the narrower the temperature sealing range under normal sealing conditions. The thicker film does not allow heat to flow so easily to melt the sealing coating or polymer, and when heated, the film retains the heat, allowing the sealant to remain fluid, with a detrimental effect on hot seal strength. Thick film also requires more pressure to bend the film and make intimate contact, particularly with crimp jaws, as found on f/f/s machines.

Jaw design has a great influence on seal integrity and strength. While the ideal seal jaw may be flat, in practice this is only true if there are no folds or tucks in the seals. Crimp jaws are used to compensate for variations in film thickness on vertical and horizontal f/f/s machines.

Seal integrity may now be evaluated as part of the in-line quality function by using standard instruments to test the pack under pressure or vacuum and identify how quickly air or oxygen will pass through the seals. A practical judgement has to be made on the time and pressure required to change the pack integrity.

7.9.2.1 Flat jaw sealing

Sealing conditions are a compromise between dwell time and the temperature and pressure of the jaws. The requirement is to apply sufficient energy to cause the sealant to fuse together and become one medium. Conduction of heat, combined with heat flow characteristics, needs to be carefully balanced to produce a perfectly formed seal, with no temperature distortion and an even seal strength throughout the sealed area. Energy input is a function of time and temperature. With heat-sensitive films such as PE and cast PP, a low temperature applied for a long period, with high pressure to remove air from between the film surfaces, is ideal. With films having a wide temperature-sealing range, the tolerance in dwell time, sealing temperature and pressure is much wider.

If possible, heat should be applied to both sides to achieve the quickest possible polymer melting. Sealing surfaces need to have good release properties to ensure that molten polymer does not stick to the heating surfaces and pull the newly made seal apart. An alternative is to use one heated surface in the form of a constant temperature metal bar, with a flat or curved profile, sealing against a rubber-faced anvil.

With PE, there is a need to avoid stressing the seal while the polymer is still fluid, and many machines are designed to have an air cooling blast or, alternatively, clamping of the seal whilst the film cools below the sealing temperature. With OPP films where the core of the film is not being melted, an effective seal is achieved by fusing sealant polymers of coextruded film or two surface coatings which flow together. The core will give rigidity to the seal, and to avoid destroying the new seal, it is only necessary to ensure that fluid coatings are not stressed. Pulling jaws apart at a perpendicular to the film overcomes the problem in practice. Film sliding over hot metal while under pressure is to be avoided as the coating may stick to the metal surface. If it is impossible to avoid the film sliding over metal under pressure, the solution is often to ensure that only point contact is made between the heated metal and the film. A rough surface minimises or avoids *hot-sticking* on the machine. The principle is to avoid total air exclusion between the contact surfaces which may be caused by the coatings flowing freely, creating a vacuum. Highly polished sealing surfaces are to be avoided. This seems contrary to the normal practice of polishing surfaces to make them more slippery but has been found to be the case on many machines.

Good film formulation with a balance of slip agents in the coating should minimise or avoid *hot stick* problems, whilst not affecting the sealing performance.

Accurate control of jaw temperature is important, particularly where the temperature sealing range is low and close to the melting point of an oriented film.

When a plastic film such as PVdC coated OPP is being used to overwrap a carton, it is only possible to use one heated surface, and the pressure necessary during sealing is provided by the rigidity of the carton. Precise jaw temperature control is essential to ensure that the envelope-shaped end folds of the film do not shrink during sealing and thereby become wrinkled and unsightly.

7.9.2.2 Crimp jaw conditions

Specific plastic film materials should, ideally, have a unique crimp jaw specification for each thickness, but a compromise is always needed as machines have to handle a wide range of films without modification or resetting mechanical parameters. As different thickness films keep crimp jaws apart by differing amounts, the loads on the crimp jaw slopes vary, and this is shown in the distortion or variability of the seal performance. Crimp jaws should be set to the ideal distance apart and spring pressures or loadings established when the crimp jaws are hot, at temperatures close to the preferred sealing temperature. Only then should knives be set to cut through the films.

Stenter-made PP films have greater extensibility in the MD, typically greater than 150% elongation before break and 70% in the transverse direction (TD). Form/fill/sealing (f/f/s) machines perform better and give better seal integrity with transverse jaw grooves to minimise stress in the TD and allow more extension in the MD.

It is seen that opaque cavitated Stenter-made films have a greater tendency to split across the film in crimp jaws which stress the film beyond their elastic limit. The film does not elongate as well in the TD as in the MD, and hence shallower angled jaws with an angle of 120° and sinusoidal profile have been developed to minimise the stress.

These designs conflict with blown (bubble) oriented films where extensibility is closer to 100% in each direction. Conditions of high pressure and the lower heat stability of bubble-made OPP will still give the same effect at the high end of sealing conditions.

PET and nylon PA films with their superior heat stability giving very wide sealing ranges do not normally develop split seals. Using PE or cast PP as the sealant of a laminate exploits the easy flow nature of cast and low-melting point polymers. The molten polymer can flow into crevices and fill any gaps or holes in the seal. While satisfactory for many pouches, the inability of the laminates to seal inside to outside layers limits the application to f/f/s with fin seals, as compared with overlapping seals, along the length and this uses slightly more material because of the extra width of film required.

The ionomer emulsions used as coatings with low melting points of around 80°C and a high level of hot tack have extended the sealing range of coated OPP to over 70°C, where the upper sealing limit is set by the shrinkage of the film, judged to be 150°C. Formally, acrylic-coated films had the widest range of 50°C with the starting point at 100°C, and this enabled linear packaging speeds of 50 m min^{-1} to be achieved.

With high packaging speeds, it is normal to have high temperatures to melt the sealant in a very short dwell time. When the machine speed varies, the high temperature of the sealing jaws damages the film. With the LTS (low-temperature sealing) coating, lower heat settings are possible, thus avoiding film damage at slower speeds. Such low sealing threshold temperatures mean that a very short dwell sealing time is possible at lower temperatures with crimp jaws, thus avoiding film shrinkage. In effect, the amount of energy required to make a seal is much lower than with other coatings, and film speeds of 100 m min^{-1} may be achieved. LTS coatings will not seal to other mediums and so only f/f/s applications can be utilised, which seal inside to inside, i.e. fin seals. Most seals are considered to be strong enough if the film tears when the seal is stressed. Seals provide built-in evidence of tampering, but the packs may still be opened easily, especially in the case of oriented films with their easy tear propagation properties. However, there is a school of thought that argues that if the seal peels open slightly and absorbs the stress without tearing, then the pack is still intact and continues to function. Tamper evidence in this case is less obvious.

In all cases of packaging small and low weight products using films/coatings which do not flow too readily during sealing, a minimum seal strength requirement of 300 g/25 mm is typical. Heavier product weights and free-flowing products such as nuts, rice, pulses and frozen vegetables may have to have seal strengths in excess of 1000 g/25 mm.

7.9.2.3 *Impulse sealing*
With impulse sealing, jaws are heated to fusion temperature by a short powerful electric impulse. The seal area remains clamped and is cooled under pressure. Impulse seals are generally narrower than hot bar seals but can be doubled up. When minor contamination is present, the impulse method may give a better seal. Voltage and duration are varied according to the material.

PE films may be sealed with impulse-heated wires or strips to make welded seals. If the seal is not to be cut through the web, the heating strip has to be covered to protect the molten polymer from sticking to the heated metal strip and destroying the seal. This is achieved by covering the strip with a release sheet such as PTFE-covered glass fibre woven cloth. The resultant seals achieve 100% film strength. It is possible to make the same type of seals with coextruded OPP using PE sealing equipment, but the seals are more sensitive to tearing close to the seal due to the normal easy tear propagation caused by high stress orientation.

7.9.2.4 *Hot wheel sealing*
In this form of heat sealing, the material to be sealed is drawn past a hot wheel. The seal area is kept under pressure until it cools and a seal has developed.

7.9.2.5 Hot air sealers

This uses hot air, heated by gas or electricity, to melt the plastic in the seal area. It is used for the sealing of plastic-coated paperboard.

7.9.2.6 Gas flame sealers

This form of sealing uses gas flames to melt the plastic in the heat seal area. It has a lower noise level and is more heat efficient than hot air sealing.

7.9.2.7 Induction sealing

A common form of induction sealing is that which is used to heat seal a diaphragm, incorporating a plastic or plastic-based heat-sealing layer, laminated or coated onto aluminium foil, already in place in the closure, to the rim of a plastic, or glass, jar or bottle. The closure is applied to the container and passed under a high-frequency induction sealing head which generates heat in the aluminium foil, which then melts the plastic and heat seals it to the perimeter of the container.

7.9.2.8 Ultrasonic sealing

This is similar to high-frequency induction heating, except that the heat is generated by molecular friction in the plastic material itself. This principle has been used to seal the corners of plastic-coated paperboard trays.

7.9.3 Cold seal

As already stated, sealing conditions are a balance of time, temperature and pressure. Where high-speed packing is required and the product is heat sensitive, such as a chocolate countline bar or chocolate-coated ice cream, the first choice sealant is cold seal latex. The adhesive is converter applied in a pattern on the reverse side where the seals are to be made, accurately registered with the print on the outside. This specification requires a release lacquer over the print on a single web film, or a release film laminated as the outer layer of a laminate.

7.9.4 Plastic closures for bottles, jars and tubs

In food packaging, the most common form of screw cap is injection moulded using PP. Where a flexible snap-on feature is required, as for instance, with an ice cream tub or as a reclosure after opening a long shelf life pack, PE is preferred. PE is used for plastic wine bottle corks.

The *hinge* property of PP has been made use of as a closure, which remains in contact with the container. A wide variety of designs are applied to containers for products which are dispensed from the container such as salt, pepper, spices and herbs.

Another thermoplastic used for closures is PS, which is harder and glossier than PP. The tightest tolerance, dimensionally, is provided by thermosetting plastic closures, though these are more commonly used for pharmaceutical and cosmetic closures. Most plastic closures can have a tamper-evident feature incorporated in the design.

7.9.5 Adhesive systems used with plastics

Most forms of adhesive can be used with plastics, e.g.

- a tie or grafting layer of plastic used to promote adhesion in extrusion coating and coextrusions
- dry bond adhesives used for laminations involving plastic substrates from which solvent is evaporated prior to bonding the surfaces together
- heat curing adhesives used for lamination, which are 100% solids, operate by cross linking to the solid state once the lamination has been completed
- hot melt adhesives, which include plastic components, for applying labels
- hot melt adhesives used to erect and close folding cartons on packing lines
- PVA water-based adhesives for side-seam-sealing folding cartons during conversion, including cartons made from one-side PE-coated paperboard, where the PE has been corona discharge treated
- pressure-sensitive and heat set label systems.

7.10 How to choose

The key to successful food packaging is to identify the packaging needs of the product. These relate to the nature of the product, the intended market, shelf life, distribution and storage, point of sale to the ultimate consumer and the use and eventual disposal of the packaging. The choice should take account of environmental and waste management issues. Ensuring food safety with respect to biological risks and needs relating to flavour, colour and texture is essential.

Packaging needs can be considered in terms of:

- protection of the product – quality, safety etc.
- appearance – sales promotion, pack design etc.
- production – extrusion, forming, printing, packing etc.

Having decided that a type of plastic pack selected from the range of possible choices, such as a film sachet, lidded tray, bottle etc., the next decision concerns the type of plastic or combination of plastics necessary to meet the functional needs. Performance is related to the structural design of the pack and whether it is made from film, sheet, moulding or expanded plastic. As we have seen, there are many plastics, each offering a range of properties, and within each packaging type there are differences.

All plastics provide barriers to the ingress of gaseous and volatile materials from the external environment into a hermetically sealed pack and from the food product both into and through the pack into the external environment. The extent to which these effects occur will depend on the food product and on the type of plastic(s), its thickness and on the temperature and RH ranges to be experienced during the life of the product.

Some plastics are heat sealable so that packs can be sealed; some are also heat resistant to meet defined needs, e.g. reheating by microwave, radiant heat and retort sterilization. Some are suitable for storage in deep freeze. Many specific needs can be met within the defined conditions of use.

In a chapter of this type, we can make readers aware of the choices and provide a basis for meaningful discussions between technologists whether they be suppliers or users of plastic packaging. The following Tables 7.1 to 7.3 give some guidance in terms of ranking for moisture vapour permeability,

Table 7.1 Ranking of various films with respect to specified properties

Polymer	Water vapour transmission rate (WVTR)	Gas permeability	Optics	Machine performance	Sealing
LDPE	3	4	4	4	1
Cast PP	3	4	2	4	2
OPP	2	2	2	2	2
OPP coated	1	1	1	2	1
PET	2	2	1	1	4
PVC (Plasticised)	3	2	2	4	2

1 = Excellent, 2 = Very Good, 3 = Good, 4 = Poor.

Table 7.2 General gas and moisture barrier properties

Film (25 µm thickness)	Water vapour transmission rate (WVTR)	Oxygen transmission rate
LDPE	10–20	6500–8500
HDPE	7–10	1600–2000
OPP	5–7	2000–2500
Cast PP	10–12	3500–4500
EVOH	1000	0.5
PVdC	0.5–1.0	2–4
PA	300–400	50–75
PS	70–150	4500–6000
PET	15–20	100–150
Aluminium	0	0

Units: WVTR in g m^{-2}/24 h at tropical conditions of 90% RH at 38°C and gas permeability in cm^3 m^{-2}/24 hrs.

Table 7.3 Examples of suitability of various films for packing the products named

Product	LDPE	OPP	OPP (metallised)	OPP (coated)	Laminate (no Al)	Laminate (+ Al)	Package type
Fresh bread	***	***	0	0	0	0	HFF
Long life bread	0	0	*	*(MAP)	**(MAP)	**(MAP)	HFF
Snacks/crisps (chips)	0	*	***	***	**	***	VFF
Biscuits	0	0	**	***	**	***	HFF
Nuts	0	0	**(MAP)	*(MAP)	**(MAP)	***(MAP)	VFF
Cooked meat	0	0	*	**	**(MAP)	***(MAP)	Pouch
Frozen food	**	*	*	0	***	***	Various

0 = Not suitable, * = short life, ** = medium life, *** = long life, MAP = modified atmosphere pack.

gas permeability, optical properties, packing machine performance and heat sealability.

The commercial consideration of cost must also be considered. Run lengths and lead times are also important. It is not unknown for there to be run length cost differences, where at one point a particular solution is cost effective relative to an alternative solution and for the position to be reversed at a different run length.

7.11 Retort pouch

7.11.1 Packaging innovation

The retort pouch is discussed here as a demonstration, case study, of the integrated approach involving packaging materials, their conversion, forming, filling and sealing together with the processing and machinery which is necessary to establish any new form of product/packaging presentation. The retort pouch is a rectangular, flexible, laminated plastic, four-side hermetically sealed pouch in which food is thermally processed. It is a lightweight, high-quality, durable, convenient and shelf stable pack. Foods packed and processed in retort pouches are in successful commercial use for a wide variety of foodstuffs in several countries, particularly Japan. They were originally developed in the 1950s and 1960s in America through research and encouragement from the US Army.

The materials from which retort pouches are made are either aluminium foil bearing/plastic laminates or foil-free plastic laminate films. These must be inert, heat sealable, dimensionally stable and heat resistant to at least 121°C for typical process times. They should have low oxygen and water vapour permeability, be physically strong and have good ageing properties (Table 7.4).

Table 7.4 Typical current examples

Types	Properties
Aluminium type	
PET/Al/CPP	Used for small pouches sold in decorated boxes for curry, sauces, household dishes
PET/Oriented PA/Al/CPP	Strong pouches widely used (small to 3 kg)
Transparent type	
PET/Oriented PA/CPP	Transparent type pouch rice, chilled hamburger steak vegetable, fish, dumplings
Transparent barrier type	
SiO_xPET/Oriented PA/CPP	Highest barrier for transparent pouch
PET metallised Al/Oriented PA/CPP	Good transparency type (strip metallised)
Oriented PA/PVdC/CPP	Vacuum packaging
Oriented PA/EVOH/CPP	High barrier, appropriate for vacuum pkg.

Source: Dai Nippon Printing Co. Ltd. 2002.

These laminates demonstrate a range of barrier properties against oxygen and moisture vapour. Those without aluminium foil are transparent. The outer polyester film provides strength and toughness, while the inner PP provides good heat-sealing properties, strength, flexibility and compatibility with all foodstuffs. The additional incorporation of an oriented PA (nylon) film layer further increases the durability of the pouch, especially where the individual pouches are not, subsequently, packed individually in cartons. AlO_x (metallised) coatings are being used to improve barrier, and SiO_x provides both high barrier and pack microwaveability. The layers are laminated to each other using specially developed thermo-stable, food compatible, adhesives.

Typical thicknesses: PET 12 μm, PA 15–25 μm, Al foil 7–9 μm and CPP at 70–100 μm

Pouches are reverse printed in a wide range of graphics on the PET film before lamination, so that the print cannot come into contact with the food. All laminates are required to meet very stringent requirements to ensure no undesirable substances can be extracted into the packaged food.

7.11.2 *Applications*

Retort pouches are used in several countries for a wide range of processed shelf stable products, from solid meat packs such as polonies to sliced meat in gravy, high-quality entrees, fish, sauces, soups, vegetables, fruits, drinks and baked items. Current markets for pouches are:

- Retail packs up to 450 g for home use and outdoor activities. Foil-free pouches have been utilised particularly for vegetables where high visibility is desirable and a short shelf life from 4 weeks to 6 months is acceptable. In these instances, oxygen permeability is the overriding factor in determining shelf life, although light is also important with regard to product browning and onset of rancidity.

- Self-standing pouches have been used for fruit juices and other drinks, soups and sauces.
- Large catering size pouches for the institutional trade up to a capacity of 3.5 kg, approximately equivalent to the A 10 can, have found ready application for prepared vegetable products such as carrots, peeled potatoes and potato chips. The relatively easier disposability of the pouch after use is also an advantage in the catering and institutional markets.
- Provision of military field rations.

Reduced heat exposure offers an opportunity for using retort pouches to process heat-sensitive products not currently suited to canning, especially in high-temperature/short-time processing where opportunities exist for optimum nutrient and flavour retention.

By far, the biggest producer is Japan where production is approximately 1 billion pouches per annum. A wide variety of products are packed; curries, stews, hashes, prepared meats, fish in sauce, mixed vegetables, all being popular dishes. Several factors which contributed to the success of pouches in the Far East are:

- limited refrigeration facilities when these packs were introduced, particularly in homes, resulting in demand for ambient shelf stable products. With increased use of refrigerators lower barrier pouches are now being used for shorter shelf life products in refrigerated storage
- social changes causing working housewives to look for convenience
- the popularity of foods such as sauces which are pumpable and ideal for pouches.

In Europe and North America, by contrast, the present market is relatively small. Main applications are for products such as rosti (fried grated potatoes), prepared meats, smoked sausage (frankfurters), smoked salmon, fish, petfood, entree dishes, vegetables and diced and sliced apples. Lack of market expansion is attributed to a highly developed frozen-food chain, the competitiveness of frozen foods of a similar type and highly automated, cost effective, canning facilities.

7.11.3 Advantages and disadvantages

The following advantages are claimed:

- less energy is required to manufacture pouches compared with cans
- transport of empty containers is cheaper (85% less space required than cans)
- packaging is cheaper than equivalent can and with carton cost is about the same

- filling lines are easily changed to a different size
- rapid heat penetration and faster process results in better nutrition/flavour
- contents are ambient shelf stable – no refrigeration is required
- packed pouch is more compact requiring about 10% less shelf space
- less brine or syrup used, pouches are lower in mass and cheaper to transport
- fast reheating of contents by immersion of pack in hot water. No pots to clean
- opens easily by tearing or cutting
- ideal for single portion packaging and serving size control
- retort pouch materials are non-corrosive
- convenient for outdoor leisure and military rations use.

There are also some disadvantages, such as:

- to achieve equivalent cannery production efficiency, a major investment in new capital equipment for filling and processing is required
- production speed on single filler/sealer is usually less than half that of common can seamers
- new handling techniques have to be adopted and may be difficult to introduce
- heat processing is more critical and more complex
- to retain rapid heat penetration there are limitations on pouch dimensions
- some form of individual outer wrapping is usually required, adding to cost
- being non-rigid products such as some fruits lose their shape
- being a new concept, education of the consumer as to correct storage and use is required during marketing.

7.11.4 *Production of pouches*

Pouches can either be formed from reels of laminated material either on in-line form/fill/seal machines in the packer's plant or they may be obtained as preformed individual pouches sealed on three sides, cut and notched. Forming consists of folding the laminate material in the middle, polyester (or PA) side out, heat sealing the bottom and side seals and cutting to present a completed pouch. Alternatively two webs can be joined, heat seal surfaces face to face, sealed, cut and separated. Hot bar sealing is the most common practice. Notches are made in the side seal at the top or bottom to facilitate opening by the consumer. Modern pouches have cut rounded corners which reduce the possibility of perforation caused by pouch to pouch contact. Rounded corner seals can also be incorporated.

The four-seal flat shape and thin cross section of the pouch is designed to take advantage of rapid heat penetration during sterilization and on reheating, prior to consumption, saving energy and providing convenience. The flat shape also enables ease of heat sealing and promotes high seal integrity. From a military point of view, the flat section is compatible with combat clothing without restricting the physical movements of the soldier.

Fin seal design and certain gusset features permit the design of upright standing pouches although they create multiple seal junctions with increased possibility of seal defects. Several of these upstanding pouches are, however, available commercially. A wide range is possible in the size and capacity of pouches.

Nominal thickness after filling varies from approximately 12 mm for a 200 g pouch to 33 mm for a 1 kg size. Some unused package volume must be allowed for, as good practice dictates no void/headspace within 40 mm of the pouch opening.

7.11.5 Filling and sealing

In-line and premade pouches are filled vertically. Vertical form/fill/seal machines can be used for liquid products. Another method employs a web of pouch material which is formed on a horizontal bed into several adjacent cavities. The cavities are filled whilst the seal areas are shielded. This method is especially useful for filling placeable products. Thereafter the filled cavities are simultaneously sealed from the top using a second web fed from the reel. The essential requirements for filling are:

- the pouch should be cleanly presented, positively opened, to the filling station, solids are filled first, followed by the liquid portion, usually at a second station
- matching fill-nozzle design and filler proportioning to the product
- non-drip nozzles
- shielding of the sealing surfaces
- bottom to top filling
- specification and control of weight consistent with the maximum pouch thickness requirement
- product consistency in formulation, temperature and viscosity
- deaeration prior to filling.

Seals and sealing machines, like fillers, are constantly being refined and speed has improved from 30 to 60 pouches per minute to the current production rate of 120–150 pouches per minute. Sealers incorporate either one of two common satisfactory sealing methods, namely, hot bar and impulse sealing. Both methods create a fused seal whilst the pouch material is clamped between opposing jaws, thereby welding the opposing seal surfaces by applying heat and pressure. Exact pouch-sealing conditions depend on the materials and machinery used, but monitoring of seal temperature, jaw pressure and dwell time is essential.

Pouch closure is normally accompanied by some means of air removal, either by steam flushing or by drawing a vacuum in a sealed chamber or simply, in the case of liquid food products, flattening the pouch by squeezing between two vertical plates. Efficient air removal prevents ballooning and rupturing during retorting. Excess air can also adversely affect heat penetration. While some very limited condensate moisture may be tolerated, a seal area clear of contamination is essential. Irrespective of the method of pouch presentation to the sealing, station grippers engage on each side, stretching the pouch opening and preventing wrinkles. The closure sealing is then carried out. Cooling after the sealing is essential to prevent wrinkling of the seal area.

All seals, whether side, bottom or closure seals, must be regularly tested. Performance is the ultimate measure of a good seal and the performance standard is the hermetically sealed can. Seals can be examined visually and sample pouches should routinely be subjected to internal pressure resistance tests (280 kPa for 30 seconds) in a suitable test jig. Seals made in this way should not yield significantly. Satisfactory seal tensile properties should also be confirmed on 13 mm sections, regularly cut from the various seals. Visual inspections at best are never wholly successful. However, inspection of all pouches before and after retorting can ensure a low rate of defects. Channel leaks, product contamination and weak seals can be detected using an ultrasonic technique (Ozguler *et al.*, 1999).

7.11.6 Processing

Processing takes place in steam heated pressure vessels or retorts. Special precautions are required to prevent unnecessary straining of the pouch seals. These involve the use of super-imposed air pressure and trays which control pouch thickness. Overpressure counter balances internal pressure build up in the pouch during processing. This is particularly essential towards the end of the cycle when cooling commences and product is at its hottest. Overpressure also provides some restraint on the pouch preventing agitation and movement of the pouch walls which could strain the seals and limits, but cannot prevent, expansion of vapour bubbles in the product. The heating system is provided by either of the following:

- steam-heated water with compressed air overpressure
- mixtures of steam and air.

Limiting the amount of air in the pouch at the time of closure to the practical minimum is essential as it can affect heat penetration during processing. Instrumentation and control valve systems are vitally important to accurately control and record both pressure and temperature ($+1°C$ to $-0.5°C$) during the retorting of pouches. Automatic process cycle control is preferred. Vertical retorts may be used but horizontal batch retorts are the most commonly used.

Fully automatic units for steam/air processing have been developed in Japan to facilitate high temperature/short time processing at 135°C, and higher. This short time high temperature treatment offers opportunities for milk and dairy-based specialities.

Trays or racks should be constructed of non-corrosive material without sharp projections or rough surfaces. Whilst heat penetration into pouches is more rapid compared to similar capacity cans, small changes in pouch thickness can have a profound effect on the lethal value achieved during the thermal process. For example, a change in thickness of only 2 mm can result in a change of Fo value of 1.5 min. For this reason, pouch dimension (thickness) is positively controlled by specially designed trays or racks which enable the easy placement of pouches in individual compartments while providing, on stacking, predictable maximum pouch thickness. Tray design usually incorporates a false bottom and sufficient void area (40%) in the supporting surface to ensure maximum exposure of each pouch to the heating medium. The maximum diameter of voids in the supporting surface should be less than the size of solid product portions which could cause slumping of the pouch surface into the holes, thus altering the maximum thickness of the pouch.

Horizontal pouch orientation is the most common as it allows the least strain on seals and favours a uniform section across the pouch surface. Vertical pouch orientation in racks is, however, also utilised. The only stipulation is that the system allows thickness control and unrestricted movement of the heating medium around each pouch. In batch systems, the trays are stacked on top of one another on trolleys. These are then pushed on rails into horizontal retorts. Several trolley loads are pushed into the retort before it is sealed. In continuous retort systems, pouch carriers or compartments are attached to conveyor chains which move through locks into and out of the processing section in much the same way as applies to cans. These carriers provide the same thickness control and exposure to the heating medium as mentioned above for batch retorts.

7.11.7 *Process determination*

Heat transfer is highly dependent on the conductivity of the food and the geometric shape of the container. Therefore, the well, known General Method and the Formula Method of Ball (and subsequent modifications) for process determination for conventional cans apply equally to retort pouches. Consequently Fo values suitable for canned products are adequate for the same product in pouches. The mathematical approach to process determination of heat transfer into the retort pouch is that of transfer into a thin slab rather than a finite cylinder, as in the case of the can. Whilst these standard mathematical approaches are of assistance in process design, they are not a substitute for full process determination by proper heat penetration or innoculated pack tests.

The process used for the retort pouch should be based on the maximum pouch thickness a particular racking system will accommodate, and deliberately include overfilled units of a degree likely to be encountered. It is always necessary in designing heat penetration tests to ensure that account is taken of the worst case and that test pouches are located in previously established slow heating points in any stack of trays. Information as to the uniformity of heat distribution in a particular retort must be established through heat distribution studies beforehand.

Ideally, temperature variations from point to point in a retort should not be greater than 1°C. Several repeats of the heat penetration determination are necessary to ensure that all variations of critical parameters likely to occur in production are taken into account. In addition to the above, it is a recommended practice to add a 10% safety factor to all process recommended settings.

7.11.8 Post retort handling

Following pressure cooling and removal in racks or trays from the retort, the pouches must be dried, inspected and placed in some form of outer packaging. Drying of pouches is achieved through a combination of pack residual temperature to encourage evaporation and a system of high velocity air knives in a drier to drive off the remaining water. When dry, pouch seals may once again be visually inspected for leaks, ruptures or weak points that have been shown up during retorting. This should not involve manual handling of the individual pouches. Systems are available for the transfer of the pouches from the retort racks to conveyor belts, thence to the pouch driers and onto inspection conveyors prior to secondary packaging.

7.11.9 Outer packaging

The secondary packaging of retort pouches for storage and distribution may either involve packing each pouch in a printed carton or, alternatively packing a number of pouches in a transit case, possibly incorporating vertical dividers. The recommendation of individual pouches in cartons is made to avoid the dangers of leaker spoilage due to external microbial contamination from the environment, workers or consumers. The practice in Japan and Europe suggests that the retail marketing of unwrapped or *naked* pouches is nonetheless possible without any apparent practical increase in spoilage. For US military field rations, a paperboard folder, or envelope, in which the individual pouch is glued, has been used. This allows for non-destructive visual inspection and reclosure, while the pouch exhibits greatly improved abuse resistance even under severe military use.

7.11.10 Quality assurance

A successful pouch packaging quality system requires:

- selection and continued monitoring of the most suitable laminate materials
- regular testing of formed pouches for seal strength, product resistance and freedom from taint
- careful selection, maintenance and control of filling, sealing, processing and handling machinery
- specifications for the control of product formulation, preparation (viscosity, aeration, fill temperature etc.) and filling (ingoing mass and absence of seal contamination)
- post sealing inspection and testing of closure seals to confirm fusion, absence of defects and contamination
- control of critical parameters influencing processing lethality such as maximum pouch thickness and residual air content
- standardised retorting procedures applying only recommended process times and temperatures confirmed to achieve adequate lethality
- regular inspection and testing of retort equipment and controls to ensure uniform heat distribution
- visual inspection of all pouches to check sealing after processing
- handling only of dry pouches and packing into collective or individual outer packaging specially tested to provide adequate, subsequent, abuse resistance
- it should be routine that all stocks are held 10–14 days prior to distribution and these should be free of blown spoilage on despatch
- careful staff selection and training at all levels.

7.11.11 Shelf life

Whilst shelf life is determined by many factors such as storage temperature and the barrier properties of the particular film used, in general, satisfactory shelf stability in excess of two years is easily obtained for a wide range of products in foil bearing pouches. US military rations tested over two years at 20°C showed no significant change in product quality ratings. Some products have been successfully stored for as long as seven years and found to be safe and edible.

Foil-free laminates will demonstrate shelf stability commensurate with oxygen permeability of the particular laminate used and the sensitivity of the product. Commercial experience confirms, however, that product stability from four weeks to six months is obtainable. Nitrogen flushing of the outer container has been successful in extending the shelf life of product in foil-free pouches.

Extensive testing under combat conditions by the US Army has proved that retort pouches if correctly packed are well able to stand up to rough conditions including being carried on a soldier's person through tough obstacle courses. Commercial experience in Europe and Japan over many years confirms that pouches can safely withstand distribution through normal trade channels and with a performance equal to that of the rigid metal can.

The retort pouch is probably the most thoroughly tested food packaging system. Its acceptance as the sole form of field rations for the US Army confirms it has fulfilled all that was expected when it was first conceived.

This short review of the integrated activities needed to market the retort pouch indicates the complexity involved and is typical of any major food processing and packaging innovation. Similar principles have been followed in other major food processing and packaging projects, e.g. aseptic packaging, frozen food packaging etc.

7.12 Environmental and waste management issues

7.12.1 Environmental benefit

About 50% of food is packaged in plastics or plastic-based packaging and the main environmental benefit of plastics food packaging is that it saves food from wastage. There are other benefits, such as significant reductions in the weight of packaging waste when plastic packaging is used in preference to alternative forms of packaging, but reducing the waste of resources is the most important environmental benefit. On the subsidiary issues, concerning sustainable development, use of resources and the consequences for manufacture and waste management, the use of plastics for food packaging has a sound environment position.

7.12.2 Sustainable development

The plastics industry overall contributes to achieving the aims of sustainable development. This subject is beyond the scope of this discussion but reference to, for example, the website of the Association of Plastics Manufacturers in Europe at www.apme.org will indicate the many areas where plastics saves resources, provides possibilities for economic development, social progress and protection of the environment. In Europe, 37% of plastics usage is for packaging, most of which is used in food packaging. Plastics in food packaging preserves food, provides choice and convenience.

7.12.3 Resource minimisation – lightweighting

Resource minimisation, or lightweighting, refers to the achievement of a similar or better performance with less packaging material. Examples of lightweighting plastic packaging include:

- In 1970, the average plastics yoghurt pot weighed 11.8 g compared with 5.0 g in 1990. (*Source*: British Plastics Federation.)
- Reductions in the thickness or gauge of plastic film used for the same purpose, 180 μm down to 80 μm has been achieved due to strength improvement. (*Source*: British Plastics Federation.)
- An example of resource minimisation is the Ecolean material which comprises PE with 40% chalk (Ecolean International A/S). This material extends the use of PE and reduces the use of energy to achieve similar performance in conventional PE packaging applications.
- The average weight of stretch film used for pallet wrap is now 350 g replacing the 1400 g used 10 years ago (British Plastics Federation).

Further examples are quoted in an INCPEN publication 'Packaging reduction doing more with less'.

The fact that plastics packaging is light in weight reduces the cost of transport of packaging material and packed product, and hence the associated fuel usage and emissions, compared with alternative forms of packaging.

7.12.4 *Plastics manufacturing and life cycle assessment (LCA)*

In manufacturing, the plastics industry claims that the energy to manufacture compares favourably with, for example, metal ore smelting and glass manufacture and that the processes used are *clean*. The conversion energy used to make plastic products from pellets is also low in relation to metal and glass processing.

Flexible packaging is energy efficient compared with premade packaging, such as glass or metal containers. This is because:

- flexible packaging is transported to the packer either flat or in reel form
- the gross weight of packed product in non-plastic packaging and managing the resulting packaging waste involves, relatively, the use of more energy.

These aspects can be evaluated quantitatively by LCA. LCA has been undertaken using internationally agreed methodology based on ISO Standards. It is conducted in two parts. Firstly, an audit, or eco-profile, is made of all resources in terms of raw materials and energy entering a previously defined system and the emissions in terms of products, waste heat, emissions to air, water and solid waste leaving the system. The plastics industry has been active in this area and many studies have been completed. The second stage of LCA comprises an assessment of the environmental impact of the process or system. Environmental impact can have local, regional and global implications, and our knowledge and understanding is still developing.

7.12.5 Plastics waste management

7.12.5.1 Introduction to plastics waste management

Returnable, refillable and reusable plastic products are in current use. In Sweden, PET drinks bottles are returnable. Plastic pallets, plastic trays and plastic boxes (totes) used in distribution are returnable and reusable. Further development of this concept will reduce the amount of plastic in the waste stream.

Plastic waste arising in manufacture is minimal as thermoplastics can be melted and reused.

The main issue concerns the 40% of the total plastic materials market which is used in packaging, and in particular the proportion which arises in domestic waste or trash. In the UK, studies have indicated that plastic waste comprises around 5–7% of household waste by weight.

The recovery of domestic plastic waste is a logistical challenge due to there being so many different types of plastic. Additional factors affecting the commercial viability of plastics waste recovery relate to the cost of virgin plastics, the low weight to volume ratio, which increases the handling cost and the fact that the waste arises over a large area geographically. Plastic packaging waste recovery rates in Europe are rising year on year. The overall average (2000) was 46%, with eight countries achieving rates of over 50%. The highest rates were Holland 100%, Switzerland 98%, Denmark 88% and Germany 74%.

Recovery itself is *not* recycling. Recycling can either comprise reuse of the material, energy recovery or composting. The European average for the reuse of plastics in all forms of waste, including packaging, is 36%. This is made up from recycling at 13% and incineration with energy recovery at 23%. The balance, 64%, is either sent to landfill or is incinerated without energy recovery.

If plastics are to be recycled as material, they must be segregated from other plastics. The most widely recycled items of plastic food packaging are PET bottles and HDPE milk bottles. Both arise in significant volumes and are easily sorted, making the process commercially viable. The plastic is reground for reuse. This process is also referred to as mechanical recycling. In the UK, about 50 000 tonnes of PE film is recovered and used to make film for the building trade and black sacks for refuse (trash) collection.

To assist segregation and sorting, a numerical code inside a recycling symbol (Mobius strip or loop) together with initial letters relating to the plastic concerned is being displayed on the base of moulded items, i.e.

- Code 1   PET or PETE – polyethylene terephthalate
- Code 2   HDPE – high density polyethylene
- Code 3   V – vinyl or polyvinyl chloride
- Code 4   LDPE – low density polyethylene
- Code 5   PP – polypropylene
- Code 6   PS – polystyrene
- Code 7   OTHER – other resins and combined structures.

The proportion of packaging waste in Europe mechanically reground in 2000 was 17.9%, i.e. a slightly higher proportion than for plastics in total (13%). However, the possibility of contamination in mixed plastic packaging waste limits the use of sorted material for food packaging.

7.12.5.2   Energy recovery
The thermal content of used plastic is relatively high. An average typical value for polymers found in household waste is $38\,MJ\,kg^{-1}$, compared with coal at $31\,MJ\,kg^{-1}$. Incineration with energy recovery produces steam, which can be used to heat buildings and generate electricity. It has the benefit that the plastics do not have to be sorted from other waste. Plastic waste is also used as fuel in cement production. Another form of energy recovery for mixed plastic waste is to convert it into fuel pellets, along with other combustible material such as waste paper and board. This material is also known as refuse derived fuel (RDF).

As stated above, of the plastics recovered in Europe (2000), the average proportion incinerated with energy recovery was 23%, but there were large variations in achievement, with Switzerland at 73%, Denmark at 75% and Germany at 26%. In tonnage terms, however, France, 32%, processed the most plastics, by this route.

There has been concern expressed about possible pollutants arising from the incineration of municipal waste. Technology is available to meet the rigorous mandatory internationally agreed safety limits and several countries, Sweden, Germany and Holland, have recently announced plans to expand the existing capacity.

7.12.5.3   Feedstock recycling
The feedstock recycling and chemical recycling of petroleum based plastics is also known as advanced recycling technology. These terms cover a range of processes which convert plastics through the use of heat into smaller molecules which are suitable for use as feedstock for the production of new petrochemicals and plastics. Process names include pyrolysis, glycolysis, hydrolysis and methanolysis. The depolymerization of PE and PP is similar to thermal cracking which is a common oil refinery process. It can only occur in the absence of oxygen.

The subject has created worldwide interest. The techniques are designed to handle contaminated plastic waste materials and are seen as being complimentary to mechanical recycling. According to the American Plastics Council (June 1999) 'feedstock recycling represents a significant technological advancement that in the case of some polymers is already supplementing existing mechanical recycling processes.'

Progress in developing feedstock recycling is slow due to the need for a reliable constant supply of waste and commercial considerations. In Europe, facilities are currently limited to Germany, where 329 000 tonnes were treated in 2000. The process uses household plastics waste collected by the DSD (German waste management organisation). Two applications are being used. Firstly, as a reductant in a blast furnace in iron and steel manufacture, and, secondly, at the SVZ gasification plant for the production of methanol. A feedstock recycling plant in France was expected to start in 2002 to handle 15 kt PET bottles, converting it to a polyolefin suitable for use in polyurethane production.

A number of processes have been successful on a pilot scale. These include Texaco (gasification of plastics waste and conversion to alcohols), BASF (conversion of packaging waste to naphtha cracker feed) and a BP consortium (conversion of household plastics into hydrocarbon feedstock for catalytic or naphtha cracker feed).

In order to invest in commercial units, long term supply contracts with an appropriate gate-fee are necessary. These logistical and commercial issues have so far prevented full scale development. New commercial investments are likely to be focussed on the feedstock recycling of PET bottles, to make monomer suitable for the production of bottle grade plastic. The mechanical recycling of PET bottles is currently widespread, but because of food contact considerations, the product is mainly used for textile applications. There is some specially authorised processing for bottles. There is a wish by some drinks manufacturers to have a recycled content in their bottles together with the recycling of more coloured and complex bottle structures, such as coextruded PET beer bottles. The challenge remains for the development of a committed supply of waste streams. (Discussion of feedstock recycling based on personal communication from Mayne, APME, 2002.)

7.12.5.4 Biodegradable plastics

Biodegradable plastics are commercially available. The question of their role in food packaging and in plastic usage as a whole is, however, debatable. Some see their use as an answer to the litter problem – but litter is not caused by packaging, it is caused by people. The idea that their use will solve the problem of persistence in landfill goes against the preferred approach for the reuse, recovery and recycling of plastic waste as a more sustainable environmental solution. There may however be niche markets, such as the packaging of organically grown fruits and vegetables, where the use of biodegradable plastics may be preferred.

Initially, the approach with plastics was to introduce materials, such as starch, into conventional plastics which could be broken down by microorganisms and thereby cause the disintegration of the polymer. In the last few years, however, two concepts have come together which have led to considerable additional interest in biodegradable plastics. Firstly, the concept of disposing of plastic materials by microbiological activity as compost is accepted as an environmentally sound method of packaging waste disposal. This is particularly attractive where the plastics are difficult to recover by reason of the places and/or ways in which they are used. In the case of food packaging, it is possible that some packaging is contaminated with food residues and this would constitute a health hazard in sorting and mechanical recycling. Secondly, there is interest in developing plastics based on naturally renewable raw materials, and these products are inherently biodegradable.

A number of polymers have been developed, the website of the Biological Plastics Society lists 12, including cellulose acetate, from both natural sources and synthetic raw materials. In recent years, several new plastics have been developed from natural raw materials, such as starch, maize (corn) and sugar. A polymer currently being used in several ways, including food packaging, is a polylactide polymer (PLA) (see website for Cargill Dow LLC). This is being used in the form of thermoformable trays, biaxially oriented coated film and extrusion coatings. Trays used for fresh produce are said to be stronger than similar trays made from PVC and PP.

Another approach is based on PET technology, e.g. DuPont Biomax, a hydro, biodegradable polyester. In the right conditions, this material is degraded by hydrolysis, making the resulting material suitable for microbiological consumption and resulting in conversion to carbon dioxide and water. A range of plastics can be made which are similar to PE and PP. They can be used as film, thermoformable sheet, blow mouldings and injection mouldings.

It should also be remembered that RCF film is a biodegradable plastic-like film derived from cellulose which has been commercially available for a long time. As already noted, however, many of the packaging applications of RCF have been lost to OPP, mainly on commercial grounds.

Appendices

Appendix 1 Simple physical tests for polyolefin film identification

Action	Observation	Conclusion
Pull in both directions	Stretches easily in both directions	Cast PE or PP
	Stretches easily in MD and splits, but not in TD	Mono axial PE
	Glass clear, becoming white in stressed areas	Cast PP
	Cloudy to milky white in stressed areas	MDPE or HDPE
	Stretches more easily in MD than TD	Oriented by Stenter
	Stretches easily in both directions	Biaxial HDPE
	Difficult to stretch in both Directions	Bubble blown OPP
	Extreme force required to stretch	PET or PA
If white or pearlescent	With cavitation	Stentered OPP
	Solid uniform film	Bubble blown OPP
Surface scratches	Coated film	OPP most likely

Appendix 2 Resistance to heat

Film	Manner of burning	Colour of flame	Odour after extinguishing
OPP	Melts, shrivels, drips	Blue	Acrid
PE	Melts easily, drips	Blue	Burning candle
PET	Softens, burns steadily	Yellow	Pleasant
PVC	Will not burn	Yellow with green	Acrid, choking
PS	Burns easily in drips	Yellow	Acrid

Appendix 3 Identification of coating

Coating	Methanol (apply 1 drop)	Blue methanol colour change	Copper wire + flame	Surface appearance
Acrylic	White after drying	Dark blue	No colour	Glossy
PVdC	No colour after drying	None	Green flame	Very slightly yellow
LTSC (low temp. sealing coating)	None	Dark blue	None	Glossy
PVOH	None	None	None	Glossy

References

Anon (2000a) Plasma (DLC) coating technology from Japan, PET Planet Insider, **1**(6), 18–19

Anon (2000b) Spendrups is the first into Glaskin high barrier PET, *Brew. Distill. Int.*, **31**(4), 48.

Anon (2000c) Beer in PET, *Verpak. Berat.*, **11**, 18–19.

Ayshford, H. (1998) Bottle coating suits beer (DLC), *Packaging Magazine*, **1**(12), 5.

Frank, M., Ulmer, H., Ruiz, J., Visani, P. and Weimar, U. (2001) Complimentary analytical measurements based on gas chromatography-mass spectrometry, sensor system and human sensory panel: a case study dealing with packaging materials, *Analytica Chimica Acta*, **431**(1), 11–29 (in English).

Grieg, S., Sherman, P.B., Pitman, R. and Barley, C. (2000) Adhesion promoters: corona, flame and ozone: a technology update, 2000 Polymers, laminations and coatings conference, Chicago, IL, USA, 27–31 August, **1**.

Haas, D. (1996) Lamination vs coextrusion: a technical and economical analysis, LatinPack '96 Conference in Columbia, 1–2 October, p. 11.

Lohwasser, W. (2001) SiO_x PET manufacture/properties, see web reference below.

Matsuoka, K., Kakemura, T., Kshima, H., Seki, T. and Tsujino (2002) Development of high barrier plastic bottles, Worldpack 2002, Improving the quality of life through packaging innovation, East lancing, MI, USA, 23–28 June, **1**, pp. 393–399.

Naegeli, H.R. and Lowhesser, W. (2001) Processing and converting of plastic films vacuum coated with inorganic barriers, 8th European Polymers, Films, Laminations and Extrusion Coatings Conference, Barcelona, Spain, 28–30 May, p. 18.

Ozguler, A., Morris, S.A. and O'Brien, W.D. (1999) Ultrasonic seal tester (retort pouches). *Packag. Technol. Sci.*, **12**(4), 161–171.

Potente, H., Heil, M. and Korte, J. (1995) Laser lamination of thermoplastics, *Plastverarbeiter* (in German), **46**(9), 42–44, 46.

Further reading

'Fundamentals of Packaging Technology' by Walter Soroka revised UK edition by Anne Emblem and Henry Emblem, published by The Institute of Packaging, ISBN 0 9464 6700 5.
'Packaging reduction – doing more with less', published by INCPEN (www.incpen.org).

Websites

Plastics industry statistics, Environmental and Waste management, see Association of Plastic Manufacturers in Europe at www.apme.
British Plastics Federation at www.bpf.co.uk.
American Plastics Council at www.americanplasticscouncil.org.
Biodegradable Plastics Society at www.bpsweb.net/02_english.
Cargill Dow at www.cdpoly/home.asp.
Biomax resins at www.dupont.com/polyester/resins/products/biomax.html.
Large scale electron beam web coating (SiO_x PET), see Lohwasser, W., Lawson Mardon Packaging Services, Neuhausen a Rhf, Switzerland at www.appliedfilms.com/Precision2/12_packaging/packaging_print.htm.
Website for European Plastics in Contact with Food regulations, see www.efpa.com/laws.html.
For general information on plastics, search websites of major plastic resin manufacturers.

8 Paper and paperboard packaging
M.J. Kirwan

8.1 Introduction

A wide range of paper and paperboard is used in packaging today – from lightweight infusible tissues for tea and coffee bags to heavy duty boards used in distribution. Paper and paperboard are found wherever products are produced, distributed, marketed and used, and account for about one-third of the total packaging market. Approximately 10% of all paper and paperboard consumption is used for packaging and over 50% of the paper and paperboard used for packaging is used by the food industry.

One of the earliest references to the use of paper for packaging food products is a patent taken out by Charles Hildeyerd on 16th February 1665 for 'The way and art of making blew paper used by sugar-bakers and others' (Hills, 1988).

The use of paper and paperboard for packaging purposes accelerated during the latter part of the 19th century to meet the needs of manufacturing industry. The manufacture of paper had progressed from a laborious manual operation, one sheet at a time, to continuous high speed production with wood pulp replacing rags as the main raw material. There were also developments in the techniques for printing and converting these materials into packaging containers.

Today, examples of the use of paper and paperboard packaging for food can be found in many places such as supermarkets, traditional markets and retail stores, mail order, fast food, dispensing machines, pharmacies, and in hospital, catering and leisure situations.

Uses can be found in packaging all the main categories of food such as:

- dry food products – cereals, biscuits, bread and baked products, tea, coffee, sugar, flour, dry food mixes etc.
- frozen foods, chilled foods and ice cream
- liquid foods and beverages – juice drinks, milk and milk derived products
- chocolate and sugar confectionery
- fast foods
- fresh produce – fruit, vegetables, meat and fish.

Packaging made from paper and paperboard is found at the point of sale (primary packs), in storage and for distribution (secondary packaging).

Paper and paperboard are sheet materials made up from an interlaced network of cellulose fibres. These materials are printable and have physical properties which enable them to be made into flexible and rigid packaging by cutting, creasing, folding, forming, gluing etc. There are many different types of paper and paperboard. They vary in appearance, strength and many other properties depending on the type(s) and amount of fibre used and how the fibres are processed in paper and paperboard manufacture.

The amount of fibre is expressed by the weight per unit area (grams per square metre, g/m^2, or lbs. per 1000 sq. ft), thickness (microns, µm or 0.001 mm, and thou (0.001 inch), also referred to as *points*) and appearance (colour and surface finish).

Paperboard is thicker than paper and has a higher weight per unit area. Paper over $200 g/m^2$ is defined by ISO (International Organisation for Standardization) as paperboard or board. However, some products are known as paperboard even though they are manufactured in grammages less than $200 g/m^2$.

Papers and paperboards used for packaging range from thin tissues to thick boards. The main examples of paper and paperboard based packaging are:

- paper bags, wrapping, packaging papers and infusible tissues, e.g. tea and coffee bags, sachets, pouches, overwrapping paper, sugar and flour bags, carrier bags
- multiwall paper sacks
- folding cartons and rigid boxes
- corrugated and solid fibreboard boxes (shipping cases)
- paper based tubes, tubs and composite containers
- fibre drums
- liquid packaging
- moulded pulp containers
- labels
- sealing tapes
- cushioning materials
- cap liners (sealing wads) and diaphragms (membranes).

Paper and paperboard packaging is used over a wide temperature range, from frozen food storage to the high temperatures of boiling water and heating in microwave and conventional radiant heat ovens.

Whilst it is approved for direct contact with many food products, packaging made solely from paper and paperboard is permeable to water, water vapour, aqueous solutions and emulsions, organic solvents, fatty substances (except grease resistant paper grades), gases, such as oxygen, carbon dioxide and nitrogen, aggressive chemicals and to volatile flavours and aromas. Whilst it can be sealed with several types of adhesive, it is not, itself, heat sealable.

Paper and paperboard, however, can acquire barrier properties and extended functional performance, such as heat sealability for leak-proof liquid packaging, through coating and lamination with plastics, such as polyethylene (PE), polypropylene (PP), polyethylene terephthalate (PET or PETE) and ethylene vinyl alcohol (EVOH), and with aluminium foil, wax, and other treatments. Packaging made solely from paperboard can provide a wide range of barrier properties by being overwrapped with a heat sealable plastic film such as polyvinylidene chloride (PVdC) coated oriented polypropylene (OPP or BOPP).

8.2 Paper and paperboard – fibre sources and fibre separation (pulping)

Paper and paperboard are sheet materials comprising an interlaced network of cellulose fibres derived from wood. Cellulose fibres are capable of developing physico-chemical bonds at their points of contact within the fibre network, thus forming a sheet. The strength of the sheet depends on the origin and type of fibre, how the fibre has been processed, the weight per unit area, and thickness. The type of fibre also influences the colour. Most paperboards have a multilayered construction which provides specific performance advantages and gives the manufacturer flexibility of choice, depending on the packaging end use, of the type of fibre used in the individual layers. Virgin, or primary, fibre is derived directly from wood by a process known as pulping. This can be done mechanically (Fig. 8.1) or with the help of chemicals which dissolve most of the non-cellulose components of the wood, which are subsequently used to provide energy (Fig. 8.2).

The terms sulphate and sulphite refer to the chemical processes used to separate the fibres from wood, sulphate being the more dominant process today. Mechanically separated fibre retains the colour of the wood though this can be made lighter by mild chemical treatment. Chemically separated fibre is brown but it can be bleached to remove all traces of non-cellulosic material. Pure cellulose fibres are translucent individually but appear white when bulked together (Fig. 8.2).

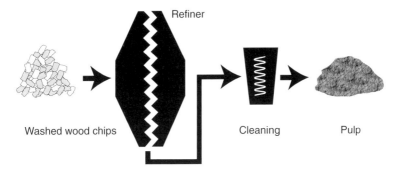

Figure 8.1 Production of mechanical pulp (courtesy of Pro Carton).

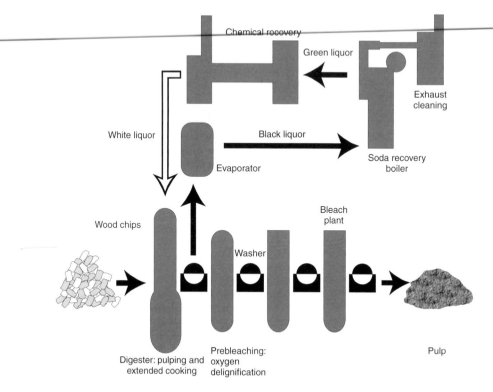

Figure 8.2 Production of chemical pulp (courtesy of Pro Carton).

Fibre recovered (secondary fibre) from waste paper and board, which is not de-inked and bleached, is grey or brown. Fibre recovered from brown packaging will be brown in colour. When mixed printed waste recovered paper and board is processed the colour of the pulp is grey. Pulp can be dyed during processing to meet a specific colour specification. The process whereby recovered fibre is made into paper and paperboard is an example of material recycling (Fig. 8.3).

Fibres can withstand multiple recycling but the process of recycling reduces fibre length and inter-fibre bonding, features related to sheet strength properties. Furthermore, it must also be appreciated that some papers and boards cannot be recovered by nature of their use and hence there is a constant need for virgin fibre to maintain the amount and strength of fibres. In practice the proportion of fibre that is recovered and recycled in various countries is between 40 and 60%. Another important factor relevant to sheet properties is the species of tree from which the fibres are derived. In general terms the industry uses long fibres for strength and short fibres for surface smoothness and efficient sheet

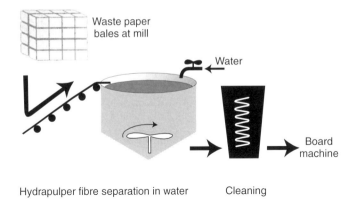

Figure 8.3 Production of recycled pulp (courtesy of Pro Carton).

forming in manufacture. Long fibres are derived from coniferous softwoods such as Spruce, Pine and Douglas Fir and have average lengths of 3–4.5 mm. Short fibres, such as those derived from Birch have average lengths of 1–1.5 mm.

8.3 Paper and paperboard manufacture

8.3.1 Stock preparation

If pulp is bought in bales it is first dispersed in water in a hydrapulper. All pulp, including pulp which comes straight from the pulpmill without drying, is then treated in various ways to prepare it for use on the paper or paperboard machine. Inter-fibre bonding can be increased by mechanical refining, in which the surface structure of the fibre is modified by swelling the fibre in water and increasing the surface area. The degree of refining, which also influences the drainage rate at the next stage in manufacture, is adjusted to suit the properties of the intended paper or paperboard product.

Additives such as alum or synthetic resins are used to increase the water repellancy of the fibres. Wet strength resins can be added to increase the strength of the product when saturated with water. Fluorescent whitening agents (FWAs), also known as optical brightening agents (OBAs), can be added at this stage to increase whiteness and brightness.

8.3.2 Sheet forming

The fibre in water suspension, roughly 2% fibre and 98% water, is *formed* in an even layer. This is achieved by depositing the suspension of fibre at a constant rate onto a moving plastic mesh, known as the wire (Fig. 8.4). On some machines, forming is carried out on a wire mesh covered cylinder. Forming results in

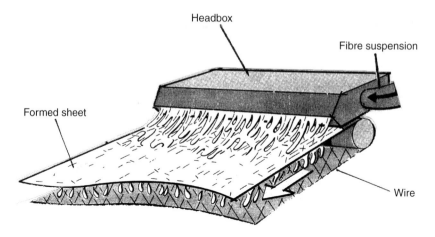

Figure 8.4 Simplified diagram of the forming process (courtesy of Iggesund Paperboard).

a layer of entangled fibres from which water is then removed by drainage, which may be assisted by vacuum. Tissue, paper, and thin boards can be formed in one layer. Thicker and heavier higher grammage paperboards require several layers of pulp, either the same type, or different pulps, depending on the board type, being brought together successively in the wet state.

Forming on a wire mesh in this way has two important consequences. Firstly, there is a slight difference in appearance between the wire side of the sheet and the other side (top side). This effect is eliminated if the sheet is subsequently coated with a white mineral (china clay) coating or if a specific type of twin wire former is used where both outer sides of the sheet are in contact with identical wire mesh surfaces. Secondly, the method of forming influences the orientation of the fibres in the sheet. Since fibres are long and thin they tend to line up in the direction of motion on the machine. This is called the machine direction (MD). Strength properties such as tensile and stiffness are higher in the MD. One of the aims of successful forming is to randomise the orientation of fibres in the sheet. Nevertheless the orientation occurs and it is normal to measure strength properties in both the MD and in the direction at right angles to the MD, known as the cross direction (CD). The fibres form an entangled network which is assisted by creating turbulence in the headbox immediately prior to forming and, on some paper machines, by shaking the wire from side to side.

8.3.3 Pressing

At the end of the forming section, or wet end of the machine, the sheet is sufficiently consolidated by the removal of water to support it's own weight to transfer into the press section (Fig. 8.5). Here it is held between absorbent blankets and gently pressed using steel rolls so that with vacuum assistance more water is removed, reducing the moisture content to about 60–65%.

8.3.4 Drying

The moisture is reduced to less than 10%, depending on grade, by passing the sheet over steam heated cylinders. Some machines include in their drying section a very large heated cylinder with a polished steel surface. This is a MG (machine glazing) cylinder – also known as a Yankee cylinder. Paper can be produced with a glazed surface on one side and on some board machines the MG cylinder is used to produce a smooth surface, whilst preserving thickness, thereby giving higher stiffness for a given grammage. A starch solution is sometimes applied towards the end of the drying section to one or both sides of

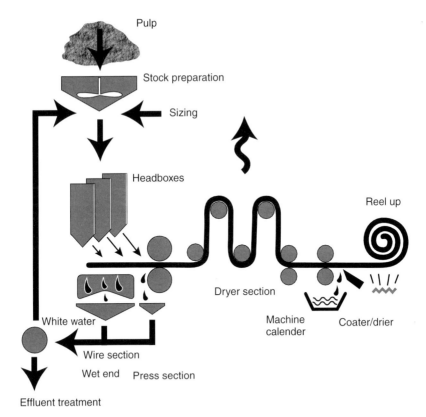

Figure 8.5 Principal features in manufacture of paper/paperboard – the number of headboxes will vary depending on the product (courtesy of Pro Carton).

the sheet. This is known as surface sizing. It improves the strength and finish of the sheet and anchors the fibres firmly in the sheet. Squeezing the sheet through a series of steel/composition rolls can enhance smoothness and thickness uniformity. This is known as calendering. Paper may be calendered at high speed in a separate process, known as supercalendering.

8.3.5 Coating

White pigmented coatings are applied to one or both sides of many types of paper and board on-machine. The coatings comprise mineral pigments, such as china clay and calcium carbonate, and synthetic binders (adhesives), dispersed in water. Excess coating is usually applied, it is smoothed and the excess removed. A number of techniques may be used – metering bar, air knife or blade coating.

One, two or three layers of coating may be applied. Coatings are dried by radiant heat and by passing the sheet over steam heated drying cylinders. They may be burnished (polished) depending on the required appearance, colour, smoothness, gloss and printing properties. Coatings can be applied off-line. In the cast coating process the smooth wet coating is cast against a highly polished chromium-plated heated cylinder. When dry, the coating separates from the metal surface leaving the coating with high smoothness and gloss.

8.3.6 Reel-up

Finally, the paper or board is reeled up prior to *finishing*.

8.3.7 Finishing

The large diameter, full machine width reels of paper and board are then slit into narrower reels of the same or smaller diameter or cut into sheets to meet customer and market needs. Sheets may be guillotined, pile turned, counted, ream wrapped, palletised, labelled and wrapped securely, usually with moisture resistant material such as PE coated paper or PE film.

8.4 Packaging papers and paperboards

A wide range of papers and paperboards are commercially available to meet market needs resulting from the choice of fibres available, bleached and unbleached, virgin and recycled, and because fibres can be modified at the stock preparation stage.

Paper and board based products can be made in a wide range of grammages and thicknesses. The surface finish (appearance) can be varied mechanically. Additives introduced at the stock preparation stage provide special properties.

PAPER AND PAPERBOARD PACKAGING   249

Coatings applied, smoothed and dried, to either one or both surfaces, offer a variety of appearance and performance features which are enhanced by subsequent printing and conversion, thereby resulting in various types of packaging.

To illustrate these features of paper and paperboard some examples are described below.

8.4.1 Wet strength paper

Paper sacks used in wet conditions need to retain at least 30% of their dry strength when saturated with water. To achieve wet strength, urea formaldehyde and melamine formaldehyde are added to the stock. These chemicals cross link during drying and are deposited on the surface of the cellulose fibres making them resistant to water absorption.

8.4.2 Microcreping

Microcreping, e.g. as achieved by the Clupak process, builds an almost invisible crimp into paper during drying enabling paper to stretch up to 7% in the MD compared to a more normal 2%. When used in paper sacks this feature improves the ability of the paper to withstand dynamic stresses, such as occur when sacks are dropped.

8.4.3 Greaseproof

The hydration (refining) of fibres at the stock preparation stage, already described, is taken much further than normal. It is carried out as a batch process and is known as *beating*. The fibres are treated (hydrated) so that they become almost gelatinous.

8.4.4 Glassine

This is a supercalendered (SC) greaseproof paper. The calendering produces a very dense sheet with a high (smooth and glossy) finish. It is non-porous, greaseproof, can be laminated to board and can be silicone coated to facilitate product release. Glassine is also available in several colours.

8.4.5 Vegetable parchment

Bleached chemical pulp is made into paper conventionally and then passed through a bath of sulphuric acid. Some of the surface cellulose is gelatinised on passing into water and redeposited between the surface fibres forming an impervious layer. This paper has high grease resistance and wet strength.

8.4.6 Tissues

Neutral pH grades with low chloride and sulphate residues are laminated to aluminium foil. The grammages range from 17–30 g/m². Tea and coffee bag tissue is a special light weight tissue available either as a heat sealable product (containing a proportion of Polypropylene fibres), or as a non-heat sealable product, in grammages from 12–17 g/m². It incorporates long fibres, such as those derived from manilla hemp which give a strong permeable sheet at the low grammages used.

8.4.7 Paper labels

These may be MG (machine glazed), MF (machine finished) or calendered kraft papers (100% sulphate chemical pulp) in the grammage range 70–90 g/m². The paper may be coated on-machine or cast coated for the highest gloss in an off-machine or secondary process.

The term *finish* in the paper industry refers to the *surface* appearance. This may be:

- MF – machine finish, smooth but not glazed
- WF – water finish where one or both sides are dampened and smoothed, to be smoother and glossier than MF
- MG – machine glazed with high gloss on one side only
- SC – supercalendered, i.e. dampened and polished off-machine to produce a high gloss on both sides.

8.4.8 Bag papers

For sugar or flour, coated or uncoated bleached kraft in the range 90–100 g/m² is used. Imitation kraft is a term on which there is no universally agreed definition, it can be either a blend of kraft with recycled fibre or it can be 100% recycled. It is usually dyed brown. It has many uses for wrapping and for bags where it may have an MG and a ribbed finish. Thinner grades may be used for lamination with aluminium foil and PE for use on form/fill/seal machines.

8.4.9 Sack kraft

Usually this is unbleached kraft (100% sulphate chemical) pulp, though there is some use of bleached kraft. The grammage range is 70–100 g/m².

8.4.10 Impregnated papers

Wax impregnated paper and fluorocarbon treatment for grease/fat resistance is produced on-machine.

8.4.11 Laminating papers

These are coated and uncoated papers (40–80 g/m^2) based on both kraft (sulphate) and sulphite pulps. These papers can be laminated to aluminium foil and extrusion coated with PE. The heavier weights can be PE laminated to plastic films and wax or glue laminated to unlined chipboard.

8.4.12 Solid bleached board (SBB)

Solid bleached board is made exclusively from bleached chemical pulp (Fig. 8.6). It usually has a mineral pigment coated top surface and some grades are also coated on the back. The term SBS (solid bleached sulphate), derived from the method of pulp production, is sometimes used to describe this product.

This paperboard has excellent surface and printing characteristics. It gives wide scope for innovative structural design and can be embossed, cut, creased, folded and glued with ease. This is a pure cellulose primary (virgin) paperboard with consistent purity for food product safety making it the best choice for the packaging of aroma and flavour sensitive products. Examples of use include chocolate confectionery, frozen foods, cheese, tea, coffee, reheatable products and as a base for liquid packaging.

8.4.13 Solid unbleached board (SUB)

Solid unbleached board is made exclusively from unbleached chemical pulp (Fig. 8.7). The base board is brown in colour. This product is also known as solid unbleached sulphate. To achieve a white surface it can be coated with a white mineral pigment coating, sometimes in combination with a layer of bleached white fibres under the coating. This board is used where there is a high strength

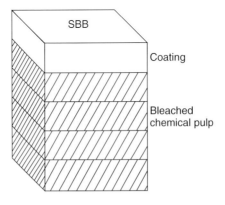

Figure 8.6 Solid bleached board (courtesy of Iggesund Paperboard).

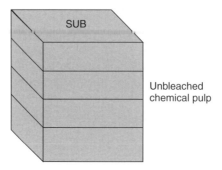

Figure 8.7 Solid unbleached board (courtesy of Iggesund Paperboard).

requirement in terms of puncture and tear resistance and/or good wet strength is required such as for bottle or can multipacks, and as a base for liquid packaging.

8.4.14 Folding boxboard (FBB)

Folding boxboard comprises middle layers of mechanical pulp sandwiched between layers of bleached chemical pulp (Fig. 8.8). The top layer of bleached chemical pulp is usually coated with a white mineral pigment coating. The back is cream (manilla) in colour. This is because the back layer of bleached chemical pulp is translucent allowing the colour of the middle layers to show through. However, if the mechanical pulp in the middle layers has been given a mild chemical treatment (bleached) it is lighter in colour and this makes the reverse side colour lighter in shade. The back layer may however be thicker or

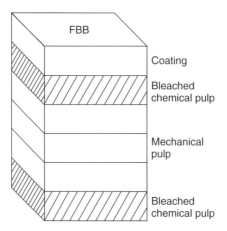

Figure 8.8 Folding boxboard (courtesy of Iggesund Paperboard).

coated with a white mineral pigment coating, thus becoming a white back folding box board. The combination of inner layers of mechanical pulp and outer layers of bleached chemical pulp creates a board with high stiffness.

Fully coated grades have a smooth surface and excellent printing characteristics. This paperboard is a primary (virgin fibre) product with consistent purity for food product safety and suitable for the packing of aroma and flavour sensitive products. It is widely used for food products such as confectionery, frozen and chilled foods, tea, coffee, bakery products and biscuits.

8.4.15 White lined chipboard (WLC)

White lined chipboard comprises middle plies of recycled pulp recovered from mixed papers or carton waste. The middle layers are grey in colour. The top layer, or liner of bleached chemical pulp is usually white mineral pigment coated. The second layer, or under liner, may also comprise bleached chemical pulp or mechanical pulp. This product is also known as newsboard. The term chipboard is also used, though this name is more likely to be associated with an unlined grade, i.e. without a white, or other colour, surface liner ply (Fig. 8.9).

The reverse-side outer layer usually comprises specially selected recycled pulp and is grey in colour. The external appearance may be white by the use of bleached chemical pulp and, possibly, a white mineral pigment coating. (White PE has also been used.)

There are additional grades of unlined chipboard and grades with special dyed liner plies for use in the manufacture of corrugated fibreboard.

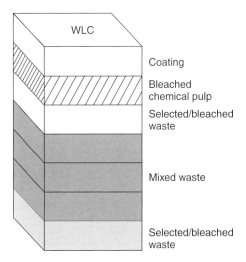

Figure 8.9 White lined chipboard (courtesy of Iggesund Paperboard).

The overall content of WLC varies from about 80–100% recovered fibre depending on the choice of fibre used in the various layers. WLC is widely used for cereals, dried foods, frozen and chilled foods, and confectionery outers.

8.5 Properties of paper and paperboard

The features of paper and paperboard which make these materials suitable for packaging relate to appearance and performance. These features are determined by the type of paper and paperboard – the raw materials used and the way they have been processed. Appearance and performance can be related to measurable properties which are controlled in the selection of raw materials and the manufacturing process.

National and international standard test procedures have been published by British Standards (UK), DIN (Germany), ISO, and in the US there is TAPPI (Technical Association for the Pulp and Paper Industry) and ASTM International (formally the American Society for Testing Materials).

8.5.1 Appearance

Appearance relates to the visual impact of the pack and can be expressed in terms of colour, smoothness and whether the surface has a high or low gloss (matte) finish.

Colour depends on the choice of fibre for the outer surface, and also, where appropriate, the reverse side. As described above, the choice is either white, brown or grey. In addition some liners for corrugated board comprise a mix of bleached and brown fibres.

Other colours are technically possible either by using fibres dyed to a specific colour or coated with a mineral pigment coloured coating.

Where paper and paperboard is required for quality printing, it is usually coated on the print side during manufacture with a mineral based coating, usually white in colour, based on china clay or calcium carbonate. The reverse side may also be coated where two side printing is required.

8.5.2 Performance

Performance properties are related to the level of efficiency achieved during the manufacture of the pack, in printing, cutting and creasing, gluing and the packing operation. Performance properties are also related to pack compression strength in storage, distribution, at the point of sale and in consumer use.

Specific measurable properties include stiffness, short span compression (rigidity) strength, tensile strength, wet strength, % stretch, tear strength, fold endurance, puncture resistance and ply bond strength. Other performance properties relate to moisture content, air permeability, water absorbency, surface friction, surface tension, ink absorbency etc. Chemical properties include pH, whilst chloride and sulphate residues are relevant for aluminium foil lamination. Flatness is easily evaluated but is a complicated issue as lack of flatness can arise from several potential causes, from the hygrosensitivity characteristics of the fibre, manufacturing variables and handling at any stage including printing and use. Neutrality with respect to odour and taint, and product safety are performance needs which are important in the context of paper and board packaging which is in direct or close proximity to food.

8.6 Additional functional properties of paper and paperboard

Additional barrier and functional performance for food packaging needs can be imparted to paper and paperboard by one or more of the following processes.

8.6.1 Treatment during manufacture

8.6.1.1 Hard sizing. Sizing is a term used to describe a treatment which delays the rate at which water is absorbed, both through the edges (wicking) and through the surface. It is achieved by the use of chemicals added during the stock, or pulp, preparation stage prior to forming in manufacture. This is known as *internal sizing*. Traditionally, alum, a natural resin, derived from wood was used. Today several synthetic sizing materials are also available. Paperboard used in packaging for frozen and chilled food and for liquid packaging needs to be hard sized.

8.6.1.2 Sizing with wax on-machine.

8.6.1.3 Acrylic resin dispersion (water based) coating – heat sealable, moderate moisture and oxygen barrier, available as one side coating on-machine.

8.6.1.4 Fluorocarbon dispersion coating, high fat resistant one side coating on-machine.

Note: The terms *on-machine* and *off-machine* are commonly used in the paper industry. The machine in question is the paper or paperboard machine. An on-machine process takes place as the paper or paperboard is being made and off-machine is a subsequent process carried out on a machine designed specially for the process concerned.

8.6.2 Lamination

This process applies another functional or decorative material, in sheet or reel form, to the paper or paperboard surface with the help of an adhesive. Examples are:

- Aluminium foil applied to one or both sides – provides a barrier to moisture, flavour, common gases such as oxygen, and UV light. Aluminium foil laminated to paper and paperboard is also used for direct contact and easy release for foods which will be cooked or reheated in radiation or convection ovens. Aluminium foil is also used to provide a decorative metallic finish as, for example, on cartons for chocolate confectionery.
- Greaseproof paper laminated to paperboard – good grease resistance for fat containing products, temperature resistance to 180°C for cooking/reheatable packs. If additionally the greaseproof paper has a release coating, this product can be used to pack sticky or tacky products.
- Glassine paper laminated to paperboard – grease resistance for products with moderate fat content such as cakes or bake-in-box applications. If the glassine is coloured the pack should not be used in reheatable applications but food contact approved grades can be used for direct contact with, for example, chocolate.

The adhesives used for lamination include PVA-type emulsions, starch based, resin/solvent based, cross linking compounds, molten wax or PE depending on the needs of the particular laminate. The presence of wax and PE also improves the barrier to water vapour. When PE is used as an adhesive the process would be described as *extrusion lamination*.

8.6.3 Plastic extrusion coating and laminating

Polyethylene (PE) heat sealable moisture barrier. Low density polyethylene (LDPE) is widely used in the plastic extrusion coating and laminating of paper and paperboard. Easier heat sealing results when PE is modified with EVA (Ethylene vinyl acetate). Medium and high density PE has a higher temperature limit, better abrasion resistance and higher barrier properties than LDPE. One and two side coatings are available (Fig. 8.10).

Polypropylene (PP) heat sealable, moisture and fat barrier. It can withstand temperatures up to 140°C and is used for packing foods to be reheated in ovens up to this temperature. One and two side coatings are available.

Polyethylene terephthalate (PET) heat sealable, moisture and fat barrier. It can withstand temperatures up to 200°C and is dual ovenable (microwave and conventional ovens). It is coated only on the non-printing side.

Polymethylpentene (PMP) moisture and fat barrier and not heat sealable. It is therefore used as flat sheets, deep drawn trays and trays with mechanically locked corners. It is coated only on the non-printing side.

Ethylene vinyl alcohol (EVOH) and polyamide (PA) heat sealable, fat, oxygen and light barrier. EVOH is moisture sensitive and needs to be sandwiched between hydrophobic materials, such as PE. It can be used as a non-metallic alternative to the aluminium foil layer.

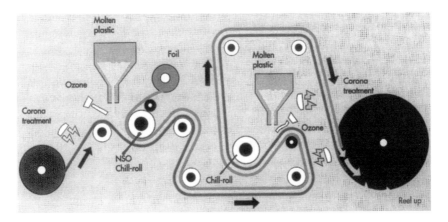

Figure 8.10 The extrusion and extrusion laminating process (courtesy of Iggesund Paperboard).

Ionomer resin (Surlyn™), a polyolefin with high resistance to fat, including essential oils in citrus fruit, and moisture with very good sealing properties, is used as a *tie* layer on aluminium foil when applying PE to foil.

The process of extrusion is often extended to include extrusion lamination so that a structure, such as paper or paperboard/PE/aluminium foil/PE, can be produced in one operation on an extruder with two extruding units.

Note: By special selection of polymers, e.g. for lids and trays, it is possible to provide easy-open peelable heat-seals.

8.6.4 *Printing and varnishing*

Usually, printing and varnishing are associated with the appearance of the pack with respect to the visual impact of the pack through colour, information, text and illustration. There are also important functional aspects of printing and varnishing which are important for food packaging.

All the main printing processes are used – Gravure, Flexographic, Letterpress, Silk Screen and Lithographic. Paper and paperboard can also be printed by the recently introduced digital process. Choice is influenced by the appearance and performance (functional) needs and commercial aspects such as order size, lead time and price.

The inks and varnishes may be those described as traditional for the process concerned, based on pigment, resin and vehicle. The vehicle, which transports the pigment and resin from the ink or varnish reservoir to the substrate via the printing plate, varnish pick-up roll etc., may be an organic solvent, water or a drying oil. For some processes, pigments are replaced by dyes. In recent years inks and varnishes cured by UV radiation have also become popular and these materials are extremely inert. They give good rub resistance in wet and dry conditions and are resistant to product absorption. The inks contain pigment, cross linking resins and a photo-initiator; they are 100% solid and are dry immediately after printing.

The functional requirements include adherence to colour standards, light fastness, rub resistance, print-to-print and print-to-pack registration and stability within the conditions of use. For some food products where the print is in close proximity to the food, e.g. chocolate confectionery, it is important that no residual solvents from the inks and varnishes, or any other interaction between print and product affects the food product.

8.6.5 Post-printing roller varnishing/coating/laminating

Post-printing roller varnishing and coating is usually associated with high gloss and can involve UV cured varnishes. The process can also be used for the application of functional varnishes to meet specific end use needs. The most common example of this is the application of heat-seal coatings for blister packs.

Another example of coating is the application of wax. This can take a variety of forms:

- Dry waxing where molten wax is applied to one or both sides of a printed paper or a printed/cut/creased carton blank. The appearance is a matte finish.
- Wet or high gloss waxing. Immediately after coating, the printed paper or carton blank is conveyed through very cold water. This causes the wax to set immediately, producing a very high gloss finish.

Waxed paperboard provides water and water vapour resistance. It can be heat sealable. The first liquid packaging cartons (~1920) were waxed. Wax is also a good gas barrier and can therefore protect food products against flavour loss or ingress of contamination. The main food applications today are for bread wrap, items of sugar confectionery (paper), frozen food and ice cream (cartons) and fresh produce (corrugated board). Cellulose acetate and OPP laminated to paperboard enhance appearance after printing.

8.7 Design for paper and paperboard packaging

Surface design concerns colour, text, illustrations, decoration, finish (gloss or matte) and surface texture. It is achieved by making use of the basic surface properties of the paper or paperboard and through lamination, coating, hot foil stamping, embossing, printing and varnishing. Surface design usually refers to the external surface of a pack but there are situations where the internal surface is important for the overall effect, e.g. the inside surfaces of chocolate and tea bag cartons.

Structural design is concerned with the shape of packages. The functional shape is determined by the needs of the pack, e.g. closure and opening features. Creative shape adds interest for promotional needs where that is appropriate. Paper and paperboard are materials which give a designer freedom to develop imaginative solutions in meeting customer needs. This is due to a number of factors:

- range of surfaces, in terms of colour and finish, available
- range of strength properties, in terms of fibre type, thickness and method of manufacture, available
- choice of functional coating, lamination, decoration, printing etc.
- ease of conversion into packages in terms of cutting, creasing, folding, gluing, locking, heat sealing etc.
- innovative machinery for conversion and packing.

8.8 Package types

8.8.1 Tea and coffee bags

These are made from very light-weight porous tissues. There are heat sealed bags where the fibre structure (grammage ~17 g/m^2) contains a heat sealable fibre such as PP. Bags may be flat, square, four side perimeter sealed or they may be round or pyramidal in shape. Another design is folded and stapled giving a larger surface area for infusion and using a lighter weight tissue (~12 g/m^2). All these bags are closely associated with the machinery which forms, fills and seals the bags – both types may have strings and tags. It is possible to link such machines with enveloping machines which can comprise paper, or paper laminated or coated, with moisture and gas barrier properties. Tea and coffee bag packing machines can include, or be linked to, cartonning or bagging machines.

8.8.2 Paper bags and wrapping paper

The paper bag is the traditional type of packaging where the product is packed at the point of sale, typically, in stores and markets where fruit and vegetables are sold and in bakeries for fresh bread and cakes. Manual wrapping using precut sheets is also widely used, e.g. in butchers shops and for fish and chips.

The paper based carrier bag with handles of various types is used for assorted items in retail shopping, and for luxury items and gifts where paper based decorative finishes are used.

8.8.3 Sachets/pouches/overwraps

These comprise paper based flexible packaging involving paper with plastics, frequently PE. Where additional barriers are required, aluminium foil or metallised PET is incorporated. This packaging requires a heat sealable layer on the inside of the packaging material. Cold seal coatings on the inside of the packaging material can be used for sealing where the product is heat sensitive.

These types of packaging are usually associated with form, fill seal machinery. Horizontal form/fill/seal machines are of two main types. There are those which form a pouch in a horizontal plane with the product filled vertically. These machines can form a base gusset (Fig. 8.11).

Vertical form/fill/seal machines are used to pack free flowing food materials and liquids. Packs made in this way are either flat, or incorporate gussets and block (flat) bottoms (Fig. 8.12).

There is also the flow wrap type which is used to pack single solid items horizontally, such as confectionery bars or multiple products already collated in trays (paperboard) (Fig. 8.13).

There are machines which form bags around mandrels, sealing being made with adhesives, so that they have a rectangular cross section and a block bottom. (This type of machine can also wrap a carton around the paper on the same mandrel to form a lined carton.)

Roll wrap machines pack rows of items, e.g. biscuits and sugar confectionery. Individual confectionery units may be wrapped in waxed paper for moisture protection and to prevent them sticking together.

Overwrapping square or rectangularly shaped cartons, with paper coated with PE or wax with neatly folded heat sealed end flaps is also used, e.g. confectionery.

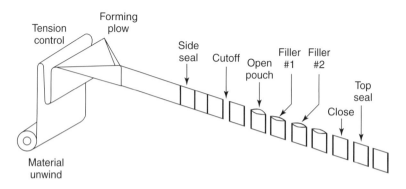

Figure 8.11 Horizontal pouch/sachet form/fill/seal machine for dry mixes (soups, sauces etc.) (courtesy of The Institute of Packaging).

PAPER AND PAPERBOARD PACKAGING   261

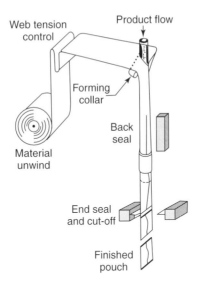

Figure 8.12 Vertical sachet form/fill/seal machine for dry, free flowing products (snack foods etc.) (courtesy of The Institute of Packaging).

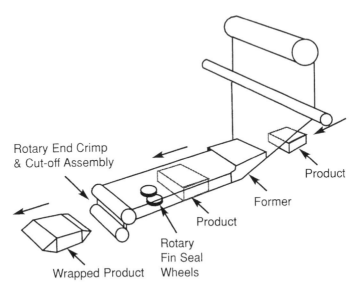

Figure 8.13 Horizontal, flowrap type, form/fill/seal machine for solid products (courtesy of The Institute of Packaging).

8.8.4 Multiwall paper sacks

Multiwall paper sacks are made from between 2 and 6 plies or layers of paper. The specifications vary according to the needs of the product and the output required. The differences concern the design of opening through which the product is filled, the design of the opposite end, the closure and the style of the sides which may be a single crease or gussetted for ease of stacking (Fig. 8.14).

The open mouth sack is closed either by sewing through a strip of creped paper folded over the edges of what was the opening, or with a metal tie. The other design of opening is the valve, a small paper tube, inserted in one corner of a pasted end, again there are several basic designs.

The main type of paper used is natural brown kraft paper which has good strength properties relating to tensile strength, % stretch, tensile energy absorption, burst strength, tear strength and where necessary, wet strength. Air permeability is important for the filling rate of powders in valved sacks and any sack filling of an aerated product. Water absorption can be important. Surface friction is relevant to pallet stacking and safety.

Where the product requires moisture protection, the moisture vapour transmission rate is important and there are various ways of achieving a low rate from the use of specially inserted PE bag liners to the use of PE or wax coated paper, PVdC and aluminium foil laminations.

The use of a bleached kraft outer ply can enhance appearance. Tougher paper in the form of extensible microcreped kraft and creped kraft is also used. Creping gives enhanced stretch properties in the MD of the paper.

Many different food products are packed in multiwall paper sacks. Examples include: sugar, dried milk, whey powder, coffee beans, flour, peanuts, potatoes and other fresh vegetable products. Traditionally this form of packing was used for the shipment of product in bulk. Smaller multiwall sacks are now used for retail packs. They can incorporate a carrying handle and a window for product visibility, e.g. for potatoes, and also for dry pet food.

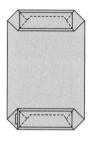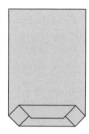

Figure 8.14 Multiwall sacks.

8.8.5 *Folding cartons*

Folding cartons are widely used in the retail packaging of food products. They are paperboard boxes which are supplied to the packaging machine either flat or folded flat. They are used to package a wide range of food products. These range from cereals, frozen and chilled foods, ice cream, chocolate and sugar confectionery, cakes and biscuits, coffee, tea, convenience food mixes (snack soups), dried food products (raisins) and food supplements in the health care market.

Products may be packed in direct contact with the inside of the paperboard or they may have already been packed in another form of packaging such as a can, bottle, sachet, bag, collapsible tube, plastic trays or pots.

The choice of paperboard used for folding cartons depends on the needs of the product in packing, distribution, storage and use, and on the surface and structural design. The basic choice is between solid bleached board (SBB), solid unbleached board (SUB), folding box board (FBB) and white lined chip board (WLC).

The protective properties of the paperboard may have been enhanced by the laminations, dispersion coatings, plastic extrusion coatings and other treatments, already discussed, in order to meet specific product needs.

Folding cartons meet many packaging needs and can be made in a wide variety of designs (Fig. 8.15). Most cartons are rectangular or square in cross section. The type of product to be packed, the method of filling and the way the cartons will be distributed, displayed and used will influence the dimensions and design in general. Rectangular shapes are easy to handle mechanically, especially when packed in large volumes at high speeds. The design may be for end loading, e.g. cereals, or top loading, e.g. tea bags.

Paperboard may be formed into trays either by heat sealing, locking tabs and slots or by gluing with hot melt adhesive depending on the application. PET lined paperboard can be deep drawn to a depth of 25 mm, or 45–50 mm in

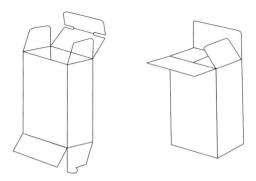

Figure 8.15 Folding cartons.

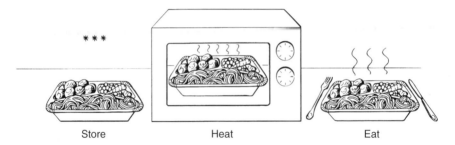

Figure 8.16 PET lined paperboard trays (courtesy of Iggesund Paperboard).

two stages. This type of tray can be used for a ready meal for frozen or chilled distribution and reheating at up to 200°C in either a microwave or a radiant heat oven. It could have a peelable heat sealable printed paperboard or plastic lid and be packed in a paperboard sleeve or carton (Fig. 8.16).

The early developments in microwave foods provided convenience and rapid heating. It was not suitable for products where browning or a degree of crispness was expected. Developments in microwave food formulations, which improved their performance in the microwave oven, and the use of susceptors in the packs has widened the range of foods suitable for this application. Susceptors absorb microwave energy and heat food rapidly, mainly by contact, and induce crispness and localised browning. Susceptors usually comprise aluminium metallised plastic film, such as PET, laminated to paper or paperboard. Inconel, nickel/chromium, susceptors can be used to induce even higher temperatures. The use of susceptors in this way is an example of *active packaging*.

Cartons may be lined by the carton maker with a flat tube of a flexible barrier material which is inserted during carton manufacture. The flexible material is usually heat sealable – examples include paper/aluminium foil/PE and laminations of plastic films. The lined cartons are supplied folded flat to the packing/filling machine. One end of the liner is heat sealed and after filling, the other end is sealed and the carton flaps closed. This type of carton is used for ground coffee, dry foods and liquids. A lined carton may be fitted with a plastic hinged lid incorporating a tamper evident diaphragm.

In another type of lined carton, flat carton blanks and a roll of the material to be used as the liner, frequently bleached kraft paper, are supplied to the packing machine. Firstly, the paper is formed around a solid mandrel. The side seam and base is either heat sealed or glued with adhesive, depending on the specification. The carton is then wrapped around the liner with the side seam and base sealed with adhesive. The product is filled and both liner and carton sealed/closed. This type of pack is suitable for the vertical filling of powders, granules and products such as loose filled tea. Folding cartons can have windows or plastic panels for product display, e.g. spirits.

Cartons may have separate lids and bases, flanged or hinged lids. A display outer is a carton which performs two functions. At the packing stage it is used as a transit pack or *outer*. When it arrives at the point of sale the specially designed lid flaps are opened and folded down inside the carton and the transit pack becomes a display pack or *display outer*. This form of pack is frequently used to pack a number of smaller items which are sold separately, e.g. confectionery products also known as *countlines*. On other designs of folding carton, lid panels, or flaps, may close as a tuck-in-flap, flip top, locked, glued or be heat sealed. Closures may be made tamper evident. Lid flaps which are repeatedly opened and closed during the life of the contents require folding endurance strength to withstand repeated opening and reclosure.

In addition, cartons can have internal display fitments or platforms, sleeves can be used for trays of chilled ready meals and multipacks for drinks cans, bottles and plastic yoghurt pots. Cartons can incorporate dispensing devices, carrying handles and easy opening tear-strip features for convenience in handling and use. Cartons can be made into non-rectangular, innovative shapes such as packaging for Easter Eggs.

Cartons can be produced in creatively designed shapes and printed, varnished, laminated or otherwise finished for luxury food products such as expensive spirits and chocolate confectionery.

Once a specific type of paperboard has been selected it is necessary to choose a grammage and thickness which will ensure adequate carton strength at each stage of the packing chain from packing through to use by the consumer.

Folding cartons are made as follows: firstly, the surface design is printed on paperboard sheets or reels; secondly, the outline profile of each carton is cut and creased. The flat carton blank which results may be supplied directly to the packer. Alternatively, the flat blank is glued, usually on the side seam and, sometimes, on the base, crashlock style, as well, and folded flat. Both approaches ensure the most efficient use of storage and transit space between the manufacturer and the packer.

There are other processes used in making cartons depending on the surface and structural design. These include varnishing, either in-line with printing or off-line in a separate operation, heat seal coating for blister packaging, embossing, hot foil stamping, window patching and many more.

8.8.6 *Liquid packaging cartons*

The concept of a liquid food package based on paperboard became a reality when it became technically possible to combine paperboard with an additional moisture and product resistant heat sealable material. This led to the development of leak-proof liquid tight packaging (Fig. 8.17). The first successful package was the gable-topped Pure-Pak patented in 1915 where wax provided the heatsealable and protective barrier properties.

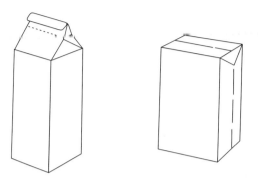

Figure 8.17 Liquid packaging cartons.

Several styles of liquid pack were subsequently developed. Most were filled through a full aperture top which was subsequently closed and sealed. The cross sections were square, rectangular or circular with the sides tapered. They were supplied to the packaging machine either as flat blanks for erection or as nested premade containers ready for filling. PE replaced wax from the 1950s and an additional development led to a reel fed form/fill/seal approach with the tetrahedral shaped Tetrapak.

A range of shelf life times are possible depending on how the product is processed prior to filling, the filling conditions, and the conditions of distribution. Products could, for example, be pasteurised prior to filling, the product could be filled hot or the product could be UHT sterilized and filled into a sterilised pack, i.e. aseptically packed. An appropriate paperboard laminate would be selected to maintain quality for the period of the required shelf life.

For fresh products, e.g. milk, with a short shelf life in chilled distribution, i.e. 0–4°C, a two side PE coated paperboard is used. For a long shelf life in ambient temperature distribution for hot-filled products and for fresh juices in chilled distribution, the barrier of the laminate is extended by a thin layer of aluminium foil. In this case the lamination would comprise PE/paperboard/PE/aluminium foil/PE.

The aseptic packaging process whereby a sterile product is filled into sterile containers and sealed under sterile conditions has been described by the Institute of Food Technologists as 'the most significant food science innovation of the last 50 years'.

This process has been successfully applied to paperboard liquid-packaging to extend shelf life at ambient temperatures. The pack requires the use of the PE/paperboard/PE/aluminium foil/PE laminate. Aluminium foil may be replaced by EVOH, an excellent oxygen barrier, and more easily handled in the domestic waste stream.

The overall result of these packaging, processing and distribution alternatives is that a wide range of liquid-food products are now available in paperboard based packaging. Examples include milk and milk derived products, juices, soups, non-carbonated water and wine. Liquid products containing particulates are usually filled into open top cartons to eliminate the possibility of product interfering with the sealing of form/fill/seal packs.

A wide range of pack sizes is available from several suppliers. Pack sizes range from the single portion tetrahedral Tetra Classic packs with volumes from 20–65 ml, through the popular volume range of packs with square or rectangular cross sections of 0.2 l, 0.33 l, 0.5 l, 1.0 l, 1.5 l and 2.0 l cartons. Several designs of single portion pack are available with straws attached and a 250 ml cylindrical container with an easy-open tab has been introduced by Walki. Even larger 4.0 l and 5.0 l cartons are available in the Pure-Pak range.

Whilst pack shapes are dominated by the gable top and brick designs based on square and rectangular cross sections, alternatives are available with shapes based on hexagons, tetrahedrons, wedges, pillow pouch and square cross sections with rounded corners. A major area of design innovation, as a result of consumer demand, in recent years has concerned ease of opening, reclosure and tamper evidence. Many convenience-in-use design features are now available from plastic straws for use with packs having an ease-of-pack-entry feature, plastic screw action closures with a tamper evident feature, peelable foil based tabs and push-fit plastic reclosures etc.

The production and marketing of liquids in cartons is one of the best examples of the integrated or systems approach to packaging whereby all aspects of the pack, filling and distribution are engineered by the manufacturer of the cartons working closely with the dairy, food processor or in-house own label retail organisation.

The paperboard used in liquid packaging is usually solid bleached, or unbleached board. This is used because it has an efficient performance in printing, cutting, creasing and folding, and particularly, as it is based on pure cellulose fibre, to protect the product from any packaging related effect on the flavour and taste of the product. Milk and milk derived products, wine and juices are flavour sensitive products requiring careful handling and packaging.

Careful attention is given to printing and extrusion coating and laminating to ensure that the materials and processes used do not have an effect on the flavour of the product. (The large gable-top carton design has also been extended to the packaging of free flowing dry foods such as rice and freeze dried vegetables for the catering market.)

8.8.7 Rigid cartons or boxes

Rigid cartons, as distinct from folding cartons, are erected before being delivered to the packer. The use of rigid cartons for food packaging is virtually confined to the luxury/gift market such as for chocolate confectionery, preserves and the more expensive bottled wines and spirits.

Rigid boxes typically comprise a baseboard, the type and thickness of which is chosen to meet the customers needs and which is cut and scored. This is corner stayed, in which gummed paper is fixed around the made up corners of the box providing rigidity. The box is covered with a decorative sheet of paper, or paper, film or aluminium foil based laminate which is also cut to a precise profile to produce a neat finish. Adhesive secures the lining material, which may be printed, to the board.

A wide range of lining materials can be used to create specific visual effects such as embossing, hot foil stamping, use of fabric materials etc. Many features can be incorporated in the design such as hinges, handles, thumb holes, domes, windows, and plastic and paperboard fitments. Most of the operations are manual or machine assisted and this together with the wide range of lining materials and design elements makes it possible for a wide range of distinctive box designs to be constructed.

8.8.8 Paper based tubes, tubs and composite containers

8.8.8.1 Tubes
Small diameter paper based tubes are used for confectionery; they may be designed with paperboard or plastic ends.

8.8.8.2 Tubs
Typically the ice cream tub must be leak-proof, resistant to the product and suitable for low temperature distribution. PE extrusion or wax coated paperboard meets these needs. Additionally, small tubs are used to pack single portion cream and yoghurt based desserts. Cross sections are circular, elliptical or square with rounded corners. Sides may be straight or tapered.

8.8.8.3 Composite containers
These containers are typically of circular cross section though designs with square and rectangular cross sections with rounded corners can also be made (Fig. 8.18). They are used for both dry food products, such as tea, powdered or granular mixes and savoury snack products, and for liquids, e.g. non-carbonated drinks. The container bodies comprise paperboard and paperboard laminates with plastic and seamed-on metal ends with either lever lid, snap-on or seamed ring-pull lids.

8.8.9 Fibre drums

Fibre drums are used to transport food products and ingredients in dry, paste or liquid form. They are usually circular in cross section with parallel sided walls which are made by winding paper, or thin paperboard, on mandrels. The winding may be either spiral or straight. The paper, or thin paperboard, is usually

PAPER AND PAPERBOARD PACKAGING

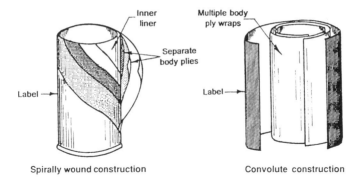

Figure 8.18 Construction of sidewalls of composite containers (courtesy of The Institute of Packaging).

unbleached kraft (brown) and the layers are adhesive bonded to provide stacking and handling strength.

The drum ends and closures can be based on fibre, metal or plastic depending on the product and distribution needs. The closure method can be by tape, metal lugs or locking metal bands and, depending on the closure, the top rim of the drum may be metal reinforced.

PE inserts or a fully laminated plastic interior with caulked bottom seals may be incorporated. Again, depending on the product, functional barrier materials can be incorporated in the construction as well as special product release coatings.

Drums can be made strong enough to allow four high stacking. They can be made suitable to provide moisture protection in outside storage. Tapered drums and drums with square cross sections incorporating rounded corners can be made.

A wide range of drum capacities are available depending on the product and the method of distribution from small drums up to as high a capacity as 280 l (75 US gal or 62 Imperial gals). Drums can be printed by silk screen, labelled or ink jet printed.

8.8.10 Corrugated fibreboard packaging

This is by far the largest paper and paperboard based sector in terms of the tonnage used. This type of packaging is synonymous with packaging for transportation and storage.

In the retail sector, boxes and trays made from corrugated fibreboard are used as secondary packaging. In the food industry, they are used to pack multiple numbers, e.g. 6, 12 etc., of primary containers for storage and distribution.

Corrugated fibreboard typically comprises three layers of paper based material and this is known as single wall material (Fig. 8.19). There are two outer

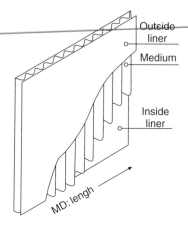

Figure 8.19 Components of single wall corrugated fibreboard (courtesy of The Institute of Packaging).

layers or liners separated by a corrugated inner ply known as fluting. The liner plies are glued to the tips of the fluting. The resulting material has high bending stiffness in relation to the weight, and high compression strength when made up in the shape of a box with glued or taped side seams and end panels. (Side seams and closures using metal staples are not normally permitted for food packaging applications.)

Double wall corrugated fibreboard comprising three liner and two fluting plies is produced but this degree of strength is not normally necessary for multiples of primary food packs.

Triple wall corrugated fibreboard is a thicker and therefore stronger material and this is used with protective inner lining bags, usually made from PE film, for the bulk packaging of free flowing food products and ingredients.

The most common lining ply material is brown kraft liner. This may be unbleached virgin kraft liner, 100% recycled fibre, also known as test liner, or mixtures of both types of fibre, the colour is brown. Bleached, white, liner plies are possible with the use of bleached kraft, and mottled white/brown liners are based on mixtures of bleached and unbleached fibres. The weights range from 115 to 400 g/m^2, though the typical values for food packaging are 125, 150 and 175 g/m^2.

The fluting medium, also known as corrugating medium, may use any of several types of fibre such as mechanical, chemical or recovered recycled fibre. Several grammages are available in the range (approximately) 100–220 g/m^2. If mechanical fibre is used it is usually of the semi-chemical type, i.e. mechanical pulp subjected to partial chemical treatment which increases the yield compared with chemical pulp but with strength characteristics which are higher than that of mechanical or recycled pulp of the same weight (grammage). The

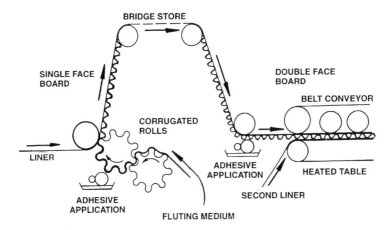

Figure 8.20 Production of corrugated fibreboard (courtesy of The Institute of Packaging).

paper is conditioned with heat and steam, and pressed between large rolls, with a gear wheel shaped surface, to produce the corrugation (Fig. 8.20).

Several standard flute configurations are available varying in the pitch height and number of corrugations per unit length, characterised by letters A (coarse), B (fine), C (medium) and E (finer than B). B-flute has a high flat crush resistance and is used for packing cans and bottles where the contents themselves contribute to the stacking strength. C-flute is used where the contents do not support the case because C-flute has a higher compression strength at the same board weight. It is also used for glass bottles where its higher flute height may provide more cushioning and higher puncture resistance.

In addition to box compression, cushioning, flat crush resistance and puncture resistance other performance features which have to be taken into consideration are print quality, efficiency of cutting and the scoring and bending characteristics. Printing is carried out either after corrugating or, where higher print quality is required, before corrugating. The latter is referred to as *pre print*.

It is sometimes more appropriate for the packer to purchase an unprinted standard sized case and print, or label, on demand. This approach may be applied where seasonally cropped fresh fruit and vegetables are being canned and it is difficult to estimate the eventual size of the crop and therefore the number of printed cases required. If the estimate is above the eventual requirement, printed cases are left in stock until the next packing season occurs, and if underestimated there is a need for urgent deliveries of additional printed cases.

Box compression strength can be calculated from the weight of contents, stacking geometry, and atmospheric conditions of storage. The manufacturers of corrugated packaging have mathematical models based on their standard

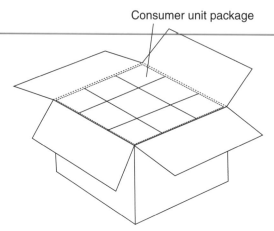

Figure 8.21 Transit package can be a corrugated fibreboard case containing a fixed number of unit packages (courtesy of Iggesund Paperboard).

materials, type of fluting, dimensions and weight of contents which can predict the compression strength of cases. Hence it is possible to estimate the weight of material and type of fluting which should provide adequate compression strength after an appropriate safety factor has been taken into consideration.

The most common design of case used in the food industry is the RSC, regular slotted container. In this design all the perimeter cutting, cutting of slots, which enable the flaps to fold neatly despite the thickness of the material, and the scoring or indenting to provide creasing and folding are carried out in straight lines in both MD and CD. Diecutting is necessary for special designs incorporating cutouts, curves and angles. Designs with these features are sometimes used in the food industry in conjunction with stretch or shrink wrapping to create more visual impact in cash-and-carry sales outlets.

Other designs of transit pack using corrugated fibreboard are typically the wrap around case erected on the packing line and the tray type packs of which there are many versions (Fig. 8.21). Some are subsequently stretch or shrink wrapped and some are erected around the primary packs to provide stacking strength and partial visibility of the primary packs without any overwrapping plastic film.

8.8.11 Moulded pulp containers

The most common food packaging applications for moulded pulp containers are the egg packs and the trays used for apples and other fresh produce.

Moulded pulp containers are made directly from a suspension of fibre in water using a mould in the form of a screen through which water is removed.

Usually recovered mixed fibres are grey in colour. If a lighter colour is required, bleached pulp, which can also be recovered fibre, of an appropriate grade is used. If other colours are required the pulp can be dyed.

There are two main processes of manufacture. The pressure injection process uses hot air under pressure to form the container which is further heated to remove excess moisture and sterilise the pack. In the other process, vacuum is applied to remove water from the mould. This process has a higher residual moisture content which has to be reduced by drying but the surface is smoother making for an improved printing result, where the packs are printed. Labelling is an alternative to printing.

8.8.12 Labels

Labels in the context of food packaging comprise the labels which are applied to:

- primary packaging in the form of cans, bottles, jars, pots, tubs, cartons, corrugated fibreboard cases, fibre drums and moulded pulp containers
- transit packs (secondary packaging) such as cases and stretch/shrink wrapped packs
- palletised loads (tertiary packaging)
- the food product directly (promotional labels), e.g. fruit.

Labels for food packaging identify, promote, inform, offer advice on the use of the products concerned and, where, for instance, a label is applied over a closure, provide security as a tamper evident feature.

Labels are characterised by their substrate, adhesion and method of application. The substrate may be paper, paperboard and laminates thereof to aluminium foil and plastic films. There is a wide choice of paper based substrate depending on the appearance and finish required. The paper may be uncoated, on-machine white mineral pigment coated in matte, satin or gloss finishes, or cast coated off-machine with white or coloured mineral pigment coatings. Where laminates to aluminium foil are used they are often embossed.

Labels may use any of the conventional print processes, the choice being influenced by the order quantity and finished appearance required. Digital printing is also used, as is ink jet printing, on the packing line. Varnishing is applied to provide protection, e.g. wet rub resistance, and gloss. Hot foil stamping is used to enhance appearance.

Dextrine adhesives are used with ungummed labels for the high speed labelling of metal and glass containers. Processed food cans and beer in glass bottles may be labelled at speeds up to 80 000 per hour. This process is known as *wet labelling*. Hot melts are used with ungummed labels on plastic containers.

Ungummed labels are usually supplied stacked in small bundles already cut to the required dimensions and shape. Where labels are picked up, held or transferred by vacuum it is essential that the substrate is not too porous to air. Where the substrate is wetted as with dextrine adhesives, care in the choice of paper is necessary to ensure that an excessive amount of curl does not develop. When moisture is applied to paper it causes the fibres to swell more significantly in the CD and as more fibres tend to be oriented in the MD the paper would tend to form a cylinder with its axis parallel to the MD – this type of curl is known as CD curl. It is necessary to use paper with a low MD/CD ratio, i.e. less MD orientation bias. As with all paper and board based packaging products, it is also important to ensure that flatness is maintained in printing, storage and end use at the packaging stage.

Some ungummed labels are applied to packaging without the use of adhesives, such as those tied on tags, and labels slipped on to the necks of bottles or otherwise clipped in place. These labels are often used with luxury food products, confectionery and drinks.

Labels supplied on reels with the adhesive already in place are referred to as being self-adhesive or pressure sensitive. As the adhesive is tacky the label stock is combined with a backing or carrier web during manufacture. The backing web comprises either glassine or bleached kraft with a siliconised surface in contact with the adhesive. The label profile is cut on the backing web; this requires a very precise control of the cutting process, since whilst the label perimeter must be cleanly cut the backing web must remain undamaged. At the point of application the label leaves the backing web, and the skeletal waste label stock and the carrier web are reeled up.

In terms of packing line speeds, self adhesive labels can be used over a very wide range from semi-automatic manually assisted lines running up to 30 units per minute to automatic lines that can be designed to run at speeds from 60 to 600 per minute. Another advantage of these labels is that changing over from one label to another on the packing line is easy.

The adhesive coating on self adhesive labels must be chosen to meet functional needs, such as whether the label is to be permanent or removable and whether there are extremes of temperature involved, e.g. frozen food storage.

Paper or paperboard in-mould labels are associated with plastic packs where the label is inserted into the tooling of an injection moulding, blow moulding or thermoforming. In-mould labels require a heat sealable coating on the reverse side which is compatible with the plastic being used for the container so that the label fuses with the container during the forming process.

There are several advantages possible with in-mould labelling. Firstly, a high quality printed image can be achieved more cost effectively than can be achieved by direct printing on round straight sided, tapered or otherwise shaped containers. Secondly, where the product requires high barrier properties, labels based on laminates of paper or paperboard to aluminium foil can give the required protection. Thirdly, with some designs of in-mould labelled container the weight of plastic used can be reduced whilst maintaining product protection and container compression strength.

A printed thin paperboard label may be side seam glued so that it is tightly applied to a tapered plastic pot in such a way that after use the label and plastic components can be easily segregated for recycling (Sandherr K3 tub from Greiner Packaging).

Heat transfer labelling, e.g. by the Dennison Therimage process, is based on a paper carrier web with a wax coating on one side. The image is reverse printed on the wax coating which is then coated with a heat sensitive adhesive. At the point of application to a plastic container the image is transferred from the paper carrier web by heat and pressure.

8.8.13 Sealing tapes

Sealing tapes are narrow width reels comprising a substrate and a sealing medium which can be dispensed and used to close and seal corrugated fibreboard cases, fibre drums, rigid boxes and folding cartons. Sealing tapes are also used to make the side seam manufacturers join on corrugated fibreboard cases and tape the corners of rigid boxes, thereby erecting or making-up corner stayed boxes.

A traditional and commonly used substrate is hard sized kraft paper, both unbleached (brown) and bleached (white). Where higher strength is required the kraft is reinforced with glass fibre, and up to four progressively stronger specifications are typically available from some suppliers. Reel widths start at 24 mm, though 50 mm width tape would be a typical width to seal the flaps of an average sized corrugated fibreboard case.

In the case of gummed tape, adhesion is achieved by coating the kraft paper with a modified starch adhesive; animal glue has largely been superseded. The adhesive is then dried and the reels are slit to size. In subsequent use the adhesive is automatically, and evenly, reactivated by water in a tape dispenser. Tape dispensers which can cut pre-set lengths for specific taping specifications are available.

The advantage of gummed paper tape is that it is permanent and provides evidence of tampering, it can be applied to a dusty pack surface without loss of adhesion, is not affected by extremes of heat and cold and does not deteriorate with time. Pressure sensitive tapes, on the other hand are used on all types of packaging from paper and paperboard to metal, glass and plastic containers. Pressure sensitive adhesive can be applied to several types of substrate, including moisture resistant kraft paper which has been coated on the other side with silicone to facilitate dispensing from the reel.

Heat fix tapes are based on kraft paper where the adhesive is applied as a thermoplastic emulsion which is subsequently reactivated by heat and applied to the sealing surface by pressure. Sealing tapes are used plain, preprinted or printed at the point of application.

8.8.14   Cushioning materials

Paper based cushioning comprises:

- Shredded paper used as a loose fill packing – this is a good use for clean recovered paper – not widely used in food packaging
- Interlocking dividers used to separate, for example, bottles in a case
- Corrugated fitments made up of one or several layers of corrugated fibreboard cut to special profiles to support, locate and protect vulnerable profiles of items. Not specifically used in food packaging but widely used in products associated with the cooking and storage of food
- Moulded pulp applications (already mentioned under Section 8.8.11 above for eggs and apples).

8.8.15   Cap liners (wads) and diaphragms

There are a number of ways of ensuring a good seal when a lid is applied to a jar, bottle or similar rigid container.

- *Pulpboard disk.* The simplest type of cap liner is a pulpboard disk made from mechanical pulp fitted inside a plastic cap. This cap liner or wad has to be compressible and inert with respect to the contents of the container. This liner could be faced with aluminium foil or PE where the nature of the contents require separation from direct contact with the pulpboard.
- *Induction sealed disk.* The disk comprises pulpboard/wax/alumium foil/ heat seal coating or lacquer. The cap with the disk in place is applied to the container and secured. It then passes under an induction heating coil. This heats the aluminium foil which causes the wax to melt and become absorbed in the pulpboard. It also activates the heat seal coating and seals the aluminium foil to the perimeter of the container. When the consumer removes the cap the adhesion between the pulpboard and the aluminium foil breaks leaving the foil attached to the container. This seal therefore provides product protection and tamper evidence. Where subsequent contact between the contents and the pulpboard is undesirable, the pulpboard is permanently bonded to the aluminium foil. (A simpler version dispenses with the wax and replaces the pulpboard with paper.)

8.9 Systems

There are many examples today of a total systems concept involving one packaging company acting in partnership with the food manufacturer in an integrated system from the point of packing and processing to the point of sale.

Paper and paperboard based packaging systems for food products implies consideration of:

- the functional needs of packaged preserved foods
- how these needs are met by paper and paperboard based materials and the packaging made from such materials
- packaging machinery
- integration of food processing with packaging, storage and distribution.

One of the best examples of this type of packaging system is the aseptic packaging of milk and juice products in paperboard based liquid packaging.

The term *packaging system* may also be used in a more limited way where a packaging material supplier, working in partnership with a product manufacturer, supplies packaging material, leases the packaging machinery and takes responsibility for technical support and maintenance of the machinery.

8.10 Environmental profile

Paperboard has a low environmental impact in that the main raw material, wood, is naturally sustainable (Fig. 8.22). Wood is derived from trees and in order to grow naturally trees need:

- sun (energy)
- soil
- water
- air (carbon dioxide).

Wood is derived from forests. Forests are essential for the well being of the world environment by:

- reversing the greenhouse effect
- stabilising climate and water levels
- preventing soil erosion
- storing solar energy.

The commercial use of wood for paper and board needs is met by sustainable forest management which:

- ensures replenishment of trees
- provides habitats for animals, plants and insects
- promotes biodiversity

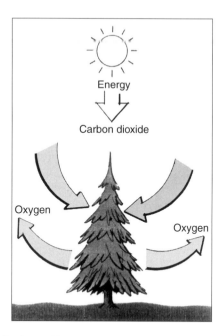

Figure 8.22 Trees grow by the combination of carbon dioxide and water, using energy from the sun. This process, which emits oxygen, is known as photosynthesis (courtesy of Iggesund Paperboard).

- protects watercourses
- preserves landscape
- maintains rural employment
- creates recreation facilities.

Forest management today meets commercial, social and environmental needs, and forests can be independently audited and certificated for environmental performance.

There is understandable public concern at the loss of forests world-wide. It is however important to differentiate between sustainable commercial forestry undertaken by the paper industry and the clearing of forests in the less well developed parts of the world where this is done to meet the needs of land hunger and where wood is the only source of fuel. Over 50% of wood cut annually world-wide is used for fuel and a large amount is used for construction.

The paper and board industry uses 10% of the wood harvested annually. It uses thinnings, the tops of large trees and saw mill waste, i.e. materials which otherwise would become waste (Fig. 8.23). Commercial forestry for the paper industry is leading to an increase in both the land area devoted to forestry and the volume of wood growing in those forests.

PAPER AND PAPERBOARD PACKAGING

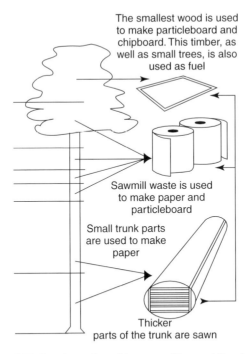

Figure 8.23 How the tree is used (courtesy of Iggesund Paperboard).

A major advantage of paper and paperboard is that it can be recycled as fibre and used to make new paper and paperboard materials (Fig. 8.24).

Pulp recovery from waste paper and board is an example of material recycling and between 40 and 60% of paper and board is recovered in Europe and North America. Commercial and industrial waste paper is relatively easy to collect and systems have been in place for 100 years or so where the driving force was based on commercial viability. In recent times attention has been focussed on domestic or post consumer waste. Systems are being developed to segregate and recover more paper and board from this source.

Packaging accounts for about 10% of all paper and board consumption and many paper and paperboard packaging products are based on recovered paper and board. The infrastructure for recovery is based on merchants and a categorisation of the various types of waste paper and board. Prices of the various grades depend on the fibre quality and the market forces of supply and demand. A quality described as 'clean white shavings arising from mills or printers trimmings' is in quality terms almost as good as virgin pulp and is high priced. Mixed unsorted waste has the lowest price.

Pulp is a world-wide commodity and a mix of recycled and virgin pulp is necessary to meet the overall needs of the market in terms of quality and

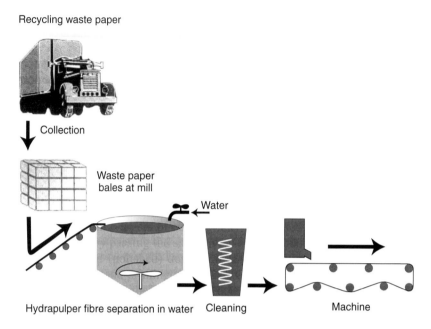

Figure 8.24 Recovery and recycling of waste paper and paperboard (courtesy of Pro Carton).

quantity. Paper consumption is rising in the Far East, especially in China, and much of this demand is being met with recovered fibre. However, 100% recovery and reuse is impossible. This is because some grades by nature of their use cannot be recovered and the fact that multiple recycling causes fibre quality and quantity to deteriorate. Hence there is an ongoing need to provide new virgin fibre.

Energy is another major resource. Pulp made by the chemical separation of fibres, and paper and paperboard mills integrated with such pulping use biomass energy, i.e. the non-cellulose components of wood are used as fuel.

Pulp derived from wood by mechanical means does use fossil fuel but in the past 10 years significant increases in efficiency have been achieved by the use of combined heat and power plants.

Other environmental aspects are:

- *Bleaching (pulp mill effluents were considered harmful).* The problem was due to the use of chlorine gas (elemental chlorine). This process has been replaced and today the by-products are harmless.
- *Use of water and subsequent contamination of water courses by paper and board mill effluent.* Such effluent was a cause of concern in the past because it reduced biological oxygen and hence affected marine life. This has been tackled by effluent treatment within mills so that water emissions are not harmful in this way.

- *Overall water consumption.* More water is now recycled within the mill. Some pulp mills today have no water emissions.
- *Paper based packaging is said to be a waste of resources.* It is necessary today to be able to demonstrate that the amount of packaging used is not excessive and that it is commensurate with the protection needs of the product thereby preventing product wastage.
- Where paper and paperboard materials are not suitable for material recycling they can still be useful as a source of energy, i.e. incineration with energy recovery, or compost.

Overall paper and board is a naturally renewable (sustainable) product which does not pollute the environment in the course of manufacture and use. It is recyclable as material, energy or compost and, if none of these processes is practical, it is biodegradable.

Reference

Hills, R.L. (1988), 'Papermaking in Britain 1488–1988', p. 49.

Further reading

Fundamentals of Packaging Technology by Walter Soroka, revised UK edition by Anne Emblem and Henry Emblem, published by The Institute of Packaging, ISBN 0 9464 6700 5.
Paperboard Reference Manual published by Iggesund Paperboard AB.
The Carton Packaging Fact File published by Pro Carton UK.
The Packaging User's Handbook by F.A. Paine, published by Blackie and Son Ltd. under the authority of The Institute of Packaging, ISBN 0 216 92975 X.

Websites

American Forest and Paper Association, www.afandpa.org.
Confederation of European Paper Industries, www.cepi.org.
Paper Federation of Great Britain, www.paper.org.
Websites of the leading manufacturers of papers and paperboards.

9 Active packaging
Brian P.F. Day

9.1 Introduction

Active packaging refers to the incorporation of certain additives into packaging film or within packaging containers with the aim of maintaining and extending product shelf life (Day, 1989). Packaging may be termed active when it performs some desired role in food preservation other than providing an inert barrier to external conditions (Rooney, 1995; Hotchkiss, 1994). Active packaging includes additives or *freshness enhancers* that are capable of scavenging oxygen; adsorbing carbon dioxide, moisture, ethylene and/or flavour/odour taints; releasing ethanol, sorbates, antioxidants and/or other preservatives; and/or maintaining temperature control. Table 9.1 lists examples of active packaging systems, some of which may offer extended shelf life opportunities for new categories of food products (Day, 1989, 2001; Rooney, 1995).

Active packaging has been used with many food products and is being tested with numerous others. Table 9.1 lists some of the food applications that have benefited from active packaging technology. It should be noted that all food products have a unique deterioration mechanism that must be understood before applying this technology. (Chapter 2 discusses biodeterioration of foods and Chapter 3 discusses food quality issues.) The shelf life of packaged food is dependent on numerous factors such as the intrinsic nature of the food, e.g. acidity (pH), water activity (a_w), nutrient content, occurrence of antimicrobial compounds, redox potential, respiration rate and biological structure, and extrinsic factors, e.g. temperature, relative humidity (RH) and the surrounding gaseous composition. These factors will directly influence the chemical, biochemical, physical and microbiological spoilage mechanisms of individual food products and their achievable shelf lives. By carefully considering all of these factors, it is possible to evaluate existing and developing active packaging technologies and apply them for maintaining the quality and extending the shelf life of different food products (Day, 1989).

Active packaging is not synonymous with intelligent or smart packaging, which refers to packaging that senses and informs (Summers, 1992; Day, 2001). Intelligent packaging devices are capable of sensing and providing information about the function and properties of packaged food and can provide assurances of pack integrity, tamper evidence, product safety and quality, and are being utilised in applications such as product authenticity, anti-theft and

Table 9.1 Selected examples of active packaging systems

Active packaging system	Mechanisms	Food applications
Oxygen scavengers	1. iron based 2. metal/acid 3. metal (e.g. platinum) catalyst 4. ascorbate/metallic salts 5. enzyme based	bread, cakes, cooked rice, biscuits, pizza, pasta, cheese, cured meats and fish, coffee, snack foods, dried foods and beverages
Carbon dioxide scavengers/emitters	1. iron oxide/calcium hydroxide 2. ferrous carbonate/metal halide 3. calcium oxide/activated charcoal 4. ascorbate/sodium bicarbonate	coffee, fresh meats and fish, nuts and other snack food products and sponge cakes
Ethylene scavengers	1. potassium permanganate 2. activated carbon 3. activated clays/zeolites	fruit, vegetables and other horticultural products
Preservative releasers	1. organic acids 2. silver zeolite 3. spice and herb extracts 4. BHA/BHT antioxidants 5. vitamin E antioxidant 6. volatile chlorine dioxide/sulphur dioxide	cereals, meats, fish, bread, cheese, snack foods, fruit and vegetables
Ethanol emitters	1. alcohol spray 2. encapsulated ethanol	pizza crusts, cakes, bread, biscuits, fish and bakery products
Moisture absorbers	1. PVA blanket 2. activated clays and minerals 3. silica gel	fish, meats, poultry, snack foods, cereals, dried foods, sandwiches, fruit and vegetables
Flavour/odour adsorbers	1. cellulose triacetate 2. acetylated paper 3. citric acid 4. ferrous salt/ascorbate 5. activated carbon/clays/zeolites	fruit juices, fried snack foods, fish, cereals, poultry, dairy products and fruit
Temperature control packaging	1. non-woven plastics 2. double walled containers 3. hydrofluorocarbon gas 4. Lime/water 5. ammonium nitrate/water	ready meals, meats, fish, poultry and beverages

product traceability (Summers, 1992; Day, 2001). Intelligent packaging devices include time-temperature indicators, gas sensing dyes, microbial growth indicators, physical shock indicators, and numerous examples of tamper proof, anti-counterfeiting and anti-theft technologies. Information on intelligent packaging technology can be obtained from other reference sources (Summers, 1992; Day, 1994, 2001).

It is not the intention of this chapter to extensively review all active-packaging technologies but rather to describe the different types of devices, the scientific principles behind them, the principal food applications and the food safety and regulatory issues that need to be considered by potential users. The major focus of this chapter is on oxygen scavengers but other active packaging technologies are also discussed and some recent developments highlighted. More detailed information on active packaging can be obtained from the numerous references listed.

9.2 Oxygen scavengers

Oxygen can have considerable detrimental effects on foods. Oxygen scavengers can therefore help maintain food product quality by decreasing food metabolism, reducing oxidative rancidity, inhibiting undesirable oxidation of labile pigments and vitamins, controlling enzymic discolouration and inhibiting the growth of aerobic microorganisms (Day, 1989, 2001; Rooney, 1995).

Oxygen scavengers are by far the most commercially important sub-category of active packaging. The global market for oxygen scavengers was estimated to exceed 10 billion units in Japan, several hundred million in the USA and tens of millions in Europe in 1996 (Anon, 1996; Rooney, 1998). The global value of this market in 1996 was estimated to exceed $200 million and is predicted to top $1 billion by 2002. The recent commercialisation of oxygen scavenging polyethylene terephthalate (PET) bottles, bottle caps and crowns for beers and other beverages has contributed to this growth (Anon, 1996; Rooney, 1998; Day, 2001).

The most well known oxygen scavengers take the form of small sachets containing various iron based powders combined with a suitable catalyst. These chemical systems often react with water supplied by the food to produce a reactive hydrated metallic reducing agent that scavenges oxygen within the food package and irreversibly converts it to a stable oxide. The iron powder is separated from the food by keeping it in a small, highly oxygen permeable sachet that is labelled *Do not eat*. The main advantage of using such oxygen scavengers is that they are capable of reducing oxygen levels to less than 0.01% which is much lower than the typical 0.3–3.0% residual oxygen levels achievable by modified atmosphere packaging (MAP). Oxygen scavengers can be used alone or in combination with MAP. Their use alone eliminates the need for MAP machinery and can increase packaging speeds. However, it is a common commercial practice to remove most of the atmospheric oxygen by MAP and then use a relatively small and inexpensive scavenger to mop up the residual oxygen remaining within the food package (Idol, 1993).

Non-metallic oxygen scavengers have also been developed to alleviate the potential for metallic taints being imparted to food products. The problem of inadvertently setting off in-line metal detectors is also alleviated even though some modern detectors can now be tuned to phase out the scavenger signal whilst retaining high sensitivity for ferrous and non-ferrous metallic contaminants (Anon, 1995). Non-metallic scavengers include those that use organic reducing agents such as ascorbic acid, ascorbate salts or catechol. They also include enzymic oxygen scavenger systems using either glucose oxidase or ethanol oxidase which could be incorporated into sachets, adhesive labels or immobilised onto packaging film surfaces (Hurme, 1996).

Oxygen scavengers were first marketed in Japan in 1976 by the Mitsubishi Gas Chemical Co. Ltd under the trade name Ageless™. Since then, several other Japanese companies including Toppan Printing Co. Ltd and Toyo Seikan Kaisha Ltd have entered the market but Mitsubishi still dominates the oxygen scavenger business in Japan with a market share of 73% (Rooney, 1995). Oxygen scavenger technology has been successful in Japan for a variety of reasons including the acceptance by Japanese consumers of innovative packaging and the hot and humid climate in Japan during the summer months, which is conducive to mould spoilage of food products. In contrast to the Japanese market, the acceptance of oxygen scavengers in North America and Europe has been slow, although several manufacturers and distributors of oxygen scavengers are now established in both these continents and sales have been estimated to be growing at a rate of 20% annually (Rooney, 1995). Table 9.2 lists selected manufacturers and trade names of oxygen scavengers, including some which are still under development or have been suspended because of regulatory controls (Rooney, 1995, 1998; Castle, 1996; Anon, 1997, 1998; Glaskin, 1997).

It should be noted that discrete oxygen scavenging sachets suffer from the disadvantage of possible accidental ingestion of the contents by the consumer and this has hampered their commercial success, particularly in North America and Europe. However, in the last few years, the development of oxygen scavenging adhesive labels that can be applied to the inside of packages and the incorporation of oxygen scavenging materials into laminated trays and plastic films have enhanced and will encourage the commercial acceptance of this technology. For example, Marks & Spencer Ltd were the first UK retailer to use oxygen scavenging adhesive labels for a range of sliced cooked and cured meat and poultry products which are particularly sensitive to deleterious light and oxygen-induced colour changes (Day, 2001). Other UK retailers, distributors and caterers are now using these labels for the above food products as well as for coffee, pizzas, speciality bakery goods and dried food ingredients (Hirst, 1998). Other common food applications for oxygen scavenger labels and sachets include cakes, breads, biscuits, croissants, fresh pastas, cured fish, tea, powdered milk, dried egg, spices, herbs, confectionery and snack food (Day, 2001). In Japan, Toyo Seikan Kaisha Ltd have marketed a laminate containing a ferrous oxygen scavenger which can be thermoformed

Table 9.2 Selected commercial oxygen scavenger systems

Manufacturer	Country	Trade name	Scavenger mechanism	Packaging form
Mitsubishi Gas Chemical Co. Ltd	Japan	Ageless	iron based	sachets and labels
Toppan Printing Co. Ltd	Japan	Freshilizer	iron based	sachets
Toagosei Chem. Industry Co. Ltd	Japan	Vitalon	iron based	sachets
Nippon Soda Co. Ltd	Japan	Seagul	iron based	sachets
Finetec Co. Ltd	Japan	Sanso-Cut	iron based	sachets
Toyo Seikan Kaisha Ltd.	Japan	Oxyguard	iron based	plastic trays
Multisorb Technologies Inc.	USA	FreshMax	iron based	labels
		FreshPax	iron based	labels
		Fresh Pack	iron based	labels
Ciba Speciality Chemicals	USA	Shelf-plus	PET copolyester	plastic film
Chevron Chemicals	USA	N/A	benzyl acrylate	plastic film
W.R. Grace Co. Ltd	USA	PureSeal	ascorbate/ metallic salts	bottle crowns
Food Science Australia/ Visy Industries	Australia	ZERO2	photosensitive dye/organic compound	plastic film
CMB Technologies	France	Oxbar	cobalt catalyst	plastic bottles
Standa Industrie	France	ATCO	iron based	sachets
		Oxycap	iron based	bottle crowns
EMCO Packaging Systems	UK	ATCO	iron based	labels
Johnson Matthey Plc	UK	N/A	platinum group metal catalyst	labels
Bioka Ltd	Finland	Bioka	enzyme based	sachets
Alcoa CSI Europe	UK	O_2-Displacer System	unknown	bottle crowns

into an Oxyguard™ tray which has been used commercially for cooked rice (see Fig. 9.1).

The use of oxygen scavengers for beer, wine and other beverages is potentially a huge market that has only recently begun to be exploited. Iron based label and sachet scavengers cannot be used for beverages or high a_w foods because, when wet, their oxygen scavenging capability is rapidly lost. Instead, various non-metallic reagents and organo-metallic compounds which have an affinity for oxygen have been incorporated into bottle closures, crowns and caps or blended into polymer materials so that oxygen is scavenged from the bottle headspace and any ingressing oxygen is also scavenged. The PureSeal™ oxygen scavenging bottle crowns (marketed by W.R. Grace Co. Ltd and ZapatA Technologies Inc., USA), oxygen scavenging plastic (PET) beer bottles (manufactured by Continental PET Technologies, Toledo, USA) and light

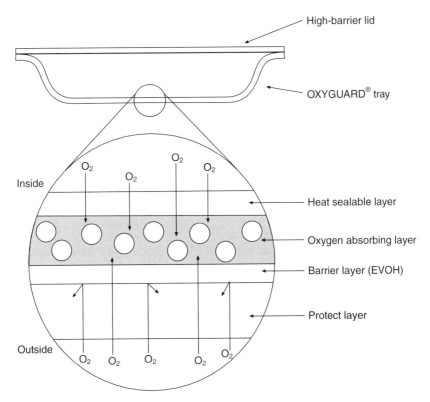

Figure 9.1 Structure of Oxyguard™ tray (courtesy of Toyo Seikan Kaisha Ltd).

activated ZERO2™ oxygen scavenger materials (developed by Food Science Australia, North Ryde, NSW, Australia and now partly owned by Visy Industries, Melbourne, Victoria, Australia) are just three of many oxygen scavenger developments aimed at the beverage market but are also applicable to other food applications (Rooney, 1995, 1998, 2000; Castle, 1996). It should be noted that the speed and capacity of oxygen scavenging plastic films and laminated trays are considerably lower compared with iron based oxygen scavenger sachets or labels (Hirst, 1998).

More detailed information on the technical requirements (i.e. for low, medium and high a_w foods and beverages; speed of reaction; storage temperature; oxygen scavenging capacity and necessary packaging criteria) of the different types of oxygen scavengers can be obtained from Labuza and Breene (1989), Idol (1993), Rooney (1995, 1998, 2000) and Hurme and Ahvenainen (1996).

9.2.1 ZERO2™ oxygen scavenging materials

As a case study, brief details of the ZERO2™ oxygen scavenging development are described here (Rooney, 2000). ZERO2™ is the trade name for a range of oxygen scavenging plastic packaging materials under development, in which the reactive components are activated by means of ultraviolet light or related high-energy processes. The plastics' oxygen scavenging properties remain inactive until activated by an appropriate stimulus and thus can be subjected to conventional extrusion-based converting processes in the manufacture of packaging such as film, sheet, coatings, adhesives, lacquers, bottles, closure liners and can coatings. This patented technology is based on research undertaken at Food Science Australia (North Ryde, NSW, Australia) and is now being developed partly with Visy Industries (Melbourne, Victoria, Australia).

Packaging problems involving the need for oxygen scavenging may be divided into two classes based on the origin of the oxygen that needs to be removed. The headspace and dissolved oxygen is present at the time of closing most packages of foods and beverages. Removal of some or all of this oxygen is required at a rate greater than that of the various food degradation processes that occur in the food. In this case, a headspace scavenger package is required. The oxygen that enters the package by permeation or leakage after closing needs to be removed, preferably before contacting the food. The scavenger required is a chemically enhanced barrier. Prototype ZERO2™ headspace scavenging polymer compositions to meet these two requirements have been synthesised from food-grade commercial polymers and extruded into film on a pilot scale. Oxygen scavenging from the gas phase can be made to occur within minutes at retort temperatures and within several hours to one or two days at room temperature. Oxygen scavenging to very low levels under refrigeration temperatures can require two or more days, as expected when gas diffusion into the polymer is slowed.

Beverages are particularly susceptible to quality degradation due to oxidation or, in some cases, due to microbial growth. Distribution can require shelf lives of up to a year under ambient conditions in some cases, resulting in a need for an enhanced barrier for plastics. The conditions found in liquid paperboard cartons, laminate pouches or multilayer barrier bottles were recently studied (Rooney, 2000) in collaboration with TNO Food Science and Nutrition (Zeist, The Netherlands). Experimental conditions were chosen using sachets of a laminate including a layer of ethylene vinyl alcohol (EVOH) with an experimental ZERO2™ layer on the inside (with an EVOH/polyethylene laminate as control). The test beverage was orange juice and, in the control packs, the dissolved oxygen concentration decreased from 8 to 0 ppm, due to reaction with the ascorbic acid (vitamin C), over one month at 25°C and 75 days at 4°C. At both temperatures, the ZERO2™ laminate removed the oxygen in less than three days and halved the loss of the vitamin C over the storage period of one year. Browning was reduced by one third after one year at 25°C. Use of lower surface-to-volume ratios found in commercial packs would be expected to result in more protection against oxidative deterioration.

Alcoholic beverages, such as beer and white wine, are also susceptible to rapid oxidative degradation. Using ZERO2™ materials, shelf life extensions of at least 33% have been demonstrated for bag-in-box wine and multilayer PET beer bottles. Cheese and processed meats are examples of refrigerated foods that are normally packaged under modified atmospheres. It is the headspace oxygen that severely limits the storage life of these products. Cheese normally requires the presence of carbon dioxide as well as an oxygen level below 1%. Results of packaging in laminates with and without a ZERO2™ layer suggest that the common spoilage moulds can be inhibited completely with little or no carbon dioxide. Sliced smoked ham can be inhibited from discolouring under refrigerated cabinet lighting conditions when the packaging laminate scavenges the initial oxygen concentration of 4% to very low levels. Development of ZERO2™ and competing processes are aimed at inhibiting the widest range of oxygen-mediated food degradation processes. Examples studied so far indicate that some fast degradation reactions can be addressed although there are more challenging cases yet to be investigated (Rooney, 2000).

9.3 Carbon dioxide scavengers/emitters

There are many commercial sachet and label devices that can be used to either scavenge or emit carbon dioxide. The use of carbon dioxide scavengers is particularly applicable for fresh roasted or ground coffees that produce significant volumes of carbon dioxide. Fresh roasted or ground coffees cannot be left unpackaged since they will absorb moisture and oxygen and lose desirable volatile aromas and flavours. However, if coffee is hermetically sealed in packs directly after roasting, the carbon dioxide released will build up within the packs and eventually cause them to burst (Subramaniam, 1998). To circumvent this problem, two solutions are currently used. The first is to use packaging with patented one-way valves that will allow excess carbon dioxide to escape. The second solution is to use a carbon dioxide scavenger or a dual-action oxygen and carbon dioxide scavenger system. A mixture of calcium oxide and activated charcoal has been used in polyethylene coffee pouches to scavenge carbon dioxide but dual-action oxygen and carbon dioxide scavenger sachets and labels are more common and are commercially used for canned and foil pouched coffees in Japan and the USA (Day, 1989; Anon, 1995; Rooney, 1995). These dual-action sachets and labels typically contain iron powder for scavenging oxygen, and calcium hydroxide which scavenges carbon dioxide when it is converted to calcium carbonate under sufficiently high humidity conditions (Rooney, 1995). Commercially available dual-action oxygen and carbon dioxide scavengers are available from Japanese manufacturers, e.g. Mitsubishi Gas Chemical Co. Ltd (Ageless™ type E and Fresh Lock™) and Toppan Printing Co. Ltd (Freshilizer™ type CV).

Carbon dioxide emitting sachet and label devices can either be used alone or combined with an oxygen scavenger. An example of the former is the Verifrais™ package that has been manufactured by SARL Codimer (Paris, France) and used for extending the shelf life of fresh meats and fish. This innovative package consists of a standard MAP tray but has a perforated false bottom under which a porous sachet containing sodium bicarbonate/ascorbate is positioned. When exudate from modified atmosphere packed meat or fish contacts the sachet's contents, carbon dioxide is emitted and this antimicrobial gas can replace the carbon dioxide already absorbed by the fresh food, so avoiding pack collapse (Rooney, 1995).

Pack collapse or the development of a partial vacuum can also be a problem for foods packed with an oxygen scavenger. To overcome this problem, dual-action oxygen scavenger/carbon dioxide emitter sachets and labels have been developed, which absorb oxygen and generate an equal volume of carbon dioxide. These sachets and labels usually contain ferrous carbonate and a metal halide catalyst although non-ferrous variants are available. Commercial manufacturers include Mitsubishi Gas Chemical Co. Ltd (Ageless™ type G), and Multisorb Technologies Inc. (Freshpax™ type M). The main food applications for these dual-action oxygen scavenger/carbon dioxide emitter sachets and labels have been with snack food products, e.g. nuts and sponge cakes (Naito *et al.*, 1991; Rooney, 1995).

9.4 Ethylene scavengers

Ethylene (C_2H_4) is a plant growth regulator which accelerates the respiration rate and subsequent senescence of horticultural products such as fruit, vegetables and flowers. Many of the effects of ethylene are necessary, e.g. induction of flowering in pineapples, colour development in citrus fruits, bananas and tomatoes, stimulation of root production in baby carrots and development of bitter flavour in bulk delivered cucumbers, but in most horticultural situations it is desirable to remove ethylene or to suppress its negative effects. Consequently, much research has been undertaken to incorporate ethylene scavengers into fresh produce packaging and storage areas. Some of this effort has met with commercial success, but much of it has not (Abeles *et al.*, 1992; Rooney, 1995).

Table 9.3 lists selected commercial ethylene scavenger systems. Effective systems utilise potassium permanganate ($KMnO_4$) immobilised on an inert

Table 9.3 Selected commercial ethylene scavenger systems

Manufacturer	Country	Trade name	Scavenger mechanism	Packaging form
Air Repair Products Inc.	USA	N/A	$KMnO_4$	sachets/blankets
Ethylene Control Inc.	USA	N/A	$KMnO_4$	sachets/blankets
Extenda Life Systems	USA	N/A	$KMnO_4$	sachets/blankets
Kes Irrigations Systems	USA	Bio-Kleen	Titanium dioxide catalyst	not known
Sekisui Jushi Ltd	Japan	Neupalon	activated carbon	sachet
Honshu Paper Ltd	Japan	Hatofresh	activated carbon	paper/board
Mitsubishi Gas Chemical Co. Ltd	Japan	Sendo-Mate	activated carbon	sachets
Cho Yang Heung San Co. Ltd	Korea	Orega	activated clays/zeolites	plastic film
Evert-Fresh Corporation	USA	Evert-Fresh	activated zeolites	plastic film
Odja Shoji Co. Ltd	Japan	BO Film	crysburite ceramic	plastic film
PEAKfresh Products Ltd	Australia	PEAKfresh	activated clays/zeolites	plastic film

mineral substrate such as alumina or silica gel. $KMnO_4$ oxidises ethylene to acetate and ethanol and in the process changes colour from purple to brown and hence indicates its remaining ethylene scavenging capacity. $KMnO_4$-based ethylene scavengers are available in sachets to be placed inside produce packages, or inside blankets or tubes that can be placed in produce storage warehouses (Labuza & Breene, 1989; Abeles *et al.*, 1992; Rooney, 1995).

Activated carbon-based scavengers with various metal catalysts can also effectively remove ethylene. They have been used to scavenge ethylene from produce warehouses or incorporated into sachets for inclusion into produce packs or embedded into paper bags or corrugated board boxes for produce storage. A dual-action ethylene scavenger and moisture absorber has been marketed in Japan by Sekisui Jushi Limited. Neupalon™ sachets contain activated carbon, a metal catalyst and silica gel and are capable of scavenging ethylene as well as acting as a moisture absorber (Abeles *et al.*, 1992; Rooney, 1995).

In recent years, numerous produce packaging films and bags have appeared in the market place which are based on the putative ability of certain finely ground minerals to adsorb ethylene and to emit antimicrobial far-infrared radiation. However, little direct evidence for these effects have been published in peer-reviewed scientific journals. Typically, these activated earth-type minerals include clays, pumice, zeolites, coral, ceramics and even Japanese Oya stone. These minerals are embedded or blended into polyethylene film bags which are then used to package fresh produce. Manufacturers of such bags claim extended shelf life for fresh produce partly due to the adsorption of ethylene by the minerals dispersed within the bags. The evidence offered in support of this claim is generally based on the extended shelf life of produce and reduction of headspace ethylene in mineral-filled bags in comparison with common polyethylene bags. However, independent research has shown that the gas permeability of mineral-filled polyethylene bags is much greater and consequently ethylene will diffuse out of these bags much faster, as is also the case for commercially available microperforated film bags. In addition, a more favourable equilibrium modified atmosphere is likely to develop within these bags compared with common polyethylene bags, especially if the produce has a high respiration rate. Therefore, these effects can improve produce shelf life and reduce headspace ethylene independently of any ethylene adsorption. In fact, almost any powdered mineral can confer such effects without relying on expensive Oya stone or other speciality minerals (Abeles *et al.*, 1992; Rooney, 1995).

9.5 Ethanol emitters

The use of ethanol as an antimicrobial agent is well documented. It is particularly effective against mould but can also inhibit the growth of yeasts and bacteria. Ethanol can be sprayed directly onto food products just prior to packaging. Several reports have demonstrated that the mould-free shelf life of bakery products can be significantly extended after spraying with 95% ethanol to give concentrations of 0.5–1.5% (w/w) in the products. However, a more practical and safer method of generating ethanol is through the use of ethanol-emitting films and sachets (Rooney, 1995).

Many applications of ethanol emitting films and sachets have been patented, primarily by Japanese manufacturers. These include Ethicap™, Antimold 102™ and Negamold™ (Freund Industrial Co. Ltd), Oitech™ (Nippon Kayaku Co. Ltd), ET Pack™ (Ueno Seiyaku Co. Ltd) and Ageless™ type SE (Mitsubishi Gas Chemical Co. Ltd). All of these films and sachets contain absorbed or encapsulated ethanol in a carrier material which allows the controlled release of ethanol vapour. For example, Ethicap™ which is the most commercially popular ethanol emitter in Japan, consists of food-grade alcohol (55%) and water (10%) adsorbed onto silicon dioxide powder (35%) and contained in a sachet made of a paper and ethyl vinyl acetate (EVA) copolymer laminate. To mask the odour of alcohol, some sachets contain traces of vanilla or other flavours. The sachets are labelled *Do not eat contents* and include a diagram illustrating this warning. Other ethanol emitters such as Negamould™ and Ageless™ type SE are dual-action sachets which scavenge oxygen as well as emit ethanol vapour (Rooney, 1995).

The size and capacity of the ethanol-emitting sachet used depends on the weight of food, the a_w of the food and the desired shelf life required. When food is packed with an ethanol-emitting sachet, moisture is absorbed by the food, and ethanol vapour is released and diffuses into the package headspace. Ethanol emitters are used extensively in Japan to extend the mould-free shelf life of high ratio cakes and other high moisture bakery products by up to 2000% (Rooney, 1995; Hebeda & Zobel, 1996). Research has also shown that such bakery products packed with ethanol-emitting sachets did not get as hard as the controls and results were better than those using an oxygen scavenger alone to inhibit mould growth. Hence, ethanol vapour also appears to exert an anti-staling effect in addition to its anti-mould properties. Ethanol-emitting sachets are also widely used in Japan for extending the shelf life of semi-moist and dry fish products (Rooney, 1995).

9.6 Preservative releasers

Recently, there has been great interest in the potential use of antimicrobial and antioxidant packaging films which have preservative properties for extending the shelf life of a wide range of food products. As with other categories of active packaging, many patents exist and some antimicrobial and antioxidant films have been marketed but the majority have so far failed to be commercialised because of doubts about their effectiveness, economic factors and/or regulatory constraints (Rooney, 1995).

Some commercial antimicrobial films and materials have been introduced, primarily in Japan. For example, one widely reported product is a synthetic silver zeolite which has been directly incorporated into food contact packaging film. The purpose of the zeolite is, apparently, to allow slow release of antimicrobial silver ions into the surface of food products. Many other synthetic and naturally occurring preservatives have been proposed and/or tested for antimicrobial activity in plastic and edible films (Anon, 1994; Rooney, 1995; Weng *et al.*, 1997; Gray, 2000). These include organic acids, e.g. propionate, benzoate and sorbate, bacteriocins, e.g. nisin, spice and herb extracts, e.g. from rosemary, cloves, horseradish, mustard, cinnamon and thyme, enzymes, e.g. peroxidase, lysozyme and glucose oxidase, chelating agents, e.g. EDTA, inorganic acids, e.g. sulphur dioxide and chlorine dioxide and antifungal agents, e.g. imazalil and benomyl. The major potential food applications for antimicrobial films include meats, fish, bread, cheese, fruit and vegetables. Figure 9.2 illustrates the controlled release of volatile chlorine dioxide that has received GRAS status for food use in the USA (Gray, 2000).

An interesting commercial development in the UK is the recent exclusive marketing of food contact approved Microban™ (Microban International, Huntersville, USA) kitchen products such as chopping boards, dish cloths and bin bags by J. Sainsbury Plc. These Microban™ products contain triclosan, an antibacterial aromatic chloro-organic compound, which is also used in soaps, shampoos, lotions, toothpaste and mouth washes (Goddard, 1995a; Jamieson,

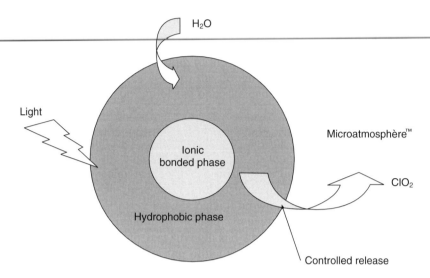

Figure 9.2 Microatmosphère™ chlorine dioxide releasing antimicrobial labels (courtesy of Bernàrd Technologies Inc., USA).

1997; Rubinstein, 2000). Another interesting development is the incorporation of methyl salicylate (a synthetic version of wintergreen oil) into RepelKote™ paperboard boxes by Tenneco Packaging (Lake Forest, Illinois, USA). Methyl salicylate has antimicrobial properties, but RepelKote™ is primarily being marketed as an insect repellent and its main food applications are dried foods which are very susceptible to insect infestations (Barlas, 1998).

Two influences have stimulated interest in the use of antioxidant packaging films. The first of these is the consumer demand for reduced antioxidants and other additives in foods. The second is the interest of plastic manufacturers in using natural approved food antioxidants, e.g. vitamin E for polymer stabilisation instead of synthetic antioxidants developed specifically for plastics (Rooney, 1995). The potential for evaporative migration of antioxidants into foods from packaging films has been extensively researched and commercialised in some instances. For example, the cereal industry in the USA has used this approach for the release of butylated hydroxytoluene (BHT) and butylated hydroxy-anisole (BHA) antioxidants from waxed paper liners into breakfast cereal and snack food products (Labuza & Breene, 1989). Recently there has been much interest in the use of α-tocopherol (vitamin E) as a viable alternative to BHT/BHA-impregnated packaging films (Newcorn, 1997). The use of packaging films incorporating natural vitamin E can confer benefits to both film manufacturers and the food industry. There have been questions raised regarding BHT and BHA's safety, and hence using vitamin E is a safer alternative. Research has shown vitamin E to be as effective as an antioxidant compared with BHT, BHA or other synthetic polymer antioxidants for inhibiting packaging film degradation during film extrusion or blow moulding. Vitamin E is also a safe and effective antioxidant for low to medium a_w cereal and snack food products where development of rancid odours and flavours is often the shelf life limited spoilage mechanism (Labuza & Breene, 1989; Rooney, 1995; Newcorn, 1997).

9.7 Moisture absorbers

Excess moisture is a major cause of food spoilage. Soaking up moisture by using various absorbers or desiccants is very effective in maintaining food quality and extending shelf life by inhibiting microbial growth and moisture related degradation of texture and flavour. Several companies manufacture moisture absorbers in the form of sachets, pads, sheets or blankets. For packaged dried food applications, desiccants such as silica gel, calcium oxide and activated clays and minerals are typically contained within Tyvek™ (Dupont Chemicals, Wilmington, Delaware, USA) tear-resistant permeable plastic sachets. For dual-action purposes, these sachets may also contain activated carbon for odour adsorption or iron powder for oxygen scavenging (Rice, 1994; Rooney, 1995). The use of moisture absorber sachets is common in Japan, where popular foods feature a number of dried products which need to be protected from humidity damage. The use of moisture absorber sachets is also quite common in the USA where the major suppliers include Multisorb Technologies Inc. (Buffalo, New York), United Desiccants (Louisville, Kentucky) and Baltimore Chemicals (Baltimore, Maryland). These sachets are not only utilised for dried snack foods and cereals but also for a wide array of pharmaceutical, electrical and electronic goods. In the UK, Marks & Spencer Plc have used silica gel-based moisture absorber sachets for maintaining the crispness of filled ciabatta bread rolls.

In addition to moisture-absorber sachets for humidity control in packaged dried foods, several companies manufacture moisture-drip absorbent pads, sheets and blankets for liquid water control in high a_w foods such as meats, fish, poultry, fruit and vegetables. Basically, they consist of two layers of a microporous non-woven plastic film, such as polyethylene or polypropylene, between which is placed a superabsorbent polymer which is capable of absorbing up to 500 times its own weight with water. Typical superabsorbent polymers include polyacrylate salts, carboxymethyl cellulose (CMC) and starch copolymers which have a very strong affinity for water. Moisture drip absorber pads are commonly placed under packaged fresh meats, fish and poultry to absorb unsightly tissue drip exudate. Larger sheets and blankets are used for absorption of melted ice from chilled seafood during air freight transportation, or for controlling transpiration of horticultural produce (Rooney,

1995). Commercial moisture absorber sheets, blankets and trays include Toppan Sheet™ (Toppan Printing Co. Ltd, Japan), Thermarite™ (Thermarite Pty Ltd, Australia) and Fresh-R-Pax™ (Maxwell Chase Inc., Douglasville, GA, USA).

Another approach for the control of excess moisture in high a_w foods is to intercept the moisture in the vapour phase. This approach allows food packers or even householders to decrease the water activity on the surface of foods by reducing in-pack RH. This can be done by placing one or more humectants between two layers of water permeable plastic film. For example, the Japanese company Showa Denko Co. Ltd has developed a Pichit™ film which consists of a layer of humectant carbohydrate and propylene glycol sandwiched between two layers of polyvinyl alcohol (PVOH) plastic film. Pichit™ film is marketed for home use in a roll or single sheet form for wrapping fresh meats, fish and poultry. After wrapping in this film, the surface of the food is dehydrated by osmotic pressure, resulting in microbial inhibition and shelf life extension of 3–4 days under chilled storage (Labuza & Breene, 1989; Rooney, 1995). Another example of this approach has been applied in the distribution of horticultural produce. In recent years, microporous sachets of desiccant inorganic salts such as sodium chloride have been used for the distribution of tomatoes in the USA (Rooney, 1995). Yet another example is an innovative fibreboard box which functions as a humidity buffer on its own without relying on a desiccant insert. It consists of an integral water vapour barrier on the inner surface of the fibreboard, a paper-like material bonded to the barrier which acts as a wick, and an unwettable, but highly permeable to water vapour, layer next to the fruit or vegetables. This multi-layered box is able to take up water in the vapour state when the temperature drops and the RH rises. Conversely, when the temperature rises, the multi-layered box can release water vapour back in response to a lowering of the RH (Patterson & Joyce, 1993).

9.8 Flavour/odour adsorbers

The interaction of packaging with food flavours and aromas has long been recognised, especially through the undesirable flavour scalping of desirable food components. For example, the scalping of a considerable proportion of desirable limonene has been demonstrated after only two weeks storage in aseptic packs of orange juice (Rooney, 1995). Commercially, very few active packaging techniques have been used to selectively remove undesirable flavours and taints, but many potential opportunities exist. An example of such an opportunity is the debittering of pasteurised orange juices. Some varieties of orange, such as Navel, are particularly prone to bitter flavours caused by limonin, a tetraterpenoid which is liberated into the juice after orange pressing and subsequent pasteurisation. Processes have been developed for debittering such juices by passing them through columns of cellulose triacetate or nylon beads (Rooney, 1995). A possible active-packaging solution would be to include limonin adsorbers (e.g. cellulose triacetate or acetylated paper) into orange juice packaging material.

Two types of taints amenable to removal by active packaging are amines, which are formed from the breakdown of fish muscle proteins, and aldehydes which are formed from the auto-oxidation of fats and oils. Unpleasant smelling volatile amines, such as trimethylamine, associated with fish protein breakdown are alkaline and can be neutralised by various acidic compounds (Franzetti *et al.*, 2001). In Japan, Anico Co. Ltd have marketed Anico™ bags that are made from film containing a ferrous salt and an organic acid such as citrate or ascorbate. These bags are claimed to oxidise amines as they are adsorbed by the polymer film (Rooney, 1995).

Removal of aldehydes such as hexanal and heptanal from package headspaces is claimed by Dupont's Odour and Taste Control (OTC) technology (Anon, 1996). This technology is based upon a molecular sieve with pore sizes of around five nanometres, and Dupont claim that their OTC removes or neutralises aldehydes although evidence for this is lacking. The claimed food applications for this technology are snack foods, cereals, dairy products, poultry and fish (Anon, 1996). A similar claim of aldehyde removal has been reported (Goddard, 1995b). Swedish company EKA Noble in co-operation with Dutch company Akzo, have developed a range of synthetic aluminosilicate zeolites which, they claim, adsorb odorous gases within their highly porous structure. Their BMH™ powder can be incorporated into packaging materials, especially those that are paper-based, and apparently odorous aldehydes are adsorbed in the pore interstices of the powder (Goddard, 1995b).

9.9 Temperature control packaging

Temperature control active packaging includes the use of innovative insulating materials, self-heating and self-cooling cans. For example, to guard against undue temperature abuse during storage and distribution of chilled foods, special insulating materials have been developed. One such material is Thinsulate™ (3M Company, USA), which is a special non-woven plastic with many air pore spaces. Another approach for maintaining chilled temperatures is to increase the thermal mass of the food package so that it is capable of withstanding temperature rises. The Adenko Company of Japan has developed and marketed a Cool Bowl™ which consists of a double walled PET container in which an insulating gel is deposited in between the walls (Labuza & Breene, 1989).

Self-heating cans and containers have been commercially available for decades and are particularly popular in Japan. Self-heating aluminium and steel cans and containers for sake, coffee, tea and ready meals are heated by an exothermic reaction which occurs when lime and water positioned in the base are mixed. In the UK, Nestlé has recently introduced a range of Nescafé coffees in self-heating insulated cans that use the lime and water exothermic reaction. Self-cooling cans have also been marketed in Japan for raw sake. The endothermic dissolution of ammonium nitrate and chloride in water is used to cool the product. Another self-cooling can that has recently been introduced is the Chill Can™ (The Joseph Company, USA) which relies on a hydrofluorocarbon (HFC) gas refrigerant. The release of HRC gas is triggered by a button set into the can's base and can cool a drink by 10°C in two minutes. However, concerns about the environmental impact of HFC's are likely to curtail the commercial success of the Chill Can™ (Anon, 1997).

9.10 Food safety, consumer acceptability and regulatory issues

At least four types of food safety and regulatory issues related to active packaging of foods need to be addressed. First, any need for food contact approval must be established before any form of active packaging is used. Second, it is important to consider environmental regulations covering active-packaging materials. Third, there may be a need for labelling in cases where active packaging may give rise to consumer confusion. Fourth, it is pertinent to consider the effects of active packaging on the microbial ecology and safety of foods (Rooney, 1995). All of these issues are addressed in an EC funded *Actipack* project which aims to evaluate the safety, effectiveness, economic and environmental impact and consumer acceptance of active and intelligent packaging (De Kruijf, 2000).

Food contact approval will often be required because active packaging may affect foods in two ways. Active packaging substances may migrate into the food or may be removed from it. Migrants may be intended or unintended. Intended migrants include antioxidants, ethanol and antimicrobial preservatives which would require regulatory approval in terms of their identity, concentration and possible toxicology effects. Unintended migrants include various metal compounds which achieve their active purpose inside packaging materials but do not need to, or should not, enter foods. Food additive regulations require identification and quantification of any such unintended migration.

Environmental regulations covering reuse, recycling, identification to assist in recycling or the recovery of energy from active packaging materials need to be addressed on a case-by-case basis. European Union companies using active packaging as well as other packaging need to meet the requirements of the

Packaging Waste Directive (1994) and consider the environmental impact of their packaging operations.

Food labelling is currently required to reduce the risk of consumers ingesting the contents of oxygen scavenger sachets or other in-pack active-packaging devices. Some active packages may look different from their passive counterparts. Therefore it may be advisable to use appropriate labelling to explain this difference to the consumer even in the absence of regulations.

Finally, it is very important for food manufacturers using certain type of active packaging to consider the effects this will have on the microbial ecology and safety of foods. For example, removing all the oxygen from within the packs of high a_w chilled perishable food products may stimulate the growth of anaerobic pathogenic bacteria such as *Clostridium botulinum*. Specific guidance is available to minimise the microbial safety risks of foods packed under reduced oxygen atmospheres (Betts, 1996). Regarding the use of antimicrobial films, it is important to consider what spectrum of microorganisms will be inhibited. Antimicrobial films which only inhibit spoilage microorganisms without affecting the growth of pathogenic bacteria will raise food safety concerns.

In the USA, Japan and Australia, active packaging concepts are already being successfully applied. In Europe, the development and application of active packaging is limited because of legislative restrictions, fear of consumer resistance, lack of knowledge about effectiveness and economic and environmental impact of concepts (Vermeiren *et al.*, 1999). No specific regulations exist on the use of active packaging in Europe. Active packaging is subjected to traditional packaging legislation, which requires that compounds are registered on positive lists and that the overall and specific migration limits are respected. This is more or less contradictory to the concept of some active packaging systems in which packaging releases substances to extend shelf life or improve quality (De Kruijf, 2000). The food industry's main concern about introducing active components to packaging seems to be that consumers may consider the components harmful and may not accept them. In Finland, a consumer survey conducted in order to determine consumer attitudes towards oxygen scavengers revealed that the new concepts would be accepted if consumers are well informed by using reliable information channels (Ahvenainen & Hurme, 1997). More information is needed on the chemical, microbiological and physiological effects of various active concepts on the packaged food, i.e. in regard to its quality and safety. So far, research has mainly concentrated on the development of various methods and their testing in a model system, but not so much on functioning in food preservation with real food products. Furthermore, the benefits of active packaging need to be considered in a holistic approach to environmental impact assessment. The environmental effect of plastics-based active packaging will vary with the nature of the product/package combination, and additional additives need to be evaluated for their environmental impact (Vermeiren *et al.*, 1999).

9.11 Conclusions

Active packaging is an emerging and exciting area of food technology that can confer many preservation benefits on a wide range of food products. Active packaging is a technology developing a new trust because of recent advances in packaging, material science, biotechnology and new consumer demands. The objectives of this technology are to maintain sensory quality and extend the shelf life of foods whilst at the same time maintaining nutritional quality and ensuring microbial safety.

Oxygen scavengers are by far the most commercially important subcategory of active packaging and the market has been growing steadily for the last ten years. It is predicted that the recent introduction of oxygen scavenging films and bottle caps will further help to stimulate the market in future years and the unit costs of oxygen scavenging technology will drop. Other active packaging technologies are also predicted to be us+ed more in the future, particularly carbon dioxide scavengers and emitters, moisture absorbers and temperature control packaging. Food safety and regulatory issues in the EU are likely to restrict the use of certain preservative releasers and flavour/odour adsorber active packaging technologies. Nevertheless, the use of active packaging is becoming increasingly popular and many new opportunities in the food and non-food industries will open up for utilising this technology in the future.

References

Abeles, F.B., Morgan, P.W. and Saltveit, M.E. (1992) *Ethylene in Plant Biology*, Academic Press Ltd, London, UK.
Ahvenainen, R. and Hurme, E. (1997) Active and smart packaging for meeting consumer demands for quality and safety. *Food Additives Contaminants*, **14**, 753–763.
Anon (1994) Fresh produce is keen as mustard to last longer. *Packaging Week*, **10**(21), 6.
Anon (1995) Scavenger solution. *Packaging News*, December edn, p. 20.
Anon (1995) Pursuit of freshness creates packaging opportunities. *Japanese Packaging News*, **12**, 14–15.
Anon (1996) Odour Eater. *Packaging News*, August edn, 3.
Anon (1996) Oxygen absorbing packaging materials near market debuts. *Packaging Strategies Supplement*, January 31 edn, Packaging Strategies Inc., West Chester, PA, USA.
Anon (1997) Labels help cooked meats to keep their colour. *Packaging Week*, **12**(40), 4.
Anon (1997) Things get hot for cool can of the year. *Packaging News*, July edn, 1.
Barlas, S. (1998) Packagers tell insects: 'Stop bugging us!'. *Packaging World*, **5**(6), 31.
Betts, G.D. (ed.) (1996) *Code of Practice for the Manufacture of Vacuum and Modified Atmosphere Packaged Chilled Foods with Particular Regards to the Risks of Botulism*. Guideline No. 11, Campden & Chorleywood Food Research Association, Chipping Campden, Glos., UK.
Castle, D. (1996) Polymer advance doubles shelf-life. *Packaging Week*, **12**(17), 1.
Day, B.P.F. (1989) Extension of shelf-life of chilled foods. *European Food and Drink Review*, **4**, 47–56.

Day, B.P.F. (1994) Intelligent packaging of foods, in *New Technologies Bulletin No. 10*, Campden & Chorleywood Food Research Association, Chipping Campden, Glos., UK, pp. 1–7.

Day, B.P.F. (2000) *Consumer Acceptability of Active and Intelligent Packaging*. Proceedings of the 'Active & Intelligent Packaging' conference, TNO, Zeist, The Netherlands, 18 April.

Day, B.P.F. (2001) Active packaging – a fresh approach. *Brand$^©$ – The Journal of Brand Technology*, **1**(1), 32–41.

De Kruijf, N. (2000) Objectives, tasks and results from the EC FAIR 'Actipak' research project, in *Proceedings of the 'International Conference on Active and Intelligent Packaging'*, Campden & Chorleywood Food Research Association, Chipping Campden, Glos., UK, 7–8 September.

Franzetti, L., Martinoli, S., Piergiovanni, L. and Galli, A. (2001) Influence of active packaging on the shelf-life of minimally processed fish products in a modified atmosphere. *Packaging Technology and Science*, **14**(6), 267–274.

Glaskin, M. (1997) Plastic bag keeps bread fresh for three years. *Financial Times*, 29 June, p. 13.

Goddard, R. (1995a) Beating bacteria. *Packaging Week*, **11**(10), 22.

Goddard, R. (1995b) Dispersing the scent. *Packaging Week*, **10**(30), 28.

Gray, P.N. (2000) Generation of active Microatmosphère™ environments from and in packages, in *Proceedings of the 'International Conference on Active and Intelligent Packaging'*, Campden & Chorleywood Food Research Association, Chipping Campden, Glos., UK, 7–8 September.

Hebeda, R.E. and Zobel, H.F. (eds) (1996) *Baked Goods Freshness – Technology, Evaluation and Inhibition of Staling*, Marcel Dekker Inc., New York, USA.

Hirst, J. (1998) *Personal Communication*, EMCO Packaging Systems Ltd, Worth, Kent, UK.

Hotchkiss, J. (1994) Recent research in MAP and active packaging systems, in *Abstracts of the 27th Annual Convention*, Australian Institute of Food Science and Technology, Canberra, Australia.

Hurme, E. and Ahvenainen, R. (1996) Active and smart packaging of ready-made foods, in *Proceedings of the International Symposium on Minimal Processing and Ready Made Foods*, SIK, Göteburg, Sweden, April edn, pp. 4–7.

Idol, R. (1993) Oxygen scavenging: Top marks. *Packaging Week*, **9**(16), 17–19.

Jamieson, D. (1997) Sainsbury's to set new standards in food hygiene. *Journal of Sainsbury plc Press Release*, 14th May, London, UK.

Labuza, T.P. and Breene, W.M. (1989) Applications of active packaging for improvement of shelf-life and nutritional quality of fresh and extended shelf-life foods. *Journal of Food Processing and Preservation*, **13**, 1–69.

Naito, S., Okada, Y. and Yamaguchi, N. (1991) Studies on the behaviour of microorganisms in sponge cake during anaerobic storage. *Packaging Technology and Science*, **4**, 4333–4344.

Newcorn, D. (1997) Not just for breakfast anymore. *Packaging World*, **4**(3), 23–24.

Patterson, B.D. and Joyce, D.C. (1993) *A Package Allowing Cooling and Preservation of Horticultural Produce without Condensation or Desiccants*, International Patent Application PCT/AU93/00398.

Rice, J. (1994) Fighting moisture – desiccant sachets poised for expanded usage. *Food Processing*, **63**(8), 46–48.

Rooney, M.L. (ed.) (1995) *Active Food Packaging*, Chapman & Hall, London, UK.

Rooney, M.L. (1998) Oxygen scavenging plastics for retention of food quality, in *Proceedings of Conference on 'Advances in Plastics – Materials and Processing Technology for Packaging'*, Pira International, Leatherhead, Surrey, UK, 25 February.

Rooney, M.L. (2000) Applications of ZERO2™ oxygen scavenging films for food and beverage products, in *Proceedings of the 'International Conference on Active and Intelligent Packaging'*, Campden & Chorleywood Food Research Association, Chipping Campden, Glos., UK, 7–8 September.

Rubinstein, W.S. (2000) Microban™ antibacterial protection for the food industry, in *Proceedings of the 'International Conference on Active and Intelligent Packaging'*, Campden & Chorleywood Food Research Association, Chipping Campden, Glos., UK, 7–8 September.

Subramaniam, P.J. (1998) Dairy foods, multi-component products, dried foods and beverages, in *Principles and Applications of Modified Atmosphere Packaging of Foods* (ed. B.A. Blakistone), 2nd edn, Blackie Academic & Professional, London, UK, pp. 158–193.

Summers, L. (1992) *Intelligent Packaging*, Centre for Exploitation of Science and Technology, London, UK.

Vermeiren, L., Devlieghere, F., van Beest, M., deKruijf, N. and Debevere, J. (1999) Developments in the active packaging of foods. *Trends in Food Science & Technology*, **10**, 77–86.

Weng, Y.M., Chen, M.J. and Chen, W. (1997) Benzoyl chloride modified ionomer films as antimicrobial food packaging materials. *International Journal of Food Science and Technology*, **32**, 229–234.

10 Modified atmosphere packaging
Michael Mullan and Derek McDowell

SECTION A MAP GASES, PACKAGING MATERIALS
AND EQUIPMENT

10.A1 Introduction

The normal gaseous composition of air is nitrogen (N_2) 78.08% (volume per volume will be used throughout this chapter), oxygen (O_2) 20.96% and carbon dioxide (CO_2) 0.03%, together with variable concentrations of water vapour and traces of inert or noble gases. Many foods spoil rapidly in air due to moisture loss or uptake, reaction with oxygen and the growth of aerobic microorganisms, i.e. bacteria and moulds. Microbial growth results in changes in texture, colour, flavour and nutritional value of the food. These changes can render food unpalatable and potentially unsafe for human consumption. Storage of foods in a modified gaseous atmosphere can maintain quality and extend product shelf life, by slowing chemical and biochemical deteriorative reactions and by slowing (or in some instances preventing) the growth of spoilage organisms.

Modified atmosphere packaging (MAP) is defined as 'the packaging of a perishable product in an atmosphere which has been modified so that its composition is other than that of air' (Hintlian & Hotchkiss, 1986). Whereas controlled atmosphere storage (CAS) involves maintaining a fixed concentration of gases surrounding the product by careful monitoring and addition of gases, the gaseous composition of fresh MAP foods is constantly changing due to chemical reactions and microbial activity. Gas exchange between the pack headspace and the external environment may also occur as a result of permeation across the package material.

Packing foods in a modified atmosphere can offer extended shelf life and improved product presentation in a convenient container, making the product more attractive to the retail customer. However, MAP cannot improve the quality of a poor quality food product. It is therefore essential that the food is of the highest quality prior to packing in order to optimise the benefits of modifying the pack atmosphere. Good hygiene practices and temperature control throughout the chill-chain for perishable products are required to maintain the quality benefits and extended shelf life of MAP foods.

10.A1.1 Historical development

The first commercial applications of the use of modified gas atmospheres were for CAS of fruits and vegetables. Fresh carcass meat was exported from New Zealand and Australia under CAS in the early 1930s. Early developments were generally for storage and transportation of bulk foods. Scientific investigations on the effect of gases on extending the shelf life of foods were conducted in 1930 on fresh meat. Killefer (1930) reported a doubling of the shelf life of refrigerated pork and lamb when these meats were stored in an atmosphere of 100% CO_2. The earliest published research on poultry products was conducted in the 1930s. Fresh poultry was stored in an atmosphere of 100% CO_2 which was found to considerably extend shelf life.

Ogilvy and Ayres (1951) conducted studies on the effect of enriched CO_2 atmospheres on the shelf life of chicken portions. They reported an increasing effect on the shelf life as the CO_2 concentration of the storage atmosphere increased up to 25% CO_2.

Commercial retailing of fresh meat in MAP tray systems was introduced in the early 1970s. European meat processing and packaging developed during the 1980s with centralised production of MAP meat in consumer packs for distribution to retail outlets. In the past few years, there has been a considerable increase in the range of foods packed in modified atmospheres for retail sale including meat, poultry, fish, bacon, bread, cakes, crisps, cheese and salad vegetables. UK retail sales of products packed under MAP grew from approximately 2 billion packs in the mid 1990s to 2.8 billion packs in 1998. Carcass meat and cooked meat and meat products accounted for 29% and 15% of the total volume of MAP retail foods (Anon, 1999).

10.A2 Gaseous environment

10.A2.1 Gases used in MAP

The three main gases used in MAP are O_2, CO_2 and N_2. The choice of gas is totally dependent upon the food product being packed. Used singly or in combination, these gases are commonly used to balance safe shelf life extension with optimal organoleptic properties of the food. Noble or *inert* gases such as argon are in commercial use for products such as coffee and snack products, however, the literature on their application and benefits is limited. Experimental use of carbon monoxide (CO) and sulphur dioxide (SO_2) has also been reported.

10.A2.1.1 Carbon dioxide

Carbon dioxide (CO_2) is a colourless gas with a slight pungent odour at very high concentrations. It is an asphyxiant and slightly corrosive in the presence of moisture. CO_2 dissolves readily in water (1.57 g kg^{-1} at 100 kPa, 20°C) to produce carbonic acid (H_2CO_3) that increases the acidity of the solution and reduces the pH. This gas is also soluble in lipids and some other organic compounds. The solubility of CO_2 increases with decreasing temperature. For this reason, the antimicrobial activity of CO_2 is markedly greater at temperatures below 10°C than at 15°C or higher. This has significant implications for MAP of foods, as will be discussed later. The high solubility of CO_2 can result in pack collapse due to the reduction of headspace volume. In some MAP applications, pack collapse is favoured, for example in flow wrapped cheese for retail sale.

10.A2.1.2 *Oxygen*
Oxygen (O_2) is a colourless, odourless gas that is highly reactive and supports combustion. It has a low solubility in water (0.040 g kg^{-1} at 100 kPa, 20°C). Oxygen promotes several types of deteriorative reactions in foods including fat oxidation, browning reactions and pigment oxidation. Most of the common spoilage bacteria and fungi require O_2 for growth. Therefore, to increase the shelf life of foods, the pack atmosphere should contain a low concentration of residual O_2. It should be noted that in some foods a low concentration of O_2 can result in quality and safety problems (for example, unfavourable colour changes in red meat pigments, senescence in fruits and vegetables and growth of food poisoning bacteria), and this must be taken into account when selecting the gaseous composition for a packaged food.

10.A2.1.3 *Nitrogen*
Nitrogen (N_2) is a relatively un-reactive gas with no odour, taste or colour. It has a lower density than air, non-flammable and has a low solubility in water (0.018 g kg^{-1} at 100 kPa, 20°C) and other food constituents. Nitrogen does not support the growth of aerobic microbes and therefore inhibits the growth of aerobic spoilage but does not prevent the growth of anaerobic bacteria. The low solubility of N_2 in foods can be used to prevent pack collapse by including sufficient N_2 in the gas mix to balance the volume decrease due to CO_2 going into solution.

10.A2.1.4 *Carbon monoxide*
Carbon monoxide (CO) is a colourless, tasteless and odourless gas that is highly reactive and very flammable. It has a low solubility in water but is relatively soluble in some organic solvents. CO has been studied in the MAP of meat and has been licensed for use in the USA to prevent browning in packed lettuce. Commercial application has been limited because of its toxicity and the formation of potentially explosive mixtures with air.

10.A2.1.5 Noble gases

The noble gases are a family of elements characterised by their lack of reactivity and include helium (He), argon (Ar), xenon (Xe) and neon (Ne). These gases are being used in a number of food applications now, e.g. potato-based snack products. While from a scientific perspective it is difficult to see how the use of noble gases would offer any preservation advantages compared with N_2 they are nevertheless being used. This would suggest that there may be, as yet unpublished, advantages for their use.

10.A2.2 Effect of the gaseous environment on the activity of bacteria, yeasts and moulds

Foods can contain a wide range of microorganisms including bacteria and their spores, yeasts, moulds, protozoa and viruses. While the packaging technologist will generally be concerned with preventing the growth of bacteria, yeasts and moulds in foods, one should be aware that certain pathogenic microorganisms, while not growing in the food, may survive during the shelf life period and cause food poisoning or disease in consumers. This section is concerned with the major microbial groups that can be controlled or affected by MAP.

10.A2.2.1 Effect of oxygen

Bacteria, yeasts and moulds have different respiratory and metabolic needs and can be grouped according to their O_2 needs (Table 10.1).

Aerobes. They require O_2 for growth and include the ubiquitous Gram-negative spoilage bacteria belonging to the *Pseudomonas* genus. This grouping also includes certain pathogenic bacteria such as *Vibrio parahaemoly-*

Table 10.1 Oxygen requirements of some microorganisms of relevance in modified atmosphere packaging

Group	Spoilage organisms	Pathogens
Aerobes	*Micrococcus* sp.	*Bacillus cereus*
	Moulds e.g. *Botrytis cinerea*	*Yersinia enterocolitica*
	Pseudomonas sp.	*Vibrio parahaemolyticus*
		Camplobacter jejuni
Microaerophiles	*Lactobacillus* sp.	*Listeria monocytogenes*
	Bacillus spp.	*Aeromonas hydrophilia*
	Enterobacteriaceae	*Escherichia coli*
Facultative anaerobes	*Brocothrix thermosphacta*	*Salmonella* spp.
	Shewanella putrefaciens	*Staphylococcus* spp.
	Yeasts	*Vibrio* sp.
Anaerobes	*Clostridium sporogenes*	*Clostridium perfringens*
	Clostridium tyrobutyricum	*Clostridium botulinum*

ticus. Note that some other *Vibrio* species are classified as facultative aerobes.

Microaerophiles. They grow under low concentrations of O_2. Thus, an environment low in O_2 may be selective for some important pathogens including *Camplobacter jejuni* and *Listeria monocyocytogenes*. Some microaerophilic bacteria, e.g. *Lactobacillus* species, may also require increased levels of CO_2 under low oxygen conditions for optimum growth.

Facultative anaerobes. They generally grow better in O_2 but are also able to grow without it. These include various important genera from the Enterobacteriacaeae including pathogenic organisms such as *Escherichia coli*, S*almonella* and *Shigella* species, *Staphylococcus aureus*, *Listeria monocytogenes*, *Brochothrix* species, *Vibrio* species, fermentative yeasts and some *Bacillus* species. *Aeromonas hydrophilia* is a new and emerging pathogen that appears to be particularly associated with fish and fish products. Many strains are psychrotrophic and some may grow between 3 and 5°C.

Anaerobes. They are inhibited or killed by the presence of O_2, e.g. the pathogenic bacterium *Clostridium botulinum*. The removal of O_2, for example in vacuum packaging, will restrict the growth of aerobic spoilage and pathogenic bacteria and therefore extend shelf life. However, as indicated above, there are other microorganisms including the pathogens *E. coli* and *A. hydrophilia* capable of growth under these conditions.

10.A2.2.2 *Effect of carbon dioxide*

The antibacterial properties of CO_2 have been known for some time (Valley & Rettger, 1927). More recent work has shown that CO_2 is effective against psychrotrophs (King & Nagel, 1975) and has potential for extending the shelf life of food stored at low temperatures.

There are several theories regarding the actual mechanism of CO_2 action. In general, CO_2 increases the lag phase and generation time of microorganisms, and this effect, as would be expected, is enhanced at lower temperatures. There appears to be an array of antimicrobial mechanisms including CO_2 lowering pH, inhibition of succinic oxidase at CO_2 concentrations greater than 10%, inhibition of certain decarboxylation enzymes and disruption of the cell membrane (Valley & Rettger, 1927; King & Nagel, 1975; Gill & Tan, 1979; Enfors & Molin, 1981). The area has also been reviewed by several workers including Daniels *et al.* (1985).

The sensitivity of selected spoilage and pathogenic bacteria to CO_2 is shown in Table 10.2. In general, the growth of Gram-negative bacteria is inhibited much more than that of Gram-positive bacteria. As indicated in Section 10.A2.1.1, the effects of CO_2 are markedly temperature dependent, and it is therefore imperative that the integrity of temperature control across the supply chain be

Table 10.2 Sensitivity of microorganisms relevant to modified atmosphere packaging to carbon dioxide

Inhibited by CO_2	CO_2 has little or no effect on growth	Growth is stimulated by CO_2
Pseudomonas spp.	*Enterococcus* spp.	*Lactobacillus* spp.
Aeromonas spp.	*Brochothrix* spp.	*Clostridium botulinum**
Bacillus spp.	*Lactobacillus* spp.	
Moulds including *Botrytis cinerea*	*Clostridium* spp.	
Enterobacteriaceae including *E. coli*	*Listeria monocytogenes*	
Staphylococcus aureus	*Aeromonas hydrophilia*	
Yersinia enterocolitica		

Effects will depend on the storage temperature and concentration of gas. For some microorganisms CO_2 concentrations close to 50% (v/v) and temperatures <10°C are required for significant effects. Growth inhibition may also be strain specific.
* Spore germination by *C. botulinum* may be enhanced in a CO_2 enriched environment (Eklund, 1982).

maintained in order to protect the health of the consumer. Of some concern is the observation that germination of spores of *C. botulinum* may be stimulated by CO_2 (Eklund, 1982).

CO_2, particularly at low temperatures, is soluble in water and lipids, and adjustment for adsorption is required. A high concentration of CO_2 can lead to defects, e.g. increased drip in fresh meats, and to container collapse. The latter can occur where CO_2 is the major gas present, and where the gas goes into solution in the water and lipid phases of the product. To counteract this effect, an insoluble gas such as nitrogen may be added to the gas mix. When CO_2 is required to control the bacterial and mould growth, a minimum of 20% is generally used. Optimal levels appear to be in the region of 20–30%. However, concentrations of 100% CO_2 may be used in bulk packs of meat and poultry.

10.A2.2.3 Effect of nitrogen

Nitrogen is a relatively un-reactive gas. It is used to displace air and, in particular, O_2 from MAP. Since air and consequently O_2 have been removed, growth of aerobic spoilage organisms is inhibited or stopped. It is also used to balance gas pressure inside packs, so as to prevent the collapse of packs containing high moisture and fat-containing foods, e.g. meat. Because of the solubility of CO_2 in water and fat, these foods tend to absorb CO_2 from the pack atmosphere.

10.A2.3 Effect of the gaseous environment on the chemical, biochemical and physical properties of foods

Food spoilage can also be caused by chemical and biochemical, including enzyme-catalysed, reactions in food. The packaging technologist should have an awareness of these effects and understand the extent to which modified atmospheres can mitigate them.

Of the gases involved in MAP, O_2, because of its reactivity, has been extensively studied. Because of the significance of O_2, this section will largely be concerned with the influence of this gas. However, CO_2, and to a lesser extent CO and ethylene (C_2H_4), have also been investigated. While reference to C_2H_4 is made in Section 10.B5 of this chapter, it is discussed in more detail in Chapter 9 of this book.

10.A2.3.1 Effect of oxygen
Apart from its effect on microorganisms, O_2 can promote oxidation of lipids, influence the colour of some food pigments, contribute to enzymic browning and promote off-flavours in some foods. It is important to note that the inclusion of O_2 in a modified atmosphere environment has the potential to have positive and or negative effects on product quality. The resultant effect is largely product dependent.

Lipid oxidation
Lipid or fat oxidation is often called oxidative rancidity and is promoted by O_2. Oxidative rancidity is a major cause of food spoilage. The reaction of O_2 with unsaturated fatty acids in fat-containing foods is a major cause of deterioration of fats or fat-containing foods. Oxidation of unsaturated fat is referred to as autoxidation, since the rate of oxidation increases as the reaction proceeds. Hydroperoxides are the predominant initial reaction products of fatty acids with oxygen. Subsequent reactions control both the rate of reaction and the nature of the products formed. Some of these products, such as acids and aldehydes, are largely responsible for the off-flavour and off-odour characteristics of rancid foods. Removal of O_2 and its replacement with N_2 or CO_2 or mixtures thereof can inhibit the development of rancidity.

Pigment colour in meat
There are three major pigments in meat, oxymyoglobin, myoglobin and metmyoglobin. Consumers value the red colour (oxymyoglobin) of fresh meat as opposed to the purple colour of myoglobin. The colour cycle in fresh meat is reversible and dynamic, with the three pigments, oxymyoglobin (red), myoglobin (purple) and metmyoglobin (brown), being constantly formed and reformed. Brown metmyoglobin, the oxidised or ferric form of the pigment, cannot bind O_2. The purple myoglobin, in the presence of O_2, may be oxygenated to the bright red pigment oxymyoglobin, producing the familiar *bloom* of fresh meat, or it may be oxidised to metmyoglobin, producing the undesirable brown colour of less acceptable fresh meat. Whether the conversion of myoglobin to oxymyoglobin or metmyoglobin is favoured depends on O_2 concentration. Under low O_2 environments, the reduced myoglobin is oxidised

to the undesirable brown metmyoglobin pigment. Conversely, high O_2 environments favour the formation of oxymyoglobin.

The red colour of raw cured meat products is due to nitrosylmyoglobin, which is formed by the reaction of myoglobin with nitric oxide (NO). During heating, red nitrosylmyoglobin is converted to pink denatured nitrosohemochrome. The red/pink colours of raw and cooked cured meat products are unstable in air and in the light.

Oxygen and light cause the dissociation of NO from the cured meat pigments, resulting in brown/grey discoloration. Hence, MAP under low O_2 levels, and in opaque packages, greatly improves the desirable red/pink colour stability of cured meat products. The use of in-pack O_2 scavenging systems can reduce and maintain residual oxygen at a level that further extends the shelf life of cured meat products. The application of O_2 scavenging technology in food packaging is discussed in Chapter 9.

Photo-oxidation of chlorophyll
The green colour of chlorophyll changes to brown/grey when oxidised to pheophytin. This is undesirable, e.g. green pasta changing in colour to brown/grey. The photo-oxidation of chlorophyll and loss of desirable green colour can be significantly reduced by MAP under low O_2 levels and in opaque packages.

Oxidative off-flavours
Oxidative off-flavours and off-odours can be caused by numerous oxidative reactions in food and drink products. Oxidative warmed-over flavour is a characteristic off-flavour primarily associated with cooked meats and poultry. Commercially, this affects mainly the chilled ready meals and other cook-chill products. In cooked meats and poultry held at chilled storage temperatures, this stale, oxidised flavour may become apparent within a short time.

Meats, fish, poultry, liquid food, beverage and dairy products, for example, are highly susceptible to oxidative processes which can initiate a chain of reactions resulting in flavour impairment. This can occur relatively quickly. MAP under low O_2 levels can delay the onset of oxidative off-flavours.

10.A2.3.2 Effects of other MAP gases
Nitrogen is un-reactive and has no direct effect on the chemical and biochemical properties of foods. Because of the high solubility of CO_2 and its reaction with water to form carbonic acid, there is potential for some adverse effects on particular foods. These are probably due to the production of localised areas of low pH on or near the food surface. These effects if they do occur, and there is debate whether they occur in practice, may result in the loss of *bloom* in some meats for example. The mechanism is likely to be associated with pH-induced protein changes including denaturation and other changes in conformation,

resulting in atypical values for light absorption and reflection from the product surface. See also Section 10.B2 of this chapter.

Carbon monoxide can combine with myogloblin to form the bright red compound carboxymyogloblin that is similar in colour to oxymyogloblin. This compound is much more stable than oxymyogloblin and is one of the reasons why CO is toxic. CO also retards fat oxidation and the formation of metmyogloblin. For further discussion of meat pigments, refer to Section 10.B3 of this chapter. Currently, CO is not approved for use in MAP.

10.A2.4 Physical spoilage

Physical or physicochemical changes in food products can cause spoilage, thereby limiting the shelf life. For example, moisture loss in cut fresh vegetables causes wilting and reduction in textural crispness, moisture migration from the filling to the pastry of bakery products can cause a soggy/sticky consistency and syneresis of dairy products results in an undesirable and unsightly separated aqueous layer. Also, some food products are sensitive to chilled temperatures. For example, certain whole tropical fruits are susceptible to chilling injury when exposed to temperatures in the range 0–10°C. Chilling injury causes loss of quality through poor ripening, pitting of the epidermal cells, rotting and development of off-flavours. With the exception of preventing moisture loss and moisture uptake, MAP does not generally directly affect physical spoilage.

10.A3 Packaging materials

Selection of the most appropriate packaging materials is essential to maintain the quality and safety of MAP foods. Flexible and semi-rigid plastics and plastic laminates are the most common materials used for MAP foods. Plastic materials account for approximately one-third of the total materials demand for food packaging applications, and their use is forecast to grow.

Relative ease of forming, light weight, good clarity, heat sealing and strength are some of the properties of plastics that make them suitable as food packaging materials. Advances in polymer processing have enabled the development of plastics that are better suited to particular food packaging applications. However, no single plastic possesses the properties that make it suited to all food packaging applications.

Plastic packaging materials may consist of a monolayer formed from a single plastic, but most, if not all, MAP films are multilayer structures formed from several layers of different plastics. Using coextrusion, lamination or coating technologies, it is possible to combine different types of plastic to form films, sheets or rigid packs. By carefully selecting each component plastic, it is possible to design a material which possesses the key properties of packaging importance to best match the requirements of the product/package system.

Plastics packaging for MAP applications is most commonly found in the form of flexible films for bags, pouches, pillow packs and top webs or as rigid and semi-rigid structures for base trays, dishes, cups and tubs. Commonly used plastic flexible laminates are produced from polyethylene (PE), polypropylene (PP), polyamide (nylons), polyethylene terephthalate (PET), polyvinyl chloride (PVC), polyvinylidene chloride (PVdC) and ethylene vinyl alcohol (EVOH). Rigid and semi-rigid structures are commonly produced from PP, PET, unplasticised PVC and expanded polystyrene.

10.A3.1 Main plastics used in MAP

The following section provides a brief overview of the commonly used plastics for MAP applications. More information on these including properties of packaging importance, definitions and terminology can be found in Chapter 7.

10.A3.1.1 Ethylene vinyl alcohol (EVOH)

Polyvinyl alcohol (PVOH) is an excellent gas barrier provided it is dry. In the presence of moisture, PVOH absorbs water, causing the plastic to swell and become plasticised. In this condition, the gas barrier properties of PVOH are greatly reduced. In order to provide greater polymer stability for commercial use, PVOH is copolymerised with ethylene to produce EVOH. The gas barrier properties of EVOH are less than those of PVOH when dry, but EVOH is less sensitive to the presence of moisture, and therefore, it is widely used as a gas barrier layer in MAP applications. This material has good processing properties and is therefore suitable for conversion into plastic films and structures. EVOH is always found laminated as a thin film, usually in the order of $5\,\mu m$ thickness, sandwiched between hydrophobic polymers, e.g. PE or PP, that protect the polymer from moisture. EVOH also possesses high mechanical strength, high resistance to oils and organic solvents and high thermal stability.

10.A3.1.2 Polyethylenes (PE)

The polyethylenes are structurally the simplest group of synthetic polymers and the most commonly used plastic materials for packaging applications. There are several types of PE classified on the basis of density. All are composed of a carbon backbone with a degree of side chain branching which influences density. Low density polyethylene (LDPE) (density, 0.910–$0.925\,\mathrm{g\,cm^{-3}}$) is generally used in film form, whereas high density polyethylene (HDPE) (density, $0.940\,\mathrm{g\,cm^{-3}}$) is commonly used for rigid and semi-rigid structures. PE are characterised as poor gas barriers, but their hydrophobic nature makes them very good barriers to water vapour. Therefore, by itself, PE cannot be used as a packaging material in MAP applications that require a high barrier to gases. PE melts at a relatively low temperature ranging from approximately 100–120°C (dependent on density and crystallinity). A less branched variant called linear low density polyethylene (LLDPE) which offers good heat sealing properties is used as a sealant layer to impart heat sealing properties on base trays and lidding films.

Modified PE-based materials that contain interchain ionic bonding are called ionomers. These have enhanced heat sealing properties that enable them to seal more effectively through meat juices, fats and powders. Ionomers also form effective heat seals with aluminium. Suryln is Dupont's trade name for its range of ionomer materials. A copolymer of ethylene and vinyl acetate, ethylene vinylacetate (EVA), offers enhanced heat sealing properties over LDPE and is found as a heat seal layer in some MAP applications.

10.A3.1.3 Polyamides (PA)

Polyamides comprise the group of plastics commonly referred to as nylons which have widespread application in food packaging. Nylons generally have high tensile strength, good puncture and abrasion resistance and good gas barrier properties. Nylons are generally moisture sensitive (hydrophilic) and will absorb water from their environment. Moisture in the nylon structure interferes with interchain bonding and adversely affects their properties, including gas barrier. Under conditions of high relative humidity, the gas transmission rate of nylon films generally increases. However, there are commercial nylons that are less affected by moisture. Their relatively high strength and toughness make them ideal as vacuum pouches for fresh meat, where hard bone ends could puncture other plastic materials. In this application, nylon is generally laminated to PE which provides the heat sealing properties.

10.A3.1.4 Polyethylene terephthalate (PET)

Polyethylene terephthalate is the most common polyester used in food packaging applications. PET is a good gas and water vapour barrier, is strong, offers good clarity and is temperature resistant. Crystalline PET (CPET) has poorer optical properties but improved heat resistance melting at temperatures in excess of 270°C. Flexible PET film is used for barrier pouches and top webs as a lidding material for tray packs. CPET is used for dual ovenable pre-formed base trays where its high temperature resistance makes it an ideal container for microwave and convection oven cooking of food.

10.A3.1.5 Polypropylene (PP)

Polypropylene is a versatile polymer that has applications in flexible, rigid and semirigid packaging structures. MAP applications are generally for rigid base trays. PP is a good water vapour barrier but a poor gas barrier. Increasing the thickness of the material somewhat compensates for the high gas transmission rate. PP melts at approximately 170°C. It can therefore be used as a container for microwaving low-fat food products. It should not be used for microwaving fatty foods, where temperatures in excess of its melting point could be reached. Foamed PP is used to provide the structural properties in laminates for MAP thermoformed base trays, where it is combined with an EVOH barrier and a PE heat sealing layer.

10.A3.1.6   Polystyrene (PS)
Pure polystyrene is a stiff, brittle material and has limited use in MAP applications. Expanded PS (EPS) which is formed from low density blown particles has been used for many years as a base tray for overwrapped fresh meat, fish and poultry products. Foamed PS has recently been used as a structural layer for pre-formed MAP base tray applications. The high gas permeability of foamed PS requires the material to be laminated to a plastic such as EVOH that provides the required gas barrier properties.

10.A3.1.7   Polyvinyl chloride (PVC)
Polyvinyl chloride has a relatively low softening temperature and good processing properties and is therefore an ideal material for producing thermoformed packaging structures. Although a poor gas barrier in its plasticised form, unplasticised PVC has improved gas and water vapour barrier properties which can at best be described as moderate. Oil and grease resistance are excellent, but PVC can be softened by certain organic solvents. It is a common structural material in MAP thermoformed base trays, where it is laminated to PE to provide the required heat sealing properties.

10.A3.1.8   Polyvinylidene chloride (PVdC)
Polyvinylidene chloride (a copolymer of vinyl chloride and vinylidene chloride) possesses excellent gas, water vapour and odour barrier properties, with good resistance to oil, grease and organic solvents. Unlike EVOH, the gas barrier properties of PVdC are not significantly affected by the presence of moisture. PVdC effectively heat seals to itself and to other materials. The high temperature resistance enables uses in packs for hot filling and sterilisation processes. Homopolymers and copolymers of PVdC are some of the best commercially available barriers for food packaging applications.

The above provides a brief introduction to the main plastic materials used in MAP applications. These and other plastics are covered in more detail in Chapter 7. It should be noted that certain desired properties can be enhanced by further processing the material. For example, coating a plastic with aluminium (metallisation) can improve the gas and vapour barrier properties and enhance the visual appearance of the material. PP is commonly metallised by passing the film through a *mist* of vapourised aluminium under vacuum. Similar treatments to improve gas and vapour barrier properties include

Table 10.3 Typical plastic-based packaging structures for MAP applications

Material	Application
UPVC/PE	Thermoformed base tray
PET/PE	Thermoformed base tray
XPP/EVOH/PE	Thermoformed base tray
PS/EVOH/PE	Thermoformed base tray
PET/EVOH/PE	Thermoformed base tray
PVdC coated PP/PE	Lidding film
PVdC coated PET/PE	Lidding film
PA/PE	Lidding film
PA/PE	Flow wrap film
PA/ionomer	Flow wrap film
PA/EVOH/PE	Flow wrap film
PET	Pre-formed base tray
PP	Pre-formed base tray
UPVC/PE	Pre-formed base tray

application of a silicon oxide (SiO_x) coating (also referred to as glass coating) to PET film and a diamond-like-carbon (DLC) coating to PET. The former has been used for MAP lidding film, with the advantages of providing excellent and stable barrier properties which are less influenced by the effects of temperature and humidity. To date, the main application of the latter has focused on non-MAP applications including a barrier coating on PET beverage bottles. Examples of MAP plastic structures are shown in Table 10.3.

10.A3.2   Selection of plastic packaging materials
Several factors must be considered when selecting package materials for MAP applications.

10.A3.2.1   Food contact approval
Packaging materials in contact with food must not transfer components from the packaging to the food product in amounts that could harm the consumer. In Europe, all food contact packaging must comply with EC Directives which are derived from the framework directive 89/109/EEC and which includes the plastics directive 90/128/EEC. Suppliers must provide evidence to demonstrate that migrant levels from plastics packaging into foods are below the recommended levels and that plastic packaging is safe in its intended use.

10.A3.2.2   Gas and vapour barrier properties
Packaging materials for MAP must have the required degree of gas and vapour barrier for the particular food application. Whereas some materials such as glass and metals (provided they are of sufficient thickness and possess no pores or other imperfections) are a total barrier to gases and vapours, plastic materials are permeable to varying degrees to gases and vapours. These molecules are

transported across a plastic package material by a mass transfer process called permeation. Permeation is defined as 'the diffusional molecular exchange of gases, vapours or liquid permeants across a plastic material which is devoid of imperfections such as cracks and perforations' (Hernandez, 1996). Essentially, the gas molecules sorb into one surface of the plastic, are transported through the material by a process of diffusion and desorb on the opposite surface. This process is shown diagrammatically in Figure 10.1.

The driving force for gas permeation through a polymer film is the difference in gas concentration between each side of the film. A concentration gradient drives a flow of permeant molecules from the high concentration side to the low concentration side of the film. In MAP, a gas concentration gradient exists between the pack headspace and the surrounding environment. In order to maintain the gas composition within the pack, the packaging material must be impermeable (a barrier) to gases.

Transmission rate is a measure of the gas or water vapour barrier of a packaging material. The transmission rate T is defined as the quantity Q of gas (or other permeant) passing through a material of area A in time t.

$$T = \frac{Q}{At}$$

Commonly used units for transmission rate are $cm^3\,m^{-2}\,day^{-1}$ (for gases) and $g\,m^{-2}\,day^{-1}$ (for vapours including water vapour).

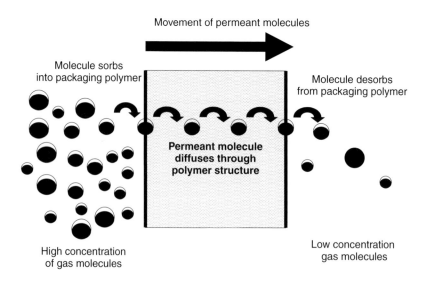

Figure 10.1 Permeability model for gases and vapours permeating through a plastic packaging film.

It is essential for packaging technologists to have details of the gas and vapour transmission rates for materials to be used for MAP. Such data should be available on the packaging specification. The lower the transmission rate value, the better the gas barrier properties of the material. Temperature has an effect on the permeability of gases and vapours and therefore should be quoted with transmission rate values on packaging specifications. Increasing temperature increases gas transmission rate across all common plastic packaging materials. The effect of temperature on the CO_2 transmission rate for a PET film is shown in Figure 10.2.

The driving force for gas transmission through a plastic film is the partial pressure difference of the gas between both sides of the material, and therefore, the gas concentration difference at the time of measurment should be quoted with the transmission rate value. Relative humidity (RH) is the driving force for water vapour transmission rate (WVTR), and therefore, WVTR values increase with increasing RH. Therefore, RH should always be quoted with WVTR values. RH can also influence the gas transmission rates of hydrophilic plastics and therefore should be quoted with gas transmission rate values (although generally gas transmission rates are measured at 0% RH). Gas transmission rates for some common food packaging plastics are summarised in Table 7.2.

Material thickness and gas concentration will affect the transmission rate. Therefore, the permeability coefficient, which compensates for material thickness and driving force, is used to compare gas barrier properties of different materials. The permeability coefficient P is defined as the quantity Q of gas under stated conditions permeating through a material of thickness l and area A, in time t, and under a partial pressure difference of Δ_p.

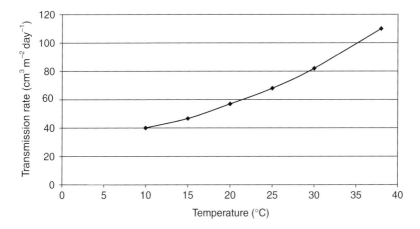

Figure 10.2 The effect of temperature on the transmission rate of CO_2 in 25μ PET film measured at 0% RH and 100% CO_2.

$$P = \frac{Ql}{At\Delta_p}$$

The SI units for P are m^3 m m^{-2} s^{-1} Pa^{-1}.

The barrier properties are primarily dependent on the type of plastic, the permeant gas or vapour and its partial pressure difference across the material and the temperature. As a general rule, the permeability of a synthetic polymer to CO_2 is approximately three to five times higher than the permeability to O_2. No single commercially available plastic provides a total barrier to gases and vapours, and therefore, the materials are selected on the basis of product type, desired shelf life, gas composition, availability and cost.

10.A3.2.3 Optical properties
Good optical properties, such as high gloss and transparency, are essential for bag, pouch and top web materials to satisfy consumer demand for a clear view of the product. To provide attractive appearance and shelf impact, some base tray materials are available in various colours. This also enhances the visual appeal of the product against the tray and helps consumers to identify product ranges and brands on the supermarket shelf. PET, PP and EPS trays are supplied in a range of colours. PVC trays are generally used in their natural form to provide a transparent pack.

10.A3.2.4 Antifogging properties
Condensation (fogging) of water vapour on the inner surface of food packs can occur when the temperature of the pack environment is reduced, resulting in a temperature differential between the pack contents and the packaging material. Fogging of the inner surface of lidding film is a result of light scattering by the small droplets of condensed moisture that leads to poor product visibility and an aesthetically unpleasing appearance of the pack. This can be overcome by applying antifogging agents to the plastic heat sealing layer, either as an internal additive or as an external coating. These chemicals decrease the surface energy of the packaging film which enables moisture to spread as a thin film across the under surface of the pack rather than collecting as visible droplets. Antifogging agents include fatty acid esters. Most lidding materials are available with antifog properties, and commonly treated plastics include LDPE, LLDPE, EVA and PET.

10.A3.2.5 Mechanical properties
Resistance to tearing and puncture and good machine handling characteristics are important in optimising the packaging operation and maintaining pack integrity during forming and subsequent handling and distribution. A further important characteristic of laminates and co-extruded films and sheets is the ability of layers to bond effectively together during the packaging operation

and during subsequent storage and handling. Certain organic compounds may have an adverse effect on bond strength to the extent that weakened bonds can result in layers peeling apart.

10.A3.2.6 *Heat sealing properties*

Effective heat seals are essential for maintaining the desired gas composition within the pack. The ability to form effective heat seals through contamination such as meat juices, powders, fats and oils is an advantage in many applications. Heat seal quality is dependent on many factors including seal material, seal width and machine settings such as temperature, pressure and dwell time.

10.A4 Modified atmosphere packaging machines

The function of MAP machines is to retain the product on a thermoformed or pre-formed base tray, or within a flexible pouch or bag, modify the atmosphere, apply a top web if required, seal the pack and cut and remove waste trim to produce the final pack. Pack format, presentation and machine performance and versatility and pack costs are essential factors for the packer filler to consider before selecting a machine for a particular product application. The following section provides an overview of the types and operation of MAP machines.

10.A4.1 *Chamber machines*

For low production throughput, chamber machines are sufficient. These are generally used with pre-formed pouches, though tray machines are available. The filled pack is loaded into the machine, the chamber closes, a vacuum is pulled on the pack and back flushed with the modified atmosphere. Heated sealing bars seal the pack, the chamber opens, packs are removed and the cycle continues. These machines are generally labour intensive and cheap, with a simple operation but are relatively slow. Some chamber machines can handle large packages and are suitable for bulk packs.

10.A4.2 *Snorkel machines*

Snorkel machines operate without a chamber and use pre-formed bags or pouches. The bags are filled and positioned in the machine. The snorkel is introduced into the bag, draws a vacuum and introduces the modified atmosphere. The snorkels withdraw and the bag is heat sealed. Bag in-box bulk products and retail packs in large MAP master packs can be produced on these machines.

10.A4.3 Form-fill-seal tray machines

Form-fill-seal (FFS) machines form pouches from a continuous sheet of roll stock (flow wrap), or form flexible or semirigid tray systems comprising a thermoformed tray with a heat sealed lid. FFS machines may be orientated in a vertical plane or a horizontal plane. Flow wrapping machines are available in both vertical and horizontal formats. The type of format is dependent on the nature of the food product being packed. FFS machines using pre-formed trays or producing thermoformed trays are almost exclusively horizontal machines. This section focuses on horizontal form-fill-seal MAP machines which are used extensively in the food industry.

Thermoformed form-fill-seal tray machines use rollstock film for base web and lidding material. Base film is carried through the machine by clamps which attach onto the edge of the web and carry it through the forming, filling, evacuation, gas modification, sealing, cutting and discharge stages.

Base trays are produced by applying heat to the base roll stock, which when softened is immediately moulded into the desired shape and size. Forming of the heated, softened sheet can be achieved by applying a vacuum, air pressure, mechanical drawing or a combination of these processes. The softened heated film is normally drawn into the forming mould under the assistance of vacuum applied through evacuation holes located along the base edges and corners of the mould. This process produces a more defined and uniform tray shape. Where deep trays are required, a more uniform distribution of plastic can be achieved by prestretching the film using mechanical devices (plugs) which prevent excessive thinning of the container walls at the base edges and corners.

The tooling for the moulds represents a significant initial capital cost of thermoformed form-fill-seal tray machines. Moulds are generally fabricated from either steel or aluminium, the latter being cheaper but less durable than the former. Inserts, called filler plates, can be placed on the base of the die to decrease the forming depth and therefore produce shallow trays. Roll stock for base tray and lidding is supplied in reels of film wound on to a core of standard diameter (usually 3 or 6″) which matches the film unwind system on the thermoforming machine. Reel diameters are usually supplied from 300 mm to 1000 mm, in increments of 50 mm.

Thermoformed trays are produced by one of the three following methods.

10.A4.3.1 Negative forming

Negative forming is suitable for flexible films but has limitations with rigid materials, unless the cavity shape is shallow with well-rounded corners. It comprises a two-stage cycle:

1. compressed air blows film up against a heating plate which softens the film
2. the softened film is blown down by compressed air through the heating plate into a mould (Fig. 10.3).

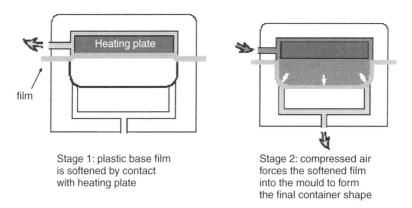

Figure 10.3 Thermoformed base tray produced by the negative forming method (courtesy of Multivac).

10.A4.3.2 Negative forming with plug assistance

Plug-assisted negative forming is used when deep or complex tray shapes are required. The plug prestretches the film to produce improved distribution of the softened plastic. This is essential to ensure tray corners have a sufficient thickness of material in order to prevent pinholes from forming or material becoming damaged during handling. This method can be used for flexible and rigid materials. It comprises a three-stage process:

1. film is softened between heated plates. Preheating can occur in one or several stages
2. plug descends and stretches the film
3. final shaping stage occurs by compressed air which pushes the film into the mould (Fig. 10.4).

10.A4.3.3 Positive forming with plug assistance

Plug-assisted positive forming is used for shaping rigid trays, where a more controlled distribution of film material is necessary to maintain the material thickness at the base and corners of the tray. It comprises three stages:

1. the film is heated between temperature-regulated plates
2. a vacuum produced at the base of the mould prestretches the film
3. the positive plug descends, and the film is blown up to it by compressed air. The plug forms the shape of the tray (Fig. 10.5).

Following the forming stage, the base tray advances to the filling station, where it is manually or automatically loaded with the food product. The pack then advances to the gassing station where the modified atmosphere is introduced and the top web heat sealed to the base tray. The sealed trays are labelled, coded and separated as necessary.

322 FOOD PACKAGING TECHNOLOGY

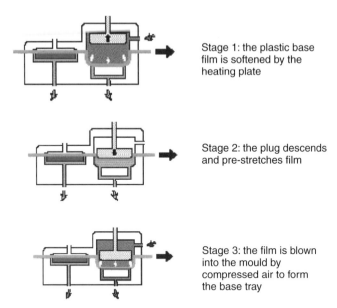

Figure 10.4 Thermoformed base tray produced by the plug assisted negative forming method (courtesy of Multivac).

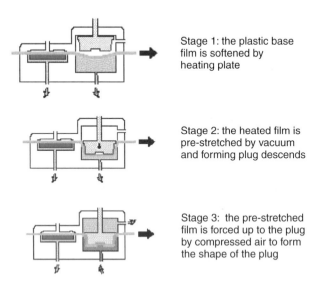

Figure 10.5 Thermoformed tray produced by plug assisted positive forming (courtesy of Multivac).

10.A4.4 Pre-formed trays

The alternative to thermoforming the base tray is to use pre-formed trays. These are loaded manually or automatically by a tray denester into the machine infeed and pass through the filling, gas flushing and sealing stages as would a thermoformed base tray. Examples of pre-formed trays are shown in Figure 10.6.

10.A4.4.1 Pre-formed trays versus thermoformed trays
Pre-formed trays have several advantages compared to thermoformed trays:

- Pre-formed trays offer more flexibility in tray design. A greater range of intricately shaped pre-formed trays are available than is currently possible to produce on thermoform machines.
- Pre-formed trays can offer enhanced appearance and presentation at point of sale.
- A greater range of tray materials can be used for pre-formed tray manufacture than would be possible with thermoform machines.
- Trays of the same shape but different colour or depth can be handled with no changeover.

Figure 10.6 Examples of plastic pre-formed trays for MAP foods.

- Greater flexibility in tray loading is possible with pre-formed trays. This operation can take place before packing and in an area separate from the packing operation.
- Generally, pre-formed trays require less downtime for changeover between different tray sizes compared to thermoform-fill-seal machines. A tooling set comprising the sealing die, frame carriages and cutting die has to be changed when a tray of different outer dimensions is used.

Thermoformed trays have the main advantage of lower packaging material costs. It is estimated that tray pack savings of between 30 and 50% are achievable mainly because costs of the thermoforming process are carried by the filler packer rather than the tray supplier. Transportation and storage costs will be higher for pre-formed trays compared to the roll stock equivalent for thermoformed trays.

10.A4.5 Modification of the pack atmosphere

MAP machines use mainly one of the two techniques to modify the pack atmosphere.

10.A4.5.1 Gas flushing

This method employs a continuous gas stream that flushes air out from the package prior to sealing. This method is less effective at flushing air out of the pack, and this results in residual oxygen levels of 2–5%. Gas flushing is therefore not suited for oxygen-sensitive food products. Generally, gas flushing machines have a simple and rapid operation and therefore a high packing rate.

10.A4.5.2 Compensated vacuum gas flushing

This method uses a two-stage process:

1. *The evacuation stage* – a vacuum is pulled on the pack to remove air. Generally, it is not possible to achieve a full vacuum, since reduced pressures will result in water to boil, at which point the vacuum cannot be improved. The vacuum achieved is generally between 5 and 10 Torr (1 Torr = 1 mmHg). As a general rule, the cooler and drier the food, the lower the achievable vacuum.
2. *Gas flushing stage* – the pack is flushed with the modified gas mix. The evacuation of air from the pack results in lower residual oxygen levels than that achieved by gas flushing, and therefore this method is better suited for packing oxygen-sensitive products.

The two-stage process employed by the compensated vacuum method results in a lower packaging rate than that possible with gas flushing.

10.A4.6 Sealing

An effective heat seal is critical to maintaining the quality and safety of the packaged product. Film factors (thickness and surface treatments) and plastic composition (resin type, molecular weight distribution and presence of additives) will determine the machine settings for the sealing operation. The correct combination of time, temperature and pressure of the seal bars is necessary to produce a good seal. Insufficient dwell time or temperature can result in ineffective seals that separate at the bond interface. Excessive dwell time or temperature can result in weakness adjacent to the seal area.

10.A4.7 Cutting

Packs are discharged as a continuous arrangement of filled and sealed packs from a thermoform-fill-seal machine, and therefore, the final operation is to separate into individual packs. This can be carried out by two methods – die cutting and longitudinal and transverse cutting.

Die cutting is achieved in one operation. A shaped blade is forced through the film which is clamped in place by a frame assembly. Transverse cutting separates packs into rows and is carried out by guillotines or punches which are driven through the film that is supported by anvils. This may be carried out in conjunction with longitudinal cutting where circular knives cut through the tray flanges parallel to the length of the film.

Regardless of the cutting method, it is important to ensure that an even flange remains around the lip of the tray in order to maximise the seal strength. Offset cutting could leave one side of the tray with a thin flange that may open during handling and distribution. Waste trim is either wound onto spindles at the machine discharge or removed by suction into collection bins.

Figures 10.7–10.11 show some of the above operations in a Multivac R230 thermoforming machine.

10.A4.8 Additional operations

Machines are generally integrated into production lines and combined with operations such as automatic filling, top web labelling, base web labelling, registration of printed top web, over printing and pack collation and case loading.

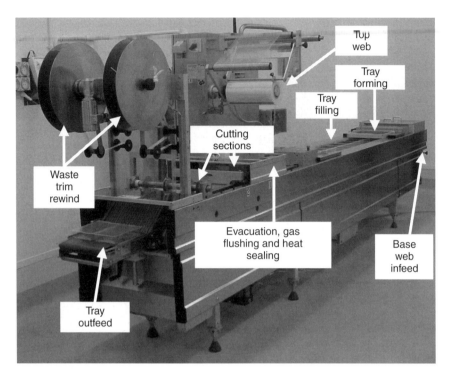

Figure 10.7 Multivac R230 thermoform fill seal machine.

10.A5 Quality assurance of MAP

Examples of instruments used in quality assurance of MAP are discussed in this section. These are provided by way of example and are not intended to be recommendations by the authors.

10.A5.1 Heat seal integrity

The majority of MAP form-fill-seal retail packs are heat sealed. The quality and safety of MAP food will be compromised if the seal integrity is lost during the required life of the pack. A breach of the heat seal will result in a rapid loss of the modified atmosphere in the pack. Therefore, the sealing operation constitutes a critical control point and must be monitored during production as part of the quality assurance procedure. It is of key importance that sealing bar temperature, pressure and dwell time are set according to machine manufacturer and packaging supplier specifications and conditions are monitored during machine operation.

Figure 10.8 Lower web unwind and infeed section on the Multivac R230.

Seal and pack integrity can be assessed by either destructive or nondestructive tests. Destructive tests are based on immersing packs in water and checking for escaping gas bubbles from around the seal. Other test methods measure seal strength by pressurising packs using compressed air until the seal fails. Nondestructive tests are based on measuring changes in pressure generated by packs under vacuum in sealed chambers. Some examples of seal integrity equipment are discussed below.

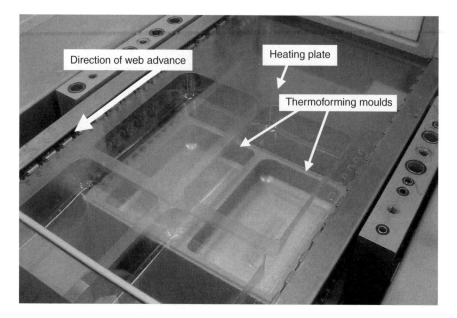

Figure 10.9 Base web thermoforming section on the Multivac R230.

10.A5.1.1 Nondestructive pack testing equipment
Ai Qualitech manufactures a range of vacuum leak testers (Q700 series) designed for the use in production or laboratory environments for MAP tray packs, pouches and pillow packs. When the vacuum is pulled, the pack expands, causing the top web to dome. A pressure sensor in contact with the top of the pack will detect pressure drop due to a leaking pack. Ease of use and objective and quantifiable measurement are possible benefits. The instrument can be supplied with pick and place equipment to enable online automatic operation. The instrument is capable of detecting holes of 10 μm or greater.

10.A5.1.2 Destructive pack testing equipment
An example of equipment which measures the heat seal strength of complete packs is LIPPKE 2500 SL Package Test System. The equipment can be used to measure seal rupture and also pack atmosphere leakage through pinholes or faulty seals. In the leak test mode, the pack is pressurised to a predetermined maximum and internal pack pressure monitored. Leaks are evident as a decrease in pack pressure. In the seal strength mode, a linear pressure increase is applied internally in the package. The pressure achieved at burst indicates the strength of the seal. Needle probe minimum penetration is 1 mm, which makes this instrument suitable for most types of pack.

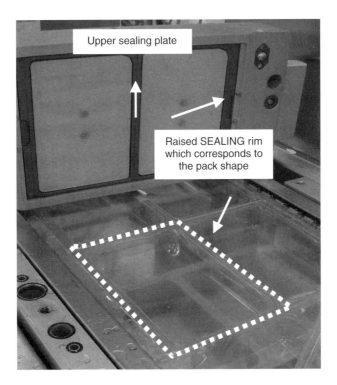

Figure 10.10 Top web heat sealing on the Multivac R230.

10.A5.2 Measurement of transmission rate and permeability in packaging films

Accurate determination of the oxygen, CO_2 and water vapour permeabilities of plastic-based film is important for MAP applications. Several methods exist for measuring transmission rate and permeability of gases and vapours across a packaging film. The most common test procedure is based on the isostatic method. In this method, both sides of the test film are maintained at the same total pressure but a constant partial pressure difference is maintained by passing test gas continuously on one side of the film while inert carrier gas continuously removes permeant from the other side of the film. This maintains a very low partial pressure of permeated test gas and establishes a constant gas concentration difference across the film. This is also referred to as the *concentration increase* method.

10.A5.2.1 Water vapour transmission rate and measurement

Water vapour transmission rate is defined as the time rate of water vapour flow, normal to the two surfaces, under steady-state conditions through unit

Figure 10.11 Cutting section on the Multivac R230.

area of a test film. There are several methods for measuring WVTR and water vapour permeability. Earlier methods were based on ASTM E96: Standard Test Methods for Water Vapor Transmission of Materials. This gravimetric procedure measured the weight increase by a desiccant sealed in an aluminium cup by the test film, with the apparatus being held in an environment of known and controlled temperature and RH. The weight increase of the dish occurs as a result of moisture uptake by the dessicant and is due to the water permeating through the test film from the surrounding environment into the sealed cup. The corresponding WVTR could be calculated from the weight increase. For high-barrier films, the test can take several days, if not weeks.

In 1990, ASTM introduced a test standard (ASTM F 1249: Standard Test Method for Water Vapor Transmission Rate Through Plastic Film and Sheeting Using a Modulated Infrared Sensor) based on the isostatic method and employing solid-state electronics with pulse-modulated infrared sensors which can detect water vapour from 1 part per million. MOCON (Modern Controls Inc.) supply equipment to measure WVTR based on this standard. Test temperatures can be controlled from 10 to 40°C (± 0.5°C) and RH from 35 to 90% RH (± 3%). The instrument can test at 100% RH by inserting water-saturated sponges in the test cell.

Water vapour permeating across the test film is transported by dry N_2 gas to an infrared detection system intended for operation over a 5–50°C temperature range. The infrared photodetector produces a low-level electrical signal in response to the change in transmitted infrared radiation. The amplifier produces a filtered DC signal in direct proportion to the water vapour in the test cell and therefore proportional to the water vapour transmission of the test film. The WVTR of high-barrier materials can be determined within 2 days. Test measurements are reported in the units: $g\,m^2\,day^{-1}$.

10.A5.2.2 Measurement of oxygen transmission rate

Oxygen transmission rate is defined as the time rate of gaseous oxygen flow, normal to the two film surfaces, under steady-state conditions through a unit area of a test film material.

MOCON Oxtran (Modern Controls, Inc.) measures the transmission rate of oxygen across flat films by a method based on the isostatic procedure. A coulometric sensor detects permeated oxygen and provides parts per billion sensitivity.

10.A5.2.3 Measurement of carbon dioxide transmission rate

The PERMATRAN C-IV, manufactured and marketed by MOCON (Modern Controls, Inc.), is an instrument for measuring the rate at which gaseous CO_2 diffuses through a flat film. Test films are clamped in a diffusion cell, and one half of the cell is flushed continuously with CO_2 gas. Permeated gas is carried to an infrared sensor where a response is generated proportional to the amount of CO_2 present. In cases where the test film offers a very low barrier to CO_2, the test film may be mounted on an aluminium foil mask to reduce the test surface area from 50 to 5 cm^2.

10.A5.3 Determination of headspace gas composition

10.A5.3.1 Oxygen determination

Oxygen promotes many food spoilage reactions as discussed in Section 10.A 2.3. Certain foods can be damaged by exposure to oxygen concentrations of 1–2%. The level of residual oxygen in the pack headspace is therefore of concern to food processors and forms part of the quality procedures for the manufacture and packing of oxygen-sensitive products. An example of the type of equipment which measures the concentration of headspace oxygen is the TORAY LC 750F Oxygen headspace analyser. This instrument determines residual oxygen by probing and sampling headspace gases in MAP packs and is suitable for production and laboratory use, with a response time of approximately 2 s.

10.A5.3.2 Carbon dioxide determination

Similar in operation to the TORAY LC 750F, the TORAY PG100 Carbon dioxide headspace analyser measures CO_2 in flexible, semi-rigid and rigid packages.

SECTION B MAIN FOOD TYPES

10.B1 Raw red meat

Microbial growth and oxidation of the red oxymyoglobin pigment are the main spoilage mechanisms that limit the shelf life of raw red meats. The packaging technologist has to maintain the desirable red colour of the oxymyoglobin pigment, by having an appropriate O_2 concentration in the pack atmosphere, and at the same time minimise the growth of aerobic microorganisms. Highly pigmented red meats, such as venison and wild boar, require higher concentrations of O_2.

Aerobic spoilage bacteria, such as *Pseudomonas* species, normally constitute the major flora on red meats. Since these bacteria are inhibited by CO_2, it is possible to achieve both red colour stability and microbial inhibition by using gas mixtures containing 20–30% CO_2 and 70–80% O_2. These mixtures can extend the chilled shelf life of red meats from 2–4 days to 5–8 days. A gas/product ratio of 2:1 is recommended.

Red meats provide an ideal medium for the growth of a wide range of spoilage and food poisoning microorganisms including *E. coli*. Because raw red meats are cooked before consumption, the risk of food poisoning can be greatly reduced by proper cooking. The maintenance of recommended chilled temperatures and good hygiene and handling practices throughout the butchery, MAP, distribution and retailing chain is of critical importance in ensuring both the safety and extended shelf life of red meat products.

10.B2   Raw poultry

Microbial growth, particularly growth of *Pseudomonas* and *Achromobacter* species, is the major factor limiting the shelf life of raw poultry. These Gram-negative aerobic spoilage bacteria are effectively inhibited by CO_2. Consequently, the inclusion of CO_2 in MAP at a concentration in excess of 20% can significantly extend the shelf life of raw poultry products. CO_2 concentrations higher than 35% in the gas mixture of retail packs are not recommended because of the risks of pack collapse and excessive drip. Nitrogen is used as an inert filler gas, and a gas/product ratio of 2:1 is recommended. Since pack collapse is not a problem for bulk MAP master packs, gas atmospheres of 100% CO_2 are frequently used.

Since poultry meat provides a good medium for the growth of pathogenic microorganisms, including some that are not inhibited by CO_2, it is critical that recommended chilled temperatures and good hygiene and handling practices throughout the supply chain are adhered to and that products are properly cooked prior to consumption.

Early research into gas mixes for MAP of poultry meat reported discolouration of the meat at CO_2 concentrations higher than 25%. Even at 15%, the authors sometimes observed a loss of *bloom* (Ogilvy & Ayres, 1951). This research is at variance with the lack of problems reported from the commercial use of relatively high levels of CO_2 with meat products, with up to 100% in some products. Gas compositions of 25–50% CO_2 and 50–75% N_2 are used routinely.

It would appear that the problems that have been occasionally encountered with high levels of CO_2, e.g. development of greyish tinges on meat, may simply be due to high residual levels of O_2 rather than the concentration of CO_2 (Gill, 1990).

It is recommended that research into the optimal gas composition and package type and size should be conducted for individual food products. Furthermore, headspace gas composition will change during storage due to microbial respiration and gas exchange between the pack headspace and the environment. Therefore, processors should conduct trials to determine the extent to which gas composition changes through the shelf life of the product. The ratio of headspace pack volume to food product volume is also important, as is the types and thickness of the package material and the package design. Shelf life evaluations must reflect the conditions from manufacture to consumption of the product. It may also be necessary to consider the effect of pack opening on the subsequent shelf life of the product.

10.B3   Cooked, cured and processed meat products

The principal spoilage mechanisms that limit the shelf life of cooked, cured and processed meat products are microbial growth, colour change and oxidative rancidity. For cooked meat products, the heating process should kill vegetative bacterial cells, inactivate degradative enzymes and fix the colour. Consequently, spoilage of cooked meat products is primarily due to post-process contamination by microorganisms, as a result of poor hygiene and handling practices. The colour of cooked meats is susceptible to oxidation, and it is important to have only low levels of residual O_2 in packs. MAP using CO_2/N_2 mixes (gas compositions of 25–50% CO_2 and 50–75% N_2) along with a gas/product ratio of 2:1 is widely used to maximise the shelf life and inhibit the development of oxidative off-flavours and rancidity. Raw cured meat products, e.g. bacon, owe their characteristic pink reddish colour to nitrosylmyoglobin. This pigment is more stable than oxymyoglobin and is unaffected by high levels of CO_2 but is slowly converted to brown metmyoglobin in air. During cooking, nitrosylmyoglobin is converted to pink denatured nitrosohemochrome pigments that are unstable in air.

Processed meat products such as sausages, frankfurters and beef burgers generally contain sodium metabisulphite, which is an effective preservative against a wide range of spoilage microorganisms and pathogens. Cooked, cured and processed meat products containing high levels of unsaturated fat are liable to be spoiled by oxidative rancidity, but MAP with CO_2/N_2 mixtures is effective at inhibiting this undesirable reaction.

Potential food poisoning hazards are primarily due to microbial contamination or growth resulting from post-cooking, curing or processing contamination.

These can be minimised by using recommended chilled temperatures, good hygiene and handling practices. The low water activity (a_w) and addition of nitrite in cooked, cured and processed meat products inhibit the growth of many food poisoning bacteria, particularly *C. botulinum*. This inhibition may be compromised in products formulated with lower concentrations of chemical preservatives than those used in traditional foods. The potential effects of any changes in product formulation on the growth and survival of pathogens should always be considered. Cooked meats stored without any added preservatives will be at risk from growth of *C. botulinum* under anaerobic MAP conditions, particularly when held at elevated storage temperatures. It should be noted that many sliced, cooked, cured and processed meat products are vacuum packed for retail sale. However, the shelf life of such products in MAP is similar to that achieved in vacuum packs, and additionally, MAP allows for easier separation of meat slices.

10.B4   Fish and fish products

There has been a very significant increase in the sale of MAP fish products in Europe and particularly in the UK. Nevertheless, packaging technologists should be aware of a major concern limiting the development of MAP, namely *C. botulinum*. There is also debate about the cost benefits of MAP, since in some applications only relatively small increases in safe shelf life have been reported. Spoilage of fish results in the production of low molecular weight volatile compounds, therefore, packaging technologists need to consider the odour barrier properties of packaging films and select appropriate high-barrier materials for packaging strong flavoured fresh, smoked and brined fish and fish products.

Spoilage of fish and shellfish results from changes caused by three major mechanisms: (i) the breakdown of tissue by the fish's own enzymes (autolysis of cells), (ii) growth of microorganisms, and (iii) oxidative reactions. MAP can be used to control mechanisms (ii) and (iii) but has no direct effect on autolysis. Because autolysis is the major cause of spoilage of fish and shellfish stored at temperatures close to 0°C compared with the activities of bacteria, this may explain the reduction in benefits achieved from MAP of fish compared to other flesh products. MAP, while potentially inhibiting oxidative reactions, may be more effective at inhibiting microbial growth.

Oxidative reactions are much more important as shelf life limiters in fish compared with other flesh meat, because seafood has a higher content of polyunsaturated lipids. Storage temperature has a major effect on fat oxidation that occurs even at frozen temperatures. Note that salt addition can accelerate oxidative processes.

Generally, the major spoilage bacteria found on processed fish are aerobes including *Pseudomonas*, *Moraxella*, *Acinetobacter*, *Flavobacterium* and

Cytophaga species. There are several microorganisms that are of particular importance when dealing with MAP fish products, these include *C. botulinum*. Use of CO_2 can effectively inhibit the growth of some of these species, see Table 10.2. The aerobic spoilage organisms tend to be replaced by slower growing, and less odour producing, bacteria, particularly lactic acid bacteria such as lactobacilli, during storage.

Because fish and shellfish contain much lower concentrations of myoglobin, the oxidation status of this pigment is less important than that in other meats. Consequently, there is potential to use higher levels of CO_2, e.g. 40%. Because of the high moisture content and the lipid content of some species, N_2 is used to prevent pack collapse.

One of the concerns about MAP of fish is that removal of O_2 and its replacement by either N_2 or N_2/CO_2 results in anaerobic conditions that are conducive to the growth of protease-negative strains of *C. botulinum*. Because these bacteria can grow at temperatures as low as $3°C$ and do not significantly alter the sensory properties of the fish, there is the potential for food poisoning that can lead to fatalities. While there is no evidence that CO_2 promotes the growth of psychotropic strains of *C. botulinum*, there are, as discussed previously, some concerns about CO_2 promoting the germination of spores of this organism.

Considerable research has been undertaken to assess, and to control, the risks associated with the growth of *C. botulinum* in MAP of fish and other products. The Advisory Committee on the Microbiological Safety of Food (ACMSF) (Anon, 1992) have recommended controlling factors that should be used singly or in combination to prevent the growth of, and toxin production in prepared chilled food by, psychotropic *C. botulinum*. As far as MAP of raw fish products is concerned, risk can be effectively eliminated if storage temperature is held at $3°C$ or below and if the shelf life is limited to no more than 10 days.

Some fish processors include O_2 in their MAP to further reduce the risk of growth of clostridia. Gas mixtures of 30% O_2, 40% CO_2 and 30% N_2 are used for white non-processed fish, i.e. nonfatty fish. While this will increase the shelf life of some fish and fish products, it would not significantly enhance the shelf life of oily or fatty fish. High, 40%, CO_2 mixes along with 60% N_2 are generally used for smoked and fatty fish. Because of the risks already discussed, it would appear reasonable to aim for a target shelf life of 10–14 days at $3°C$.

10.B5 Fruits and vegetables

Consumers now expect fresh fruit and vegetable produce throughout the year. MAP has the potential to extend the safe shelf life of many fruits and vegetables. Packaging fresh and unprocessed fruits and vegetables poses many challenges for packaging technologists. As with all products, it is essential to work with the highest quality raw materials, and this is especially true for this product group, often referred to as *fresh produce*. The quality of fresh produce is markedly dependent on growing conditions, minimising bruising and other damage during harvesting and processing, adherence to good hygienic practices, controlling humidity to prevent desiccation while avoiding condensation to prevent mould growth, and maintaining optimum storage temperatures. Unlike other chilled perishable foods, fresh produce continues to respire after harvesting. The products of aerobic respiration include CO_2 and water vapour. In addition, respiring fruits and vegetables produce C_2H_4 that promotes ripening and softening of tissues. The latter if not controlled will limit shelf life.

Respiration is affected by the intrinsic properties of fresh produce as well as various extrinsic factors, including ambient temperature. It is accepted that the potential shelf life of packed produce is inversely proportional to respiration rate. Respiration rate increases by a factor of 3–4 for every 10°C increase in temperature. Hence, the goal of MAP for fruits and vegetables is to reduce respiration to extend shelf life while maintaining quality. Respiration can be reduced by lowering the temperature, lowering the O_2 concentration, increasing the CO_2 concentration and by the combined use of O_2 depletion and CO_2 enhancement of pack atmospheres. If the O_2 concentration is reduced beyond a critical concentration, which is dependent on the species and cultivar, then anaerobic respiration will be initiated. The products of anaerobic respiration include ethanol, acetaldehyde and organic acids. Anaerobic respiration, or anaerobiosis, is usually associated with undesirable odours and flavours and a marked deterioration in product quality. While increasing the CO_2 concentration will also inhibit respiration, high concentrations may cause damage in some species and cultivars.

Reducing O_2 concentrations below 5% will slow the respiration rate of many fruits and vegetables. Kader *et al.* (1989) have tabulated the minimum O_2 concentration tolerated by a range of fresh produce; while some cultivars of apples and pears can tolerate O_2 concentrations as low as 0.5%, potatoes undergo anaerobic respiration at around 5% O_2. In general, O_2 concentrations below about 3% can induce anaerobic respiration in many species of fresh produce.

Elevated CO_2 can also inhibit respiration. If the gas concentration is too high, then anaerobic respiration is induced with consequent quality problems. CO_2 sensitivity is both species and cultivar dependant; strawberries are able to tolerate 15% CO_2 whereas celery is stressed by CO_2 concentrations above 2% (Kader *et al.*, 1989). The tolerance of strawberries to CO_2 can be used to inhibit the growth of the mould *Botrytis cinerea*.

The use of low concentrations of O_2 and elevated levels of CO_2 can have a synergistic effect on slowing down respiration and, indirectly, ripening. While the mechanisms whereby MAP can extend the shelf life of fresh produce are not fully understood, it is known that the low O_2/high CO_2 conditions reduce the conversion of chlorophyll to pheophytin, decrease the sensitivity of plant tissue to C_2H_4, inhibit the synthesis of carotenoids, reduce oxidative browning and discolouration and inhibit the growth of microorganisms. These mechanisms are all temperature dependent. The effects of MAP on the physiology of fruits and vegetables have been the subject of extensive research by many groups and have been well reviewed, e.g. Kader (1986).

Packaging technologists should be aware of several major pathogens as far as MAP fresh produce is concerned, in particular *L. monocytogenes* and *C. botulinum*. As previously discussed, *L. monocytogenes* can grow under reduced O_2 levels and is not markedly inhibited by CO_2. This combined with its ability to grow at temperatures close to 0°C helps explain the concern.

The use of MAP atmospheres containing low concentrations of O_2 and elevated CO_2 concentrations may permit the growth of psychotropic protease-negative strains of *C. botulinum*. However, provided packs are stored at 3°C or below for not more than 10 days, there is unlikely to be a problem with clostridia. Temperature control is critical, since temperature abuse could lead to pack contents becoming toxic.

The environment in which fruits and vegetables are grown may harbour pathogens including *Salmonella* species, enterotoxigenic *E. coli* and viruses. While these microorganisms may not grow in MAP packs, particularly if the storage temperature is maintained around 3°C, they may survive throughout storage and could cause food poisoning through cross-contamination in the home or due to the consumption of raw or under-processed product. Hygienic preparation, sanitation in chilled-chlorinated water, rinsing and dewatering prior to MAP are now considered as essential treatments to fruits and vegetables prior to packaging to ensure low microbial counts and assure safety. Since there is a risk of anaerobic pathogens, such as *C. botulinum*, growing in MAP packs, a minimum level of O_2 (e.g. 2–3%) is usually recommended to ensure that potentially hazardous conditions are not created.

Equilibrium MAP (refer to Chapter 2) has been used for fresh produce. Essentially, this involves using knowledge of the permeability characteristics of particular packaging films, along with the respiration characteristics of the product to balance the gas transfer rates of O_2 and CO_2 through the package with the respiration rate of the particular product.

Increasingly, gas packing fresh produce along with $CO_2/O_2/N_2$ gas mixtures is being used. This approach may have benefit in reducing enzymic browning reactions before a passively generated equilibrium modified atmosphere has been established.

10.B6 Dairy products

MAP has the potential to increase the shelf life of a number of dairy products. These include fat-filled milk powders, cheeses and fat spreads. In general, these products spoil due to the development of oxidative rancidity in the case of powders and/or the growth of microorganisms, particularly yeasts and moulds, in the case of cheese.

Whole milk powder is particularly susceptible to the development of off-flavours due to fat oxidation. Commercially, the air is removed under vacuum and replaced with 100% N_2 or N_2/CO_2 mixes and the powder is hermetically sealed in metal cans. Due to the spray drying process, air tends to be absorbed inside the powder particles and will diffuse into the container over a period of ten days or so. This typically will raise the residual headspace O_2 content to 1–5% or higher (Evans, Mullan and Pearce, unpublished results). Because some markets require product with low levels of residual O_2 (<1%), some manufacturers re-pack the cans after ten days of storage. Obviously, this is both expensive and inconvenient. We have found that use of N_2/CO_2 mixes (Evans, Mullan and Pearce, unpublished results) can be helpful. Use of O_2 scavenging may also be useful. Refer to Chapter 9 for a more detailed discussion of O_2 scavengers.

English territorial cheeses, e.g. Cheddar, have traditionally been vacuum packed. Increasingly MAP is being used with high CO_2 concentration gas mixes. This has the advantage of obtaining a low residual O_2 content and a tight pack due to the CO_2 going into solution. It is important to balance this process using the correct N_2 level in the gas mix so as to avoid excessive pressure being put on the pack seal.

Use of N_2/CO_2 atmospheres has significant potential for extending the shelf life of cottage cheese. The cottage cheese is a high-moisture, low-fat product that is susceptible to a number of spoilage organisms including *Pseudomonas* spp. Use of gas mixtures containing 40% CO_2 balanced with 60% N_2 can increase the shelf life significantly.

References

Anon (1992) Advisory Committee on the Microbiological Safety of Food. Report on vacuum packaging and associated processes, HMSO, London, ISBN 0–11–321558–4.
Anon (1999) MSI Data Report: modified atmosphere packaging: UK, MSI.
Daniels, J.A., Krishnamurthi, R. and Rizvi, S.S.H. (1985) A review of effects of carbon dioxide on microbial growth and food quality. *J. Food Protect.*, **48**, 532.
Eklund, M.W. (1982) Significance of *Clostridium botulinum* in fishery products preserved short of sterilisation. *Food Technol.*, **115**, 107–112.
Enfors, S. and Molin, G. (1981) The influence of temperature on the growth inhibitory effect of CO_2 on *Pseudomonas fragii* and *Bacillus cereus*. *Can. J. Microbiol.*, **27**, 15.
Gill, C.O. (1990) Controlled atmosphere packing of chilled meat. *Food Control*, **1**, 74–79.

Gill, C.O. and Tan, K.H. (1979) Effect of CO_2 on growth of *Pseudomonas fluorescens*. *Appl. Environ. Microbiol.*, **38**, 237.

Hernandez, R.H. (1996) Plastics in Packaging, Chapter 8 in *Handbook of Plastics, Elastomers, and Composites*, 3rd edn, New York, McGraw-Hill.

Hintlian, C.B. and Hotchkiss, J.H. (1986) The safety of modified atmosphere packaging: a review. *Food Technol.*, **40**(12), 70–76.

Kader, A.A., Zagory, D. and Kerbel, E.L. (1989) Modified atmosphere packaging of fruits and vegetables, in *CRC Critical Reviews of Food Science and Nutrition*, **28**(1), 1–30.

Kader, A.A. (1986) Biochemical and physiological basis for effects of controlled and modified atmospheres on fruits and vegetables. *Food Technol.*, **40**(5), 99–100, 102–104.

King, A.D. and Nagel, C.W. (1975) Influence of carbon dioxide up on the metabolism of *Pseudomonas aeruginosa*. *J. Food Sci.*, **40**, 362.

Ogilvy, W.S. and Ayres, J.C. (1951) Post-mordem changes in meats II. The effect of atmospheres containing carbon dioxide in prolonging the storage of cut-up chicken. *Food Technol.*, **5**, 97–102.

Valley, G. and Rettger, L.F. (1927) The influence of carbon dioxide on bacteria. *J. Bacteriol.*, **14**, 101–113.

Index

Achromobacter, 332
Acinetobacter, 334
active packaging
   Actipak, 298
   carbon dioxide scavengers/emitters, 289
   definition, 282
   flavour/odour absorbers, 296
   food applications, 285–9
   food safety, consumer and regulatory issues, 298–9
   preservative releasers, 293
adhesive lamination, plastics, 208
Advisory Committee on the Microbiological Safety of Food (ACMSF), 335
aerobes, 334
   definition, 306
Aeromonas hydrophilia, 307
air composition, 303
American Society of Testing and Materials (ASTM), 110–12
anaerobes definition, 307
anaerobiosis, 336
antifogging properties, 318
argon
   MAP applications, 304

biodegradable plastics, 237
biodeterioration, agents of
   bacteria, 35
   enzymes, 33
   fungi, 38
   non-enzymic biodeterioration, 40
blanching, 42
bleaching of cellulose fibre, 243
blow and blow process, glass container making, 157
blown plastic film, 179
Botrytis cinerea, 336
bread, modified atmosphere packaging of, 304
British Standards Institute (BSI), 113

Campylobacter jejuni, 307
cans, *see* metal cans
carbon dioxide
   Ageless™, 290
   effect on foods, 304–5

effect on microbial growth, 305
food applications, 289–90
Fresh Lock™, 290
Freshilizer™, 286, 290
carbon monoxide
   effect on meat pigments, 311
   MAP applications, 305
   properties, 305
cast plastic film, 179
chemically separated cellulose fibre, 243
cling film wrapping, 198
Clostridium botulinum, 37, 38, 43, 49, 61, 299, 307–8, 334, 335, 337
coating of plastics, 205
   acrylic coatings, 205
   DLC (Diamond Like Coating), 208
   extrusion coating with PE, 208
   low temperature sealing coatings (LTSC), 206
   metallising with aluminium, 207
   PVdC coatings, 206
   PVOH coatings, 206
   SiO_x coatings, 207
cold seal, 221
cold/chill chain, 109
Comite European de Normalisation (CEN)
   see also distribution performance tests
communication, reference logistics, 99
compression strength, 105
compression testing, 111
consumer needs, 18
controlled atmosphere storage (CAS), 304
   definition, 303
corrugated fibreboard, 112–13, 269–72
crisps, modified atmosphere packaging of, 304
critical control point (CCP), 100, 109
cube utilization, 98, 102
Cytophaga, 335

dairy products, modified atmosphere packaging of, 304, 338
diamond-like-carbon (DLC) coating, 208, 315
direct product profitability (DPP), 107
distribution centres, 104

INDEX

distribution needs, 13
distribution performance testing, 109
  compression testing, 111–12
  equipment, 110
  impact, 110–11
  shock, 110–11
  standards, *see* ASTM; ISO; ISTA
  see also transit testing
  vibration testing, 110

efficient consumer response (ECR), 107
electronic data interchange (EDI), 25, 99
environmental performance of packaging, 26–7, 171, 233–7, 277–81
equilibrium modified atmosphere, 337
equilibrium relative humidity (ERH), 36
ergonomic standard, 98
Escherichia coli, 37, 38, 53, 307, 332, 337
ethanol emitters, 292–3
ethylene, 290, 309, 336–7
  scavengers, 290–92
  scavenger/moisture absorber, Neupalon™, 291
EU Directive for Packaging Waste, 299
EU Directive for Plastics, 174
extrusion blow moulding, 186
extrusion lamination, plastics, 210

facultative anaerobes, definition, 307
fat, *see* lipid
fish modified atmosphere packaging of, 334
Flavobacterium, 334
flavour/odour adsorbers, 296–7
  types, 297
food
  contact approval (packaging materials), 315
  poisoning, 332
  quality, 303
  shelf life, *see* shelf life
food contact issues for plastics, 214
food preservation methods
  blanching, 42
  chilling and cooling, 53
  continuous thermal processing (aseptic), 47–51
  curing, 57
  drying and water activity control, 54
  fermentation, 59
  freezing, 52, 53
  high pressure processing, 61
  irradiation, 62
  membrane processing, 62
  microwave processing, 63
  modification of atmosphere (MAP), 60
  ohmic heating, 62
  pasteurisation, 51–2
  pickling, 58
  smoking, 58
  thermal processing, 42–7, 137–8, 165, 229–31
food product quality, factors affecting
  enzyme activity, 73
  flavour scalping (loss), 81
  ingress of off-flavours, prevention thereof, 81
  insect damage, 78
  microbiological processes, packaging and, 77
  moisture changes in food, 78
  oxidation, 70
  physical damage, 77
food spoilage
  chemical, 332
  enzyme, 334
  gas, *see* carbon dioxide effect on foods, oxygen effect on foods
  microbiological, 334
  physical, 311
fruit
  modified atmosphere packaging of, 331
  respiration, 336

gas permeation
  definition, 316
  gas exchange, 303
gas transmission rate
  definition, 316
  measurement, 25
gas
  barrier properties, 315–18
  flushing, 324
  flushing, compensated vacuum, 324
  headspace composition determination, 331
  gaseous composition of air, 303
  headspace composition measurement, 331
  measurement of transmission rate, 331
  properties, 304–5
glass composition, 153
  amber (brown), 154
  blue, 154
  dark green, 154
  pale green (half white), 154
  white flint (clear glass), 153
glass container closure selection, 163
  normal seals, 164
  pressure seals, 164
  vacuum seals, 164

342 INDEX

glass container manufacture, 156
   cold end treatment, 159
   container forming, 157
   design parameters, 158
   furnace (melting), 156, 157
   hot end treatment, 158
   inspection and quality, 161
   low-cost production tooling, 160
   surface treatments, 158
glass container usage
   cleaning, 169
   consumer acceptibility, 156
   due diligence in the use of, 169
   food market sectors, 153
   labelling and decoration, 165
   marketing benefit, 172
   pack design and specification, 167
   pack integrity, 156
   pack safety, 156
glass
   attributes of packaging in, 154
   definition, 152
   packaging, 152
   strength in theory and practice, 166

hard sizing of paper and board, 255
heat sealing, 217–21, 228–9
   importance to modified atmosphere packaging, 319
   integrity, 229, 326–7
   measurement, 328
helium, 306

inert gases, *see* noble gases
injection blow moulding, 186
injection moulding, 186
injection stretch blow moulding, 186
intelligent packaging, 282–3
International Organisation for Standardization (ISO), 111–12, 118
International Safe Transit Association (ISTA), 110–11

labelling of rigid plastic containers, 213
Lactobacilli, 307, 335
lamb, modified atmosphere packaging of, 304
life cycle assessment (LCA), 28
life cycle inventory analysis (LCI), 28
lipid, oxidation, 309
Listeria monocytogenes, 37, 38, 53, 307, 337

logistical packaging issues
   food processing and retailing, 100
   retail customer service, 106
   supply chain integration, 108
   transport, 101
   warehousing, 104
   waste management, 107
logistical packaging, functions, 96–9
   communication, 99
   protection, 97
   productivity, 98
   utility, 98
logistics packaging materials and systems
   corrugated fibreboard boxes, 112
   reusable totes, 115
   shrink wrapping, 115
   unitization, 116–19

manufacture of plastic packaging, 178
Marks & Spencer, 285, 295
meat
   modified atmosphere packaging of, 308, 331, 333
   oxygen, its effect on pigments, 309, 333
mechanical properties of packaging, 318–19
mechanically separated cellulose fibre, 243
mesophilic, 35
metal can manufacture
   coatings, film laminates and inks, 131
   easy-open can ends, 130–1
   plain food can ends, 130
   three-piece welded cans, 125
   two-piece drawn and ironed (DWI), 127
   two-piece single drawn and multiple drawn (DRD), 126
metal can packaging issues
   can reception at packer, 132
   filling and exhausting, 133
   handling, 139
   heat processing, 137
   post processing, cooling, drying and labelling, 138
   seaming, 135
   storage and distribution, 140
metal can shelf life issues
   aluminium, 147
   dissolution of tin, 144
   environmental stress cracking aluminium ends, 149
   external corrosion, 149
   function of tin, 142

interactions between can and contents, 142
internal corrosion, 148
iron, 146
lacquers, 147
lead, 147
stress corrosion cracking, 148
sulphur staining, 149
tin toxicity, 145
metal cans
 container designs, 121
 packaging overview, 120
 performance needs, 120
 raw materials, aluminium, 124
 raw materials, steel, 123
 recycling, 124
metallising of plastic films (OPP, PET, PA), 183, 218, 224–5, 314
microaerophiles, definition, 307
migration
 avoiding migration and taint, 89
 factors affecting, 88
 from other packaging materials, 86
 from plastics, 83
 issues for plastics, 214
 monitoring and measuring, 89
Mitsubishi Gas Chemical Company, 285, 290
modified atmosphere packaging machinery
 chamber, 319
 compensated vacuum gas flushing, 324
 form-fill-seal, 320, 322
 gas flushing, 324
 negative forming, 320–21
 negative forming with plug assistance, 321–2
 positive forming with plug assistance, 321–2
 snorkel, 319
modified atmosphere packaging
 carbon dioxide scavengers/emitters, 289–90
 carbon dioxide headspace determination, 331
 definition, 303
 ethylene scavengers, 290
 MAP gases, 304–6
 MAP packaging materials, 303
 market for foods, 304
 measurement of carbon dioxide transmission rate, 331
 measurement of oxygen transmission rate, 331
 measurement of transmission/permeability rates, 329
 oxygen headspace determination, 331

oxygen scavengers, 284–9
water vapour transmission rate, 329
modified atmosphere packaging, food applications
 cooked, cured and processed meat products, 333
 dairy products, 338
 fish and fish products, 334
 fruits and vegetables, 335
 raw poultry, 332
 raw red meat, 331
moisture absorbers, 295–6
 types, 297
moisture
 food spoilage, 303
Moraxella, 334

narrow neck press process, glass container manufacture, 157
neon, 306
nitrogen
 effect on foods, 310
 effect on microbial growth, 308
 gaseous composition of air, 303
 properties, 305
noble gases
 properties, 305
 use in modified atmosphere packaging of foods, 304, 306
nylon, *see* plastic, polyamide

optical properties of packaging, 183, 318
oriented plastic film, 179
oxygen
 effect on microbial growth, 306–7
 effect on foods, 303, 309
 gaseous composition of air, 303
 headspace composition measurement, 331
 properties, 305
 transmission rate, measurement, 331
oxygen scavengers, 284–9
 Ageless™, 285
 beer, 284, 286
 food applications, 285
 iron based scavengers, 284
 market, 284
 non-metallic scavengers, 285–6
 Oxyguard™, 286–7
 smoked ham, 289
 wine, 286
 ZERO2™, 287–9

packaging papers and paperboards
  bag papers, 250
  folding boxboard (FBB), 252
  glassine, 249
  greaseproof, 249
  impregnated papers, 250
  laminating papers, 251
  microcreping, 249
  paper labels, 250
  sack kraft, 250
  solid bleached board (SBB), 251
  solid unbleached board (SUB), 251
  tissues, 250
  vegetable parchment, 249
  wet strength paper, 249
  white lined chipboard (WLC), 253
packaging specifications and standards, 28
Packaging Waste Directive (EU) (1994), 299
packaging
  definition, 8
  design and development, 9, 29
  food service, 109
  functions of, 8
  historical perspective, 2–4
  legislation, 107
  optimisation, 11–12
  protective role of, 4
  recovery, 172, 235, 244, 279
  recycling, 107–8, 171–2, 235–7, 244, 270, 279–80
  reduction, 107–8, 172, 233–4
  reuse, 107–8, 171, 235
  strategy, 9
  total quality management (TQM), 28
  total product concept, 10
pallet, 116–18
  construction, 117
  fibreboard, 116
  load, 116
  plastic, 116
  stacking height, 118
  stacking pattern, 118
  wood, 116
paper and board based systems, 277
paper and board environmental profile, 277
paper and board types of packaging, 259
  cap liners and diaphragms, 276
  composite containers, 268
  corrugated fibreboard packaging, 269
  fibre drums, 268
  folding cartons, 263

induction sealed disc, 276
interlocking dividers, 276
labels, 276
liquid packaging cartons, 265
moulded pulp applications, 276
moulded pulp containers, 272
multiwall sacks, 262
paper bags and wrapping paper, 259
pulpboard disc, 276
rigid cartons or boxes, 267
sachets/pouches/overwraps, 260
sealing tapes, 275
shredded paper, 276
tea and coffee bags, 259
tubes, 268
tubs, 268
paper and board, added processes, 255
  Acrylic dispersion coating, 255
  Fluorocarbon dispersion coating, 255
  hard sizing, 255
  lamination, 255
  plastic extrusion/laminating, 256
  printing and varnishing, 257
  varnishing/coating/laminating, 258
  wax sizing, 255
paper and board, fibre sources and pulping, 243
paper and paperboard manufacture
  coating, 248
  drying, 247
  finishing, 248
  pressing, 246
  reel-up, 248
  sheet forming, 245
  stock preparation, 245
paper and paperboard, packaging design of, 258
paper and paperboard, properties of, 254
pasteurisation, *see* thermal processing
pathogens, 32–3, 38, 66–7, 75, 138–9
permeability coefficient, 317–18
permeability issues for plastics, 215
permeation, *see* gas permeation
pest control, 105
plastic packaging manufacture
  packs based on plastic films, laminates, 183
  plastic film and sheet for packaging, 179
  rigid plastic packaging, 186
plastic packaging, containers and components, 179, 183, 186

INDEX   345

plastics environmental issues, 233
plastics in food packaging
   acrylonitrile butadiene styrene (ABS), 201
   cellulose based materials, 203
   ethylene vinyl acetate (EVA), 197
   ethylene vinyl alcohol (EVOH), 201, 312
   fluoropolymers, 203
   gas and water vapour barrier properties, 15
   high nitrile polymers (HNP), 202
   ionomers, 196, 313
   packaging materials, 311
   polyamide (PA), 197, 313
   polycarbonate (PC), 196
   polyethylene (PE), 115, 189, 312–13
   polyethylene naphthalene dicarboxylate (PEN), 195
   polyethylene terephthalate (PET or PETE), 194, 313, 315, 317
   polymethyl pentene (TPX), 202
   polypropylene (PP), 191, 313–14
   polystyrene (PS), 200, 314
   polyvinyl acetate (PVA), 204
   polyvinyl chloride (PVC), 198, 314
   polyvinylidene chloride (PVdC), 199, 314
   styrene butadiene (SB), 201
plastics waste management, 235
plastics, sealing and closing, 217
pork, modified atmosphere packaging of, 304
poultry, modified atmosphere packaging of, 304, 308, 332
preservative releasers, 293–5
   Microban™, 293
   RepelKote™, 294
press and blow, glass container manufacture, 157
printing of plastic films
   digital, 212
   flexographic, 212
   gravure, 211
printing of rigid plastic containers
   dry offset printing, 213
   heat transfer printing, 213
product packaging needs, 13
productivity, reference logistics, 97
properties of paper and board, 254
   appearance, 254
   performance, 254
protection, reference logistics, 97
Pseudomonas, 306, 332, 334, 338

psychrotropic, 35
PureSeal™, 286

Radio frequency identification (RFID), 99, 106–7
recovered (secondary) cellulose fibre, 244
retail distribution centre (RDC), 95, 104
retail market needs, 22
retort pouch, 224
reusable totes, 115–16

Salmonella, 37, 38, 53, 62, 307, 337
self cooling cans, 298
self heating cans, 298
shelf life, 65, 232, 303, 304
   factors affecting, 68
shock testing, 110
shrink bundles, 115
shrink sleeving (labels)
shrink wrapping, 115, 118, 176, 186,199
silicon oxide (SiO_x) coating, 315
silk screen printing, 213
slip sheet (logistics), 117–18
Staphylococcus, 37, 307
stock keeping units (SKU), 99, 106
stretch blow moulding, plastics, 186
stretch wrapping, 118, 198
sulphur dioxide
   MAP applications, 304
   temperature control packaging, 297

temperature controlled packaging, 297–8
   products, 297–8
thermal lamination, plastics, 211
thermal processing
   aseptic, 47–51, 188–9, 202, 266
   canned foods, 42–7, 137–8
   glass packed foods, 165
   pasteurisation, 51–2
   retort pouches, 224–33
thermoduric, 35
thermoforming, 186
thermophilic, 35
Toppan Printing Co. Ltd., 285, 290
Toyo Seikan, Kaisha Ltd., 285
transit damage, 101
   see also distribution performance tests
transit testing, 101, 107, 109
transmission rate, definition, 316–17
   see also CO_2, O_2 and water vapour transmission rate

transport
  air, 103
  rail, 102–3
  road, 101–2
  sea, 103–4

US Fibre Box Association, 113
US Occupational Safety and Health
    Administration (OSHA), 98

vacuum packaging, 334
value of packaging to society, 7
vegetables
  modified atmosphere packaging of, 304, 335
  respiration, 336
vibration testing, 102, 110

Vibrio parahaemolyticus, 306–7
virgin (primary) cellulose fibre, 243

warehouse, 104–6
waste management issues, 26–7, 107–8, 171,
    233–7, 278–80
water activity (a_w), 36
water vapour transmission rate
    (WVTR)
  definition, 316
  effect of relative humidity (RH), 317
  measurement of, 329
  test standard ASTM E96, 330
  test standard ASTM F 1249, 330

zenon, 306

LIVERPOOL
JOHN MOORES UNIVERSITY

I.M. MARSH LRC
Tel: 0151 231 5216